NATO ASI Series

Advanced Science Institutes Series

A series presenting the results of activities sponsored by the NATO Science Committee, which aims at the dissemination of advanced scientific and technological knowledge, with a view to strengthening links between scientific communities.

The Series is published by an international board of publishers in conjunction with the NATO Scientific Affairs Division

A Life Sciences B Physics	Plenum Publishing Corporation London and New York
C Mathematical and Physical Sciences D Behavioural and Social Sciences E Applied Sciences	Kluwer Academic Publishers Dordrecht, Boston and London
F Computer and Systems Sciences G Ecological Sciences H Cell Biology I Global Environmental Change	Springer-Verlag Berlin Heidelberg New York London Paris Tokyo Hong Kong Barcelona Budapest

PARTNERSHIP SUB-SERIES

1. Disarmament Technologies	Kluwer Academic Publishers
2. Environment	Springer-Verlag/Kluwer Academic Publishers
3. High Technology	Kluwer Academic Publishers
4. Science and Technology Policy	Kluwer Academic Publishers
5. Computer Networking	Kluwer Academic Publishers

The Partnership Sub-Series incorporates activities undertaken in collaboration with NATO's Cooperation Partners, the countries of the CIS and Central and Eastern Europe, in Priority Areas of concern to those countries.

NATO-PCO DATABASE

The electronic index to the NATO ASI Series provides full bibliographical references (with keywords and/or abstracts) to about 50000 contributions from international scientists published in all sections of the NATO ASI Series. Access to the NATO-PCO DATABASE is possible via a CD-ROM "NATO Science & Technology Disk" with user-friendly retrieval software in English, French and German (© WTV GmbH and DATAWARE Technologies Inc. 1992).

The CD-ROM can be ordered through any member of the Board of Publishers or through NATO-PCO, Overijse, Belgium.

Series H: Cell Biology, Vol. 104

Springer
Berlin
Heidelberg
New York
Barcelona
Budapest
Hong Kong
London
Milan
Paris
Santa Clara
Singapore
Tokyo

Cellular Integration of Signalling Pathways in Plant Development

Edited by

Fiorella Lo Schiavo

University of Padua, Department of Biology
Via Colombo 3
I-35121 Padua, Italy

Robert L. Last

Boyce Thompson Institute for Plant Research
and Section of Genetics and Development
Cornell University, Tower Road
Ithaca, NY 14853-1801, USA

Giorgio Morelli

Unità di Nutrizione Sperimentale,
Istituto Nazionale della Nutrizione
Via Ardeatina 546
I-00178 Rome, Italy

Natasha V. Raikhel

Michigan State University,
Department of Energy and Plant Research Laboratory
East Lansing, MI 48824-1312, USA

With 68 Figures (1Colour Plate)

Springer

Published in cooperation with NATO Scientific Affairs Division

Proceedings of the NATO Advanced Study Institute "Cellular Integration of Signalling Pathways in Plant Development", held at Acqua Fredda di Maratea, Italy, May 20–30, 1997

Library of Congress Cataloging-in-Publication Data

Cellular integration of Signalling pathways in plant development / edited by Fiorella Lo Schiavo ... [et al.].
p. cm. – (NATO ASI series. Series H, Cell biology; vol. 104)
"VIIIth NATO Advanced Study Institute on Plant Cell and Molecular Biology was held in Acqua Fredda di Maratea, Italy, from May 20–30, 1997."
Includes bibliographical references.
ISBN 3-540-64014-2 (hardcover)
1. Plant cellular signal transduction–Congresses. 2. Plants-Development–Congresses.
I. Lo Schiavo, Fiorella, 1950– . II. NATO Advanced Study Institute on Plant Cell and Molecular Biology (8th : 1997 : Acquafredda di Maratea, Italy) III. Series.
QK725.C4 1998 571.8'2–dc21 97-51323 CIP

ISSN 1010-8793
ISBN-13: 978-3-642-72119-9 e-ISBN-13: 978-3-642-72117-5
DOI: 10.1007/978-3-642-72117-5

Softcover reprint of the hardcover 1st edititon 1998

Typesetting: Camera ready by authors/editors
Printed on acid-free paper
SPIN 10571809 31/3137 - 5 4 3 2 1 0

PREFACE

The VIIIth NATO Advanced Study Institute on Plant Cell and Molecular Biology was held in Acqua Fredda di Maratea, Italy from May 20-30, 1997.
The topic, "Cellular Integration of Signalling Pathways in Plant Development" brought together scientists who investigate the cellular and molecular basis of plant development. The aim of this course was for a multidisciplinary group to confront with each other diverse research approaches-genetic, biochemical, cellular and molecular-pursued in various laboratories in order to stimulate new perspectives in the plant biology field.

In the past few years there have been tremendous advances in the understanding of signals and signalling pathways that operate at the cellular level and lead to developmental processes.
The course covered a broad area of Plant Biology. It was designed to review, in a unified way, the field of basic plant biology, biochemistry and genetics focusing on emerging topics that are critical to our understanding of the unique properties of plant cells.
The course highlighted the most recent progress on signals, machinery and pathways in the plant cell. Emphasis was placed on integrating these studies with those on cell division, cell plate formation, and other aspects of plant development to elucidate the intricate relationships among them. Several presentations brought up to date our knowledge of the plant cell wall and cell plate formation, an area unique to plant biology. In the protein trafficking field, several components of the machinery that targets proteins to specific organelles have been identified and cloned. An effort was made to introduce the field of electrophysiology, a very powerful and rapidly developing field, to the plant molecular scientists. The various chapters of this book summarize lectures of invited speakers. They have not been subjected to any major cutting or editing, in order to preserve the individual nature of each contribution.

We are pleased to acknowledge the financial support of the following companies and organizations who recognize the importance of the basic plant research for the development of agricultural biotechnology. In particular we thank the NATO for the generous contribution without whose support this meeting could not possibly take place, and then the Companies Monsanto, Pioneer Hi-bred, Beckman Analytical-Italia, Du Pont and Perkin Elmer-Italia for their help.
We thank the many people who helped us to organize of the course. And finally, we thank all the participants in this meeting for their valuable contributions and for the stimulating discussions that characterized the course.

Fiorella Lo Schiavo
Padua, Italy

Natasha Raikhel
East Lansing, MI, USA

Robert Last
Ithaca, NY, USA

Giorgio Morelli
Rome, Italy

CONTENTS

Part 7 Post-embryonic Development

Part 8 Cell-Cell Communication

Mechanisms for the Transport of Soluble Proteins to the Vacuole in Plants

Sridhar Venkataraman and Natasha V. Raikhel

Department of Energy Plant Research Laboratory, Michigan State University, East Lansing, Michigan 48824-1312

Abstract. In eukaryotic cells, most proteins destined for the cell surface or for organelles of the secretory pathway are synthesized by endoplasmic-reticulum-bound ribosomes and co-translationally transported into the ER lumen. This translocation process is dependent on the presence of a signal sequence found at the N-terminus of most soluble secretory proteins. If no other signal is present, soluble proteins are transported to the cell surface by a default mechanism, with a positive sorting signal being required for retention within the endomembrane system (reviewed in Bednarek and Raikhel, 1992) [Fig. 1]. Soluble vacuolar proteins (in plant and yeast cells) or lysosomal proteins (in mammalian cells) are transported through the secretory pathway to the *trans*-Golgi network (TGN), where they are segregated from secreted proteins and carried to the vacuole or lysosome in clathrin-coated vesicles. The mechanisms for sorting proteins to these organelles differ between organisms. Sorting to mammalian lysosomes is usually mediated by oligosaccharide modifications, whereas sorting to yeast and plant vacuoles is mediated by short peptide domains that may be removed before or upon deposition in the organelle, or it may form a part of the mature protein (for review see, Bar-Peled et al., 1996). Soluble proteins are transported through the secretory system via transport vesicles that bud from one compartment and fuse specifically with the next. The specificity of these fusion events is thought to be determined by integral membrane proteins on the vesicles and the acceptor membrane and several candidates for these targeting molecules have recently been identified (Rothman, 1994).

Keywords: secretory pathway, vacuole, endoplasmic reticulum, Golgi, protein trafficking

1 Plant Vacuolar Targeting Signals

The targeting signals found on proteins take many forms, and often a single organelle will have several distinct signals corresponding to non-redundant pathways for targeting. For example, vacuolar targeting signals are found to vary considerably in sequence and location. In mammalian cells, a post-translational addition of mannose-6-phosphate is sufficient for

NATO ASI Series, Vol. H 104
Cellular Integration of Signalling Pathways in Plant Development
Edited by F. Lo Schiavo, R. L. Last, G. Morelli, and N. V. Raikhel

targeting to the lysosome (equivalent to the vacuole; Kornfeld and Mellman, 1989); however, it has been found that carbohydrate modification does not appear to be involved in vacuolar targeting in yeast or plants (Bednarek and Raikhel, 1991; Chrispeels, 1991). Rather, in plants, three distinct targeting signals have been characterized (Fig. 1). Barley lectin (BL) contains a 15-amino-acid C-terminal propeptide (CTPP) that is proteolytically removed to produce the mature protein (Lerner and Raikhel, 1989). The CTPP has been shown to be necessary and sufficient for sorting to the vacuole, and a detailed mutagenesis study has suggested that the C-terminus of the propeptide is recognized by the sorting machinery (Bednarek and Raikhel, 1991; Dombrowski et al., 1993). Other proteins have also been shown to contain a CTPP. Tobacco chitinase is found in both the vacuole and the extracellular space of tobacco cells. The vacuolar form has an extension of seven amino acids at the C-terminus, which is absent from the extracellular form (Shinshi et al., 1990). When tobacco chitinase cDNA, with and without the CTPP, was expressed in tobacco plants, chitinase protein lacking the CTPP was shown to be secreted. When the C-terminal extension of either the BL or tobacco chitinase was added to the normally secreted cucumber chitinase, the chitinase was redirected to the vacuole, showing that this extension contains all the information required for sorting to the vacuole (Bednarek and Raikhel, 1991; Neuhaus et al., 1991).

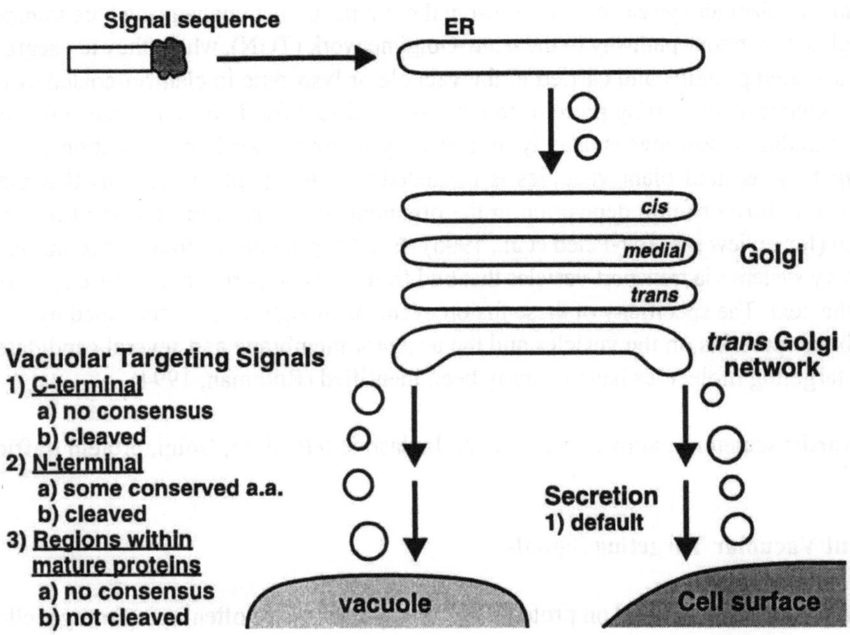

Figure 1. Targeting of cotranslationally translocated soluble proteins to plant vacuoles.

Site-directed mutagenesis analyses of CTPPs of both the BL (Dombrowski et al., 1993) and tobacco chitinase (Neuhaus et al., 1994) demonstrate that there is no specific peptide sequence required for a vacuolar sorting signal, and a wide variety of sequences are able to function in vacuolar targeting in both cases. However, the addition of two or more glycines to the BL CTPP causes secretion (Dombrowski et al., 1993). This implies that some components of the sorting machinery recognize the CTPP from the C-terminus. Other vacuolar proteins have C-terminal extensions, and these extensions contain sorting information (Melchers et al., 1993). No primary structural homologies can be identified between CTPPs, however they do have an overall hydrophobic character that may be important in recognition and sorting. Nothing is known yet about the sorting machinery for these proteins.

A second type of sorting signal is found in the vacuolar protease aleurain, which contains an N-terminal propeptide (NTPP). Regions of proaleurain important for vacuolar sorting were identified by their incorporation into proendoproteinase B (proEP-B), protease that is normally secreted. Two short stretches of amino acids in the aleurain NTPP were able to direct proEP-B to the vacuole with low efficiency. Residues adjacent to this sorting signal were found to increase the efficiency of sorting (Holwerda et al., 1992). Another protein with an N-terminal sorting domain is sporamin, a storage protein from sweet potato. Sporamin is synthesized with a 16-amino-acid NTPP, which is necessary for vacuolar sorting, as deletion of the propeptide causes secretion in transformed tobacco cells (Matsuoka and Nakamura, 1991). Unlike CTPPs, NTPPs from various vacuolar proteins contain a conserved motif.

Some plant vacuolar proteins are synthesized without a cleaved propeptide and thus the sorting information must be contained within the mature protein (von Schaewen and Chrispeels, 1993). All three plant vacuolar sorting signals have been tested for sorting in yeast, but none of them has been recognized by the yeast sorting machinery. This implies that the receptors recognizing the sorting signals are different in plants and yeast (for review see Gal and Raikhel, 1993; Gal and Raikhel, 1994).

Although the signals for sorting to the plant vacuole have been determined now for several proteins, little is known about the mechanism for sorting these proteins. A potential vacuolar sorting receptor has been identified and extracted from clathrin-coated vesicles of developing pea cotyledons (Kirsch et al., 1994). This 80-kDa integral membrane glycoprotein was also enriched in Golgi vesicles, and protease digestion studies indicated that an N-terminal lumenal domain is responsible for its NTPP-binding activity. The protein was shown to bind to an affinity column containing the proaleurain NTPP and to be subsequently eluted at pH 4.0. It did not bind to a control column containing the N-terminus of a secreted protein. A peptide consisting of the prosporamin vacuolar sorting signal competed for binding to the 80-kDa protein against the proaleurain NTPP, whereas a proBL CTPP peptide or a mutant peptide no longer competent for sorting did not. The identity of the protein responsible for the binding was confirmed by cross-linking.

These data are consistent with the binding protein being a receptor for vacuolar proteins containing the conserved NTPP motif, and suggest that vacuolar proteins containing the

CTPP may use a different receptor (Kirsch et al., 1994). The binding specificity of BP-80 has been examined using affinity columns containing peptides corresponding to vacuolar targeting signals from several proteins. In addition to binding to proaleurain NTPP, BP-80 is also retained on columns containing the prosporamin NTPP and the targeting sequence from 2S albumin of Brazil nut, but not the probarley lectin CTPP (Kirsch et al., 1996).

Recently, Ahmed et al. (1997) identified a receptor-like protein, AtELP, by computer-aided sequence analysis of the Arabidopsis EST database. The sequence of this protein shares 60% similarity with BP80 from pea. AtELP is a membrane protein highly expressed in monocot and dicot plants. Various approaches of subcellular localization have been used to show that the protein is present in two membrane fractions corresponding to a fraction enriched in clathrin- and adaptin-associated vesicles and to a novel compartment. The function of AtELP remains to be addressed.

Studies on tobacco chitinase have revealed the existence of a receptor for proteins containing a CTPP. When this chitinase was transiently expressed at high levels in tobacco protoplasts some portion of the protein was secreted, indicating that the sorting system was saturated (Neuhaus et al., 1994). This can best be explained by the presence of a receptor protein that is saturated when there is too much cargo to be transported. The presence of individual receptors for various vacuolar proteins in yeast further suggests that plant proteins with different sorting signals may use different receptors. However, in contrast to the case with an NTPP many different alterations and deletions in CTPP are tolerated by the sorting machinery (Dombrowski et al., 1993; Neuhaus et al., 1994).

One important question that arises from the differences in vacuolar targeting signals and the possibility of more than one receptor is whether there are also multiple mechanisms for the transport of proteins to the vacuole. Co-expression of barley lectin and sporamin in transgenic tobacco plants followed by electron microscopic immunolocalization and pulse-chase analysis indicate that these two proteins are localized to the same large vacuoles in leaves (Schroeder et al., 1993). However, in some plant cells two functionally distinct vacuole types have been identified (Hoh et al., 1995; Paris et al., 1996). Interestingly, Matsuoka et al. (1995) showed that the proBL CTPP and prosporamin NTPP are interchangeable in vacuolar targeting by switching the two signals. The mechanism of transport of these two proteins has been investigated in transgenic tobacco cells using the fungal metabolite wortmannin, an inhibitor of phosphatidylinositol 3-kinase (PI 3-kinase) activity. A PI 3-kinase (Vps34 protein) has been demonstrated to be essential for vacuolar protein sorting in yeast (Stack et al., 1993). Both BL and sporamin containing the proBL CTPP are missorted to the cell surface in the presence of wortmannin, whereas proteins containing the prosporamin NTPP are correctly targeted to the vacuole under these conditions (Matsuoka et al., 1995). This indicates that CTPP- and NTPP-mediated transport occurs via different mechanisms, CTPP-mediated transport being sensitive to wortmannin and NTPP-mediated transport insensitive to wortmannin. The target of wortmannin in tobacco cells has also been investigated: phospholipid synthesis as well as PI kinase activities were shown to be inhibited by this metabolite. A comparison of dose dependencies of inhibition suggests that the synthesis of phospholipids may be involved in CTPP-dependent vacuolar transport (Matsuoka et al., 1995).

2 Components of the Vacuolar Transport Machinery

Recent work in mammalian and yeast systems has begun to illuminate the trafficking machinery. For example, we now know that it is not the sorting signals themselves that lead to targeting of the protein to a given organelle. Rather, each signal appears to be recognized by a receptor in the preceding compartment that deposits the signal-containing protein into an appropriate transport vesicle. The trafficking machinery then acts upon these vesicles; specific proteins found on the vesicle surface direct them to the correct organelle in a process described by the SNARE hypothesis (Söllner et al., 1993). On the vesicle surface are found proteins called v-SNARE (alternately, termed synaptobrevins or VAMPS) which provide specificity for vesicle transport by interacting with proteins called t-SNARES (or syntaxins) on the target membrane. Following this interaction, the involvement of specific cytosolic factors has been implicated in vesicle fusion. These factors are called NSF (N-ethylmaleimide-sensitive factor) and SNAPs (soluble NSF-associated proteins) in mammalian cells, although homologues of these factors have been found in yeast and other organisms. Other members of the trafficking machinery have also been found to be conserved among eukaryotes. For example, v-SNARE- and t-SNARE-type proteins have been isolated from many organisms including yeasts, mammals, insects, and plants (Aalto et al., 1993; Protopopov et al., 1993; Bassham et al., 1995).

Although considerable research has been done on the trafficking machinery in yeast and mammalian cells, very little is known about the mechanics of trafficking in plants. Recently, a functional homologue of a yeast syntaxin involved in vacuolar transport (*PEP12*) has been isolated from *Arabidopsis thaliana* (*AtPEP12*) as a cDNA that complements a yeast *pep12* deletion mutant (Bassham et al., 1995). The cDNA isolated in this screen is predicted to encode a protein homologous to *PEP12* and mammalian syntaxins. It is likely that this protein represents an essential part of the plant vacuolar targeting machinery.

Like Pep12p and other syntaxins, AtPEP12p contains a C-terminal coiled-coil domain and a highly conserved C-terminal membrane anchor domain (59% homologous to Pep12p). Southern analysis has indicated that *atPEP12* resides at a single locus, which has been mapped to chromosome V at approximately 33.7 cM (between the nga106 and mi322 markers; D. Bouchez and N. V. Raikhel, unpublished). Northern analysis has shown that a single 1.3-kb mRNA is expressed in plants, the highest level of which is found in roots, stems and flowers, with lower levels found in leaves (Bassham et al., 1995). In rabbits, antibodies have been produced to the bacterially overexpressed hydrophilic N-terminus of atPEP12p; these antibodies are highly specific and immunoprecipitate *in vitro* translated atPEP12p (as a single band of ~35 kDa). Use of this antisera in western analysis of several plant tissues shows, similar to what was found for *atPEP12* mRNA, that the highest amount of atPEP12p is found in roots, with lower levels in leaves and stems. Biochemical fractionation has indicated that the atPEP12p is found exclusively associated with cellular membranes, as expected for an integral membrane protein. Gradient centrifugation has revealed that atPEP12p does not co-fractionate with membrane markers available for the *Arabidopsis* ER (AtSEC12p) (Bar-Peled and Raikhel, 1997), Golgi (latent-IDPase and fucosyl transferase), tonoplast (a-TIP), or PM (RD48) (Conceição et al., 1997). Immunogold electron microscopy of cryofixed *Arabidopsis* seeds and roots has localized atPEP12p to the membranes of small uncoated vesicles approximately 100 μm in diameter; these have been proposed to be of the late post-Golgi compartment (Fig.2) [Conceição et al., 1997). Precise identification of these vesicles as a late post-Golgi compartment can be made if cargo destined for the vacuole (proteins with vacuolar-targeting signals) is found associated with these AtPEP12p-vesicles. The question that should be answered is whether the vacuolar proteins, using three different targeting signals, are delivered to the vacuole via the compartment carrying AtPep12p or whether some unidentified syntaxin proteins are responsible for the delivery of a subset of vacuolar proteins.

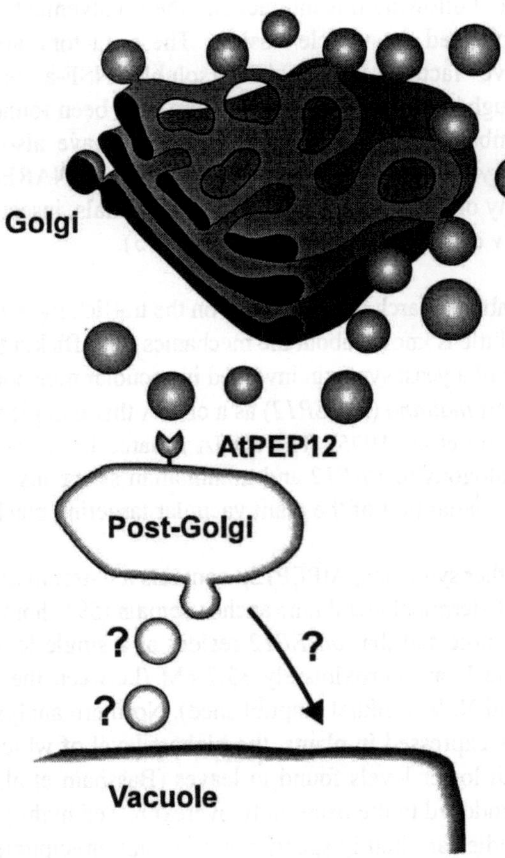

Figure 2. AtPEP12 syntaxin resides on the late post-Golgi compartment in plants.

Further characterization and localization of AtPEP12p in plant cells will likely provide insight into the mechanisms of vacuolar targeting in plants. Already, AtPEP12p is an important molecular marker for a unique class of vesicles, and perhaps for the post-Golgi compartment–a marker which has not been available in plants yet. Moreover, AtPEP12p can be used as a tool for identifying other members of the protein trafficking machinery, perhaps revealing details about how a plant cell differentiates between the many cellular destinations in protein targeting.

Other genes, presumably involved in vacuolar protein sorting, have been isolated by homology to yeast genes. A PI 3-kinase showing some sequence homology to the yeast *VPS34* gene and to the mammalian PI 3-kinase gene has been isolated from Arabidopsis (Welters et al., 1994). Although overexpression of *AtVP3S4* gene in transgenic plants indicates that the gene encodes a functional PI 3-kinase, the *AtVPS34* is not able to complement a yeast *vps34* mutant. The function of AtVPS34 *in planta* is not yet known. An Arabidopsis gene showing homology to the yeast *VPS1* gene has also been identified (Dombrowski and Raikhel, 1995) although it has not been demonstrated to function in vacuolar targeting. Several small GTP-binding proteins have been found to be associated with vesicles carrying vacuolar proteins in pumpkin cotyledon cells, and their function in protein transport has been proposed (Shimada et al., 1994). Another small GTP-binding protein, Rab6, presumably functions in the regulation of vesicle transport from the TGN in mammals. The Arabidopsis homolog of that gene has been also cloned (Bednarek et al., 1994). AtRAB6 has approximately 79% sequence identity to the mammalian (Rab6) and yeast (Ryh1 and Ypt6) protein counterparts. Although the AtRAB6 functionally complemented the *YPT6* mutant from yeast, its function *in vivo* is not yet known.

The transport of soluble proteins through the plant secretory pathway to the vacuole has been studied in some detail at the level of the targeting signals within the proteins. However, many major questions remain about the mechanisms by which the proteins are transported. Few components of the vesicle transport machinery have been isolated; however, their precise function in planta remains to be addressed. In addition, the use of genetic and biochemical approaches will be essential for isolating further components of the vesicle targeting machinery in the near future. Only when the majority of markers that allow us to differentiate various components of the secretory machinery are available, will it be possible to address the differences between the various cell types mechanistically (Bassham and Raikhel, 1997).

References

Aalto MK, Ronne H, Kerãnen S (1993) Yeast syntaxins Sso1p and Sso2p belong to a family of related membrane proteins that function in vesicular transport. EMBO J 12: 4095-4104

Ahmed SU, Bar-Peled M, Raikhel NV (1997) Cloning and subcellular location of an Arabidopsis receptor-like protein that shares common features with protein-sorting receptors of eukaryotic cells. Plant Physiol 114: 325-336

Bar-Peled M, Bassham DC, Raikhel NV (1996) Transport of proteins in eukaryotic cells: More questions ahead. Plant Mol Biol 32: 223-249

Bar-Peled M, Raikhel NV (1997) Characterization of AtSEC12 and AtSAR1. Plant Physiol 114: 315-324

Bassham DC, Raikhel NV (1997) Molecular aspects of vacuole biogenesis. Adv Bot Res 25: 43-58

Bassham DC, Gal S, Conceição AS, Raikhel NV (1995) An *Arabidopsis* syntaxin homologue isolated by functional complementation of a yeast *pep12* mutant. Proc Natl Acad Sci USA 92:7262-7266

Bednarek SY, Raikhel NV (1991) The barley lectin carboxyl-terminal propeptide is a vacuolar protein-sorting determinant in plants. Plant Cell 3: 1195-1206

Bednarek SY, Raikhel NV (1992) Intracellular trafficking of secretory proteins. Plant Mol Biol 20: 133-150

Bednarek SY, Reynolds TL, Schroeder M, Grabowski R, Hengst L, Gallwitz D, Raikhel NV (1994) A small GTP-binding protein from *Arabidopsis thaliana* functionally complements the yeast *YPT6* null mutant. Plant Physiol 104: 591-596

Chrispeels (1991) Sorting of proteins in the secretory system. Annu Rev Plant Physiol Plant Mol Biol 42: 21-53

Conceição AS, Marty-Mazars D, Bassham DC, Sanderfoot AA, Marty F, Raikhel NV (1997) The syntaxin homolog AtPEP12p resides on a late post-Golgi compartment in plants. Plant Cell 9: 571-582

Dombrowski JE, Raikhel NV (1995) Isolation of a cDNA encoding a novel GTP-binding protein of *Arabidopsis thaliana*. Plant Mol Biol 28: 1121-1126

Dombrowski JE, Schroeder MR, Bednarek SY, Raikhel NV (1993) Determination of the functional elements within the vacuolar targeting signal of barley lectin. Plant Cell 5: 587-596

Gal S, Raikhel NV (1993) Protein sorting in the endomembrane system of plant cells. Curr Opin Cell Biol 5: 636-640

Gal S, Raikhel NV (1994) A carboxy-terminal plant vacuolar targeting signal is not recognized by yeast. Plant J 6: 235-240

Hoh B, Hinz G, Jeong B-K, Robinson DG (1995) Protein storage vacuoles form de novo during pea cotyledon development. J Cell Sci 108: 299-310

Holwerda BC, Padgett HS, Rogers JC (1992) Proaleurain vacuolar targeting is mediated by short contiguous peptide interactions. Plant Cell 4: 307-318

Kirsch T, Paris N, Butler JM, Beevers L, Rogers JC (1994) Purification and initial characterization of a potential plant vacuolar targeting receptor. Proc Natl Acad Sci USA 91: 3403-3407

Kirsch T, Saalbach G, Raikhel NV, Beevers L (1996) Interaction of a potential vacuolar targeting receptor with amino- and carboxy-terminal targeting determinants. Plant Physiol 111: 469-474

Kornfeld S, Mellman I (1989) The biogenesis of lysosomes. Annu Rev Cell Biol 5: 483-525

Lerner D, Raikhel NV (1989) Cloning and characterization of root-specific barley lectin. Plant Physiol 91: 124-129

Matsuoka K, Nakamura K (1991) Propeptide of a precursor to a plant vacuolar protein required for vacuolar targeting. Proc Natl Acad Sci USA 88: 834-838

Matsuoka K, Bassham DC, Raikhel NV, Nakamura K (1995) Different sensitivity to wortmannin of two vacuolar sorting signals indicates the presence of distinct sorting machineries in tobacco cells. J Cell Biol 130: 1307-1318

Melchers LS, Sela-Buurlage MB, Vloemans SA, Woloshuk CP, Van Roekel JSC, Pen J, van den Elzen PJM, Cornelissen BJC (1993) Extracellular targeting of the vacuolar tobacco proteins AP24, chitinase and ß-1,3-glucanase in transgenic plants. Plant Mol Biol 21: 583-593

Neuhaus J-M, Sticher L, Meins F, Boller T (1991) A short C-terminal sequence is necessary and sufficient for the targeting of chitinases to the plant vacuole. Proc Natl Acad Sci USA 88: 10362-10366

Neuhaus J-M, Pietrzak M, Boller T (1994) Mutation analysis of the C-terminal vacuolar targeting peptide of tobacco chitinase: Low specificity of the sorting system, and gradual transition between intracellular retention and secretion into the extracellular space. Plant J 5: 45-54

Paris N, Stanley CM, Jones RL, Rogers JC (1996) Plant cells contain two functionally distinct vacuolar compartments. Cell 85: 563-572

Protopopov V, Govindan B, Novick P, Gerst JE (1993) Homologs of the synaptobrevin/VAMP family of synaptic vesicle proteins function on the late secretory pathway in *S. cerevisiae*. Cell 74: 855-861

Rothman JE (1994) Mechanisms of intracellular protein transport. Nature 372: 55-63

Schroeder MR, Borkhsenious ON, Matsuoka K, Nakamura K, Raikhel NV (1993) Colocalization of barley lectin and sporamin in vacuoles of transgenic tobacco plants. Plant Physiol 101: 451-458

Shimada T, Nishimura M, Hara-Nishimura I (1994) Small GTP-binding proteins are associated with the vesicles that are targeted to vacuoles in developing pumpkin cotyledons. Plant Cell Physiol 35: 995-1001

Shinshi H, Neuhaus J-M, Ryals J, Meins F Jr (1990) Structure of a tobacco endochitinase gene: Evidence that different chitinase genes can arise by transposition of sequences encoding a cysteine-rich domain. Plant Mol Biol 14: 357-368

Söllner T, Whiteheart SW, Brunner M, Erdjument-Bromage H, Geromanos S, Tempst P, Rothman JE (1993) SNAP receptors implicated in vesicle targeting and fusion. Nature 362: 318-324

Stack JH, Herman PK, Schu PV, Emr SD (1993) A membrane-associated complex containing the Vps15 protein kinase and the Vps34 PI 3-kinase is essential for protein sorting to the yeast lysosome-like vacuole. EMBO J 12: 2195-2204

von Schaewen A, Chrispeels MJ (1993) Identification of vacuolar sorting information in phytohemagglutinin, an unprocessed vacuolar protein. J Exp Bot 44(Supplement): 339-342

Welters P, Takegawa K, Emr SD, Chrispeels MJ (1994) *AtVPS34*, a phosphatidylinositol 3-kinase of *Arabidopsis thaliana*, is an essential protein with homology to a calcium-dependent lipid binding domain. Proc Natl Acad Sci USA 91: 11398-11402

Aquaporins: Their Role and Regulation in Cellular Water Movement

Vipula K. Shukla & Maarten J. Chrispeels

Department of Biology 0116, University of California San Diego, 9500 Gilman Drive, La Jolla, CA 92093-0116 USA

Keywords. aquaporins, water channels, membrane permeability, regulation

All living cells are separated from their surroundings by a surface (plasma) membrane. Eukaryotic cells are further compartmentalized internally by a variety of membrane systems, including the endoplasmic reticulum, the vacuolar membrane (tonoplast) and organellar membranes. By necessity, these membranes are selectively permeable, since cells regulate the amount, type and direction of substances which move across them. Control of the exchange of substances across membranes often depends on the physical and chemical properties of the membranes themselves, as well as proteinaceous components including specific channels, transporters and pumps. Water is by far the most important of the materials which are moved across membranes, in and out of cells and their organelles.

The recent discovery of proteinaceous water channels in plants and animals (Preston et al., 1992; Maurel et al., 1993) has added a layer of complexity to the paradigm of how water moves across cellular membranes. In plants, the primary determinants of transmembrane water movement are hydrostatic (turgor) and osmotic pressure gradients. Previously, these gradients were assumed to drive diffusion of water across the lipid bilayer in a mechanism that was not clearly understood. However, it was obvious that some specialized cellular membranes were able to transport water at a rate much higher than could be accounted for by diffusional models. The identification of a class of proteins, called aquaporins, which form channels specific for the transport of water, has provided significant insights into the molecular mechanisms of this membrane permeability. Consideration of aquaporins in models of water flow has forced a re-evaluation of water relations at the cellular, tissue and organismal levels. For the first time, data about the function of aquaporins provides support for early theories of pore-mediated bulk flow of water across biological membranes, first proposed in animal systems (Koefoed-Johnsen & Ussing, 1953) and later suggested for plants (Dainty, 1963).

A frequently asked question concerning aquaporins is: why do cells need water channels? After all, lipid bilayers exhibit an intrinsic diffusional permeability to water (P_d), calculated to be <0.01 cm.s^{-1}. This diffusion is characterized by an inability to be inhibited by pharmacological compounds, and typically shows a high Arrhenius activation energy (E_a>10 kcal.mol^{-1}). In contrast, specialized, biologically active membranes such as those found in mammalian erythrocytes show a high osmotic water permeability coefficient (P_f) of approximately 0.04 cm.s^{-1} (Solomon,

NATO ASI Series, Vol. H 104
Cellular Integration of Signalling Pathways in Plant Development
Edited by F. Lo Schiavo, R. L. Last, G. Morelli, and N. V. Raikhel

1989), which is inhibited by mercuric chloride and shows a low Arrhenius activation energy (E_a < 5 kcal.mol^{-1}) (Chrispeels & Agre, 1994). Growing plant cells require a large influx of water in a relatively short time; the $T_{1/2}$ of water exchange for elongating soybean cortical cells is approximately 3 seconds (Steudle & Boyer, 1985), and these cells show a high hydraulic conductivity. As seen in these examples, certain biologically active membranes need to be able to transport large quantities of water in short period of time; this type of water movement cannot be accounted for by the intrinsic permeability of the lipid bilayer. Therefore, one answer to the question of why water channels are necessary lies in the dynamic nature of living cells. At certain times in their life cycle, under certain environmental conditions, living cells require the rapid transport of large quantities of water; this is most easily accomodated by a channel-mediated process. Furthermore, the ability to regulate the movement of water is essential for the general maintenance of osmotic relations within any type of cell, tissue or organism. In particular, because of their sessile nature, higher plants critically need to adapt their physiology to ensure proper growth and development. For example, the absorption of water into the vacuole of a plant cell results in hydrostatic pressure (turgor), which drives a number of processes including cell expansion and closing of the stomata. It is therefore imperative that organisms be able to regulate water movement across their cellular membranes in a developmentally, temporally, spatially and environmentally controlled manner. Water channels clearly are involved in this regulatory process.

1 What is the Role of Aquaporins in Water Transport?

To demonstrate the water transport activity of aquaporin proteins in the laboratory, a heterologous expression system is used. cRNA encoding aquaporin proteins is injected into oocytes of *Xenopus laevis* under conditions of controlled osmolarity. These oocytes express the aquaporin proteins and target them to the plasma membrane, which is the default destination for membrane proteins that enter the secretory system in animal cells. It should be noted that direct evidence, such as immunolocalization, that this targeting pathway functions accurately for plant proteins in oocytes is lacking. The presence and proper insertion of aquaporin proteins into the plasma membrane of oocytes is inferred from the change in the permeability of the membrane. After injection and a 2-4 day incubation period, the oocytes are shifted to a hypoosmotic medium, which causes the cells to swell due to water influx. The rate of this swelling can be measured and quantitated using video microscopy and image capture, and the osmotically induced water permeability (P_f) can be calculated from this rate. Control (water injected) cells swell at a rate significantly slower than cells that are expressing aquaporins, which show up to a 10-20-fold increase in P_f (Chrispeels & Agre, 1994). Similar results have been obtained with both plant and animal aquaporins (Preston et al., 1992; Maurel et al., 1993). Importantly, the observed increase in P_f is not accompanied by an increase in electrogenic ion transport, suggesting that aquaporins are water-specific channels and do not transport ions or neutral metabolites (Maurel et al. 1993). However, exceptions to this classification have been demonstrated: when expressed in *Xenopus* oocytes, mammalian AQP3 is permeable to glycerol and urea (Ishibashi et al., 1994). Thus, while the term aquaporin is a functional definition, it appears that some of these channels may have multiple specificities.

These experiments demonstrate the ability of aquaporin proteins to transport water in a heterologous system. What is the evidence that water channels function as such *in vivo*? In humans, the pathophysiology of certain diseases, such as nephrogenic diabetes insipidus (NDI), a kidney ailment whose symptoms include an inability to concentrate urine, has been correlated with water channel activities (Mulders et al., 1996). Specifically, mutations in the AQP2 gene result in mistargeted water channels; it is inferred that the membranes lacking these channels are subsequently altered in their water permeability (Mulders et al., 1997). However, there are examples of human mutations that result in a deficiency of functional water channels, yet do not result in disease states (Preston et al., 1994).

In plants, evidence for *in vivo* water channel activity is emerging, and recent studies have begun to focus on this question. In tomato roots, Maggio & Joly (1995) found that the aquaporin inhibitor mercuric chloride caused a significant and rapid reduction in pressure-induced water flux through the vascular tissue. These decreases were not accompanied by a change in the K^+ concentration, suggesting that the difference in water flux was not due to imbalances in ion movement, but rather a change in the osmotic components that drive water movement. Mercuric chloride has also been used in experiments with internodal cells of the alga *Chara corallina,* a model system that is amenable to direct measurements of single-cell membrane permeability using pressure probes (Henzler & Steudle, 1995). In this study, the authors found that inhibition of aquaporins caused a 3-4 -fold decrease of the hydraulic conductivity of the membrane, while the permeability coefficients of other solutes were unchanged. This effect was reversed when the mercuric chloride was removed with 2-mercaptoethanol, supporting the hypothesis that water is transported primarily through mercury-sensitive channels. Based on their measurements of permeability and reflection coefficients, Henzler and Steudle concluded that a composite model of the membrane in which proteinaceous pores including specific water channels are arranged in parallel with lipid arrays is likely, and these aquaporin channels account for a majority of the water transport taking place in the alga (Henzler & Steudle, 1995).

Transgenic technology has enabled plant biologists to address water transport questions at the molecular level. In *Arabidopsis thaliana*, plants expressing an antisense construct of the plasma membrane protein AthH2, predicted to be an aquaporin, were used to isolate protoplasts (Kaldenhoff et al., 1995). These protoplasts were subjected to hypoosmotic stress, leading to a rapid influx of water, and the proportion of burst cells at several time points was measured. After 3 minutes, 95-100% of control (untransformed) protoplasts had burst, while approximately half of the antisense cells were still intact. These data indicate that the rate of water influx in cells with less AthH2 protein is significantly slower than in the wild type control, suggesting a role for this protein in water transport in the plant.

Combining genetic and biophysical techniques, a direct demonstration of aquaporin-mediated water transport activity has been achieved. The gene encoding the Arabidopsis plasma-membrane protein RD28 was first cloned from a drought-induced cDNA library (Yamaguchi-Shinozaki et al., 1992) and later shown to be a water channel *in vitro* using the Xenopus oocyte assay (Daniels et al., 1994). This protein was subsequently expressed in transgenic tobacco plants, under the control of the

Table I. RD28 Enhances the Hydraulic Conductivity of the Plasma Membrane in Transgenic Tobacco Plants*

Characteristic	Wild Type Control	RD28 Transgenic
Turgor Pressure (MPa)	0.82 ± 0.015 (n=56 cells)	0.86 ± 0.013 (n=64 cells)
Cell Volume (μm^3)	33.4 ± 1.8 (n=199 cells)	31.3 ± 1.4 (n=119 cells)
$T_{1/2}$ for H_2O Exchange(s)	3.36 ± 0.26 (n=56 cells)	2.37 ± 0.16 (n=64 cells)

*Pressure microprobe measurements performed by A. Thomas & T.C. Hsiao, University of California, Davis. Data are for leaf surface vein cells.

CMV 35-S promoter. When plants expressing this protein were analyzed using pressure microprobe techniques, it was found that the $T_{1/2}$ for water exchange was significantly faster than in control (wild type) plants (Table I). Importantly, the turgor pressure and cell volume of the transgenic plants was unchanged relative to wild type, leading to the conclusion that the change in water exchange rate was due to an increase in hydraulic conductivity of the cells, resulting from the presence of the RD28 protein. This experiment provides the first substantive data showing that the water transport activity of aquaporins observed in Xenopus oocytes is indicative of a similar activity in plant cells.

While it has become increasingly obvious that aquaporins function as water channels in biological membranes, additional routes for facilitated water transport cannot and should not be excluded from consideration. Ion channels, which function in membrane depolarization and are believed to be involved in early events in signal transduction (Ward et al., 1995), may play a role in regulation of the osmotic gradient across membranes. In particular, movement of ions through K^+ and anion channels is accompanied by water flux due to displacement of water molecules in the channel pore. For example, under conditions of rapid turgor change such as stomatal opening/closing, these channels are activated and their contribution to changes in guard cell turgor has been characterized (reviewed in Zeiger et al., 1987; Willmer & Fricker, 1996). Although the amount of water transported by ion channels under these conditions varies, depending on the channel, it is generally agreed that ion channels alone cannot account for the overall changes in guard cell turgor that are observed, suggesting that aquaporins may be acting concurrently.

Additionally, a recent study has suggested that cotransporters play an important role in water homeostasis. Loo and colleagues (Loo et al., 1996) have tested the hypothesis that water transport is directly linked to solute transport by the Na^+/glucose cotransporter of mammalian intestinal cells, using the Xenopus oocyte expression system. Electrophysiological measurements were carried out using a two-electrode voltage clamp system and coupled cotransport of Na^+ ions and glucose molecules was measured as a transmembrane current associated with Na^+ movement.

These measurements indicated that for every two Na^+ ions, one glucose molecule was cotransported across the membrane. Additionally, oocyte swelling rates were calculated to quantitate water influx. Surprisingly, it was found that activation of Na^+/glucose cotransport was rapidly (within 10 seconds) associated with water influx, and the two activities showed a similar high activation energy, approximately 25 kcal.mol^{-1}. Therefore, this water transport was not due to endogenous aquaporin activity. Based on a comparison of transport rates, Loo et al. determined that the stoichiometry of cotransport was 260 water molecules per 2 Na^+/1 glucose (Loo et al., 1996). The mechanism of this high water transport activity may be due to the apparent ability of cotransporter proteins to bind a large number of water molecules in the region of the sugar binding site. As these proteins convey their cargo to different membrane compartments, the bound water molecules are carried along, and it is possible that conformational changes associated with sugar binding/release may also control delivery of water. Since cotransporters are found throughout evolution, these studies are of widespread significance; the authors also mention similar preliminary results with a plant H^+/amino acid cotransporter. Clearly, further studies on this mechanism for water transport are needed to determine the relative contribution of cotransporters to the overall control of osmotic relations.

2 Are Aquaporins Involved in Cellular Osmo-homeostasis?

Current and ongoing studies point to a definite role for aquaporins in the movement of water between cells, tissues and organs of many diverse biological systems. However, data about the role of water channels at the subcellular level is only starting to emerge. Since eukaryotic cells are subdivided into many compartments that experience different pH, salt, ion and protein concentrations, and since these compartments are defined by membrane systems, one could expect that these membranes will contain water channels that participate in the maintenance of homeostasis. An obvious place to study such channels, at the single cell level, is in bacteria. *E. coli* contains a MIP homolog, AqpZ, that is a functional water channel (Calamita et al., 1995). Study of this model organism will help define the role of aquaporins within a cell, as it responds to various environmental changes in osmotic conditions. Other MIP homologs have been identified in cyanobacterial systems (Kashiwagi et al., 1995; Shukla & Chrispeels, unpublished observations); these studies are ongoing as well.

In eukaryotes, plant cells provide a good context to study the relative contributions of the different membrane systems and their water channels to cellular osmoregulation. A recent study by Maurel et al. (1997) has examined the vacuolar (tonoplast) and plasma membranes from tobacco cells by separating the membranes and performing permeability measurements on isolated vesicles (Maurel et al., 1997). This study found low P_f values (6-8 μm.s^{-1}) and high activation energies (E_a = 13-14 kcal.mol^{-1}) for the plasma membrane, implying that water transport occurs mostly by diffusion across the lipid membrane in this system. In contrast, the tonoplast exhibited a 100-fold higher P_f and a reduced E_a of approximately 2.5 kcal.mol^{-1}. Furthermore, these features of the vacuolar membrane were reversibly inhibited by mercuric chloride, an aquaporin inhibitor. These data led the authors to conclude that water transport in the tonoplast is mediated by aquaporins, and this water transport into/out of the vacuole may play a role in buffering osmotic fluctuations occuring in the cytoplasm. If this is the case, it points to a direct osmoregulatory function for aquaporins. Since many types of cells contain well defined compartments, it is possible that such a mechanism exists not only in vacuolated plant cells, but elsewhere in other biological systems.

3 Aquaporins are Members of a Large Gene Family

Aquaporins are members of an ancient superfamily of channel proteins called the MIP (major intrinsic protein) family, first identified in mammalian lens tissue (Gorin et al., 1984). In addition to water channels, this family includes glycerol facilitators, ion channels, and channels involved in the transport of small solutes. Thus, while all aquaporins are MIP proteins, not all MIPs are aquaporins; the term aquaporin is a functional definition that connotes water channel activity. To date, the MIP family consists of more than 80 members, including proteins from bacteria, yeast, animals and plants (reviewed in Park & Saier, 1996) (Fig. 1).

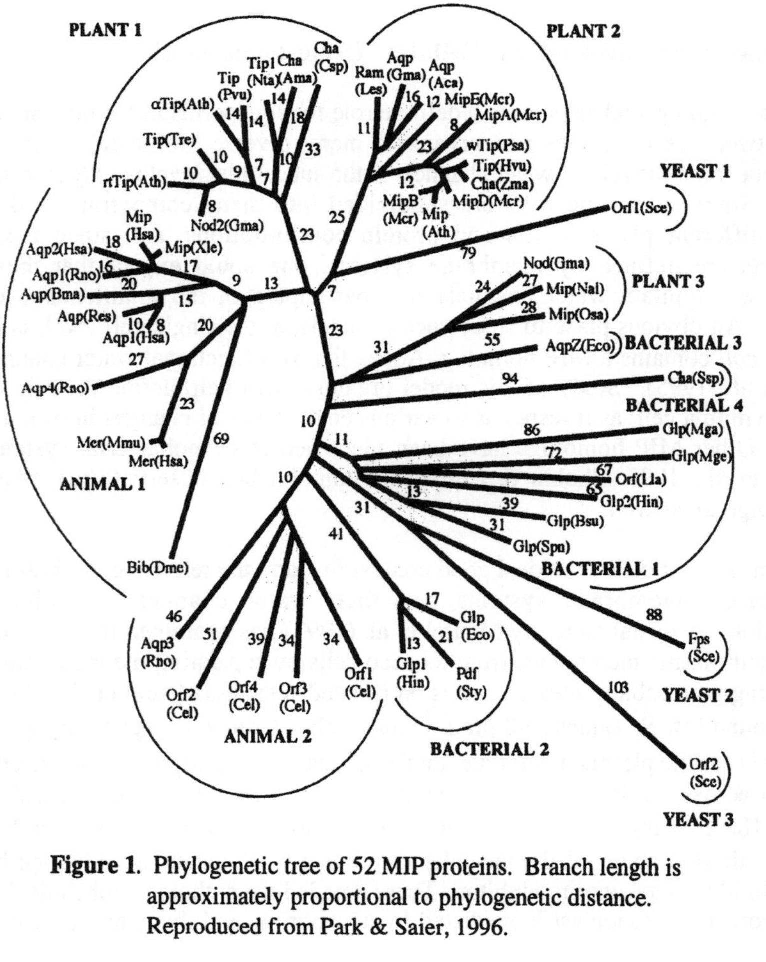

Figure 1. Phylogenetic tree of 52 MIP proteins. Branch length is approximately proportional to phylogenetic distance. Reproduced from Park & Saier, 1996.

Aquaporins, like all members of the MIP family, are 24-28 kDa polypeptides, with six membrane-spanning domains connected by short exposed loops, and cytosolic amino and carboxyl termini. The amino and carboxyl halves of aquaporin proteins are themselves sequence related, and combine to form a unique structure that has been modeled as an "hourglass" (Jung et al., 1994). The insertion and arrangement of aquaporin polypeptide monomers, in a biologically active tetramer within the membrane, is thought to create a "hole" which permits the passage of water molecules in the presence of an osmotic or physical pressure differential (Fig. 2). It should be noted that aquaporins are true channels and not active pumps; water transport through the pore is driven by osmotic differentials and does not require ATP. All aquaporins analyzed so far contain certain conserved motifs, which according to structural models and mutational analysis, are found near the pore and are probably required for water transport activity. Current studies of animal and plant aquaporins support this hourglass model (Jap & Li, 1995; Mitra et al., 1995; Daniels, 1997; Li et al., 1997); the basic structure has been described as a cylinder with a trapezoidal cross section, the wall of which is formed by six transmembrane a-helices and an additional short a-helix, with a branched rod-like substructure within the cylinder (Li et al., 1997). The rapid, ongoing elucidation of detailed structural components of aquaporins will undoubtedly enhance our understanding of the mechanisms whereby water traverses the pore.

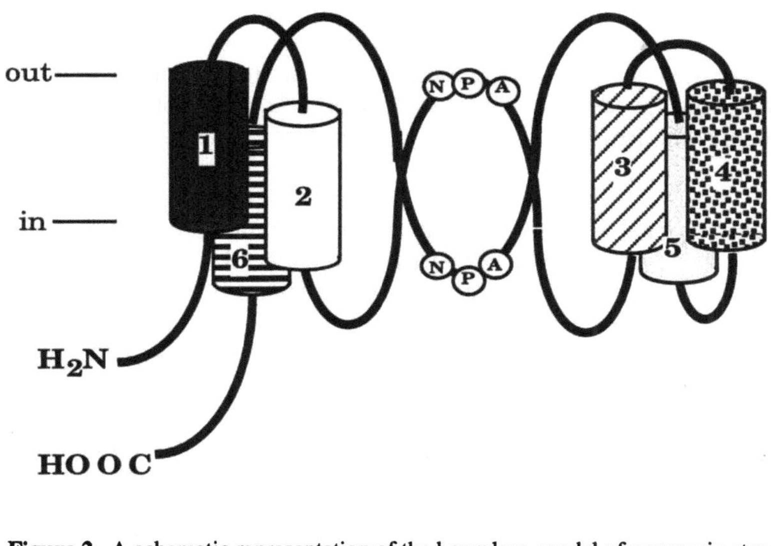

Figure 2. A schematic representation of the hourglass model of aquaporin structure, described in Jung et al., 1994.

In higher plants, aquaporins are found in both the tonoplast (vacuolar) membrane and the plasma membrane, and recent studies have identified a large number of aquaporins in different plant species (reviewed in Maurel, 1997). Different aquaporins are expressed in a developmental and tissue-specific manner; for example Arabidopsis expresses the vacuolar γ-TIP (tonoplast intrinsic protein) in expanding cells at the time when the large central vacuole is being formed, during cell enlargement (Ludevid et al., 1992). In both Arabidopsis and *Pisum sativum*, genes for an aquaporin-like protein have been found to be water-stress induced (Guerrero et al., 1990; Yamaguchi-

Shinozaki et al., 1992), and the Arabidopsis protein (RD28) has been shown to be a plasma membrane water channel (Daniels et al., 1994). Within a given plant species, there appear to be many different aquaporin genes expressed. For example, a search of the Arabidopsis EST (expressed sequence tag) database for conserved regions of MIP genes has identified at least 23 MIP homologues in this plant (Weig et al., 1997) about half of which have been shown to have water channel activity. Based on sequence comparisons, these genes fall into 3 general classes, 2 of which correlate, in general, with the membranes (tonoplast and plasma membrane) in which the protein is located. The reason for this redundancy is likely to be the temporal, developmental, spatial and environmental differences in expression between individual proteins in a single plant. Thus, the overall control of water flow in a plant involves the concerted regulation of many different water channel genes and their proteins at the transcriptional, translational and post-translational levels.

4 How are Aquaporins Regulated?

Given the complexity of expression patterns and activity of aquaporin proteins, it is clear that many regulatory mechanisms must be engaged to allow organisms to maintain osmotic homeostasis and respond to various signals. One argument in favor of extensive regulation of water channel activity lies in the tissue/cell specificity observed for many aquaporins; within an organ (e.g. roots) neighboring cells that are presumably experiencing similar osmotic conditions may show different levels of aquaporin expression and activity, resulting in different rates of water flux. This is particularly true in the case of organ development; cell expansion and elongation requires the targeted, temporal regulation of water flux leading to changes in cell turgor.

At the transcriptional level, several aquaporin genes have been shown to be regulated by salt or water stress (reviewed in Maurel, 1997). Aquaporin gene expression can also be controlled by various plant hormones (ABA, GA) (Kaldenhof et al., 1993; Phillips & Huttly, 1994) and analysis of the promoter region of one gene has recently revealed sequence elements involved in GA and ABA induction (Kaldenhoff et al., 1996). Tissue specificity has also been shown to be controlled by promoter cis-acting regulatory elements in the case of tobacco roots under both normal and nematode-infection stress-induced conditions (Yamamoto et al., 1991). In addition to induction/repression of gene expression, factors such as message stability/turnover may contribute to the transcriptional control of aquaporins, and unraveling these mechanisms will provide insights into general schemes for control of gene expression and signal transduction.

While there is currently no evidence suggesting translational control of aquaporin proteins, the role of post-translational modification is being examined closely. Protein phosphorylation has been studied in several different aquaporins. In both plants and animals, phosphorylation of some aquaporins has been shown to stimulate the water channel activity of these proteins in Xenopus oocytes (Kuwahara et al., 1995; Maurel et al., 1995; Yool et al., 1996), although the amount of increase in membrane permeability is highly variable. Johansson et al. (1996) found that the phosphorylation of a spinach plasma membrane water channel, PM28A, is negatively regulated by osmotic pressure *in vivo*, and this phosphorylation is strictly dependent on submicromolar Ca^{2+} concentrations *in vitro* (Johansson et al. 1996), implying that calcium-dependent protein kinases are involved in the regulatory process. Sequence analysis of this protein and other plant PIPs and TIPs has revealed that several of these proteins have up to six potential phosphorylation sites, some of which are highly conserved (Fig. 3).

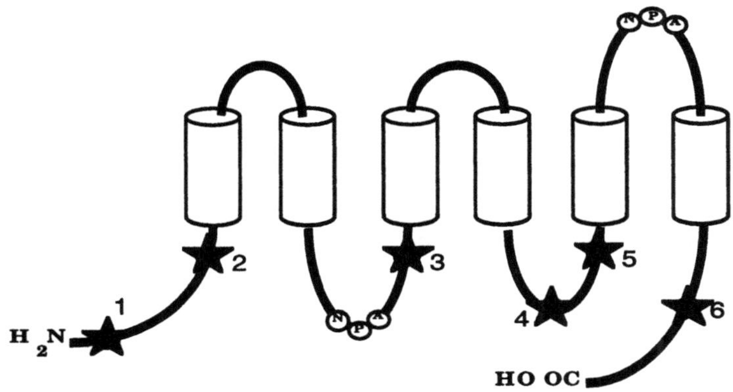

1 KEVS
2 KLWS
3 RKVS
4 PKRS
5 ARDS
6 KALGSFR

Figure 3. Potential phosphorylation sites in the spinach PIP PM28A

However, not all aquaporin phosphorylation events lead to increased water channel activity (Lande et al., 1996). Although it has been proposed that aquaporin phosphorylation may provide a mechanism for "gating" of the water channel (Maurel et al., 1995), alternate schemes have also been presented. In the case of the vasopressin-regulated mammalian aquaporin AQP2, protein phosphorylation is believed to play a role in the control of membrane localization and recycling (Sabolic et al., 1995; Katsura et al., 1996). Based on the data available, it is obvious that the role of phosphorylation in the activity and specificity of aquaporin water channels is complex and may vary from protein to protein.

In addition to phosphorylation, other post-translational modifications occur in aquaporins. Proteolytic processing of a precursor polypeptide to a mature protein may provide a tool for controlling both the activity and abundance of functional water channels; one example of this is given in the case of a 23 kDa TIP from pumpkin seeds (Inoue et al., 1995). Also, early work on the animal aquaporin CHIP28, the first protein to be identified as a water channel, revealed that this polypeptide is glycosylated (Smith & Agre, 1991); although the significance of this glycosylation for water channel activity is not known, it is possible that such modification confers structural features that affect the pore. Finally, structural work on CHIP 28 and other aquaporins has shown that they form tetrameric aggregates in the membrane (Jap & Li, 1995). Although the functional unit of aquaporins is believed to be the monomer (van Hoek et al., 1991), aggregation state may play a role in protein stability, affecting the abundance of channels in the membrane.

These observations imply that the regulation of aquaporins occurs at many levels. This is not surprising, given the abundance, redundancy and ancient origins of this gene family. The ability to control water relations at both the cellular and organismal levels requires diverse mechanisms and the ability to adapt quickly. As work on aquaporins progresses, we will undoubtedly learn more about these mechanisms and the implications they have for control of biologically critical processes in general.

Acknowledgements

Work on aquaporins in our laboratory has been supported by grants from the National Science Foundation and the United States Department of Agriculture to M.J.C. and by a grant (95-37304-2297) from the USDA to V.K.S.

References

Calamita, G., et al. (1995). Molecular Cloning and characterization of AqpZ, a Water channel from *Escherichia coli*. *J. Biol. Chem.* **270**: 29063-29066.

Chrispeels, & M.J. Agre, P. (1994). Aquaporins: water channel proteins of plant and animal cells. *TIBS* **19**: 421-425.

Dainty, J. (1963). Water relations of plant cells. *Adv. Bot. Res.* **1**: 279-326.

Daniels, M.J. (1997). Characterization of Water Channels (Aquaporins) in Plants. University of California, San Diego.

Daniels, M.J.,Mirkov, T.E.&Chrispeels, M.J. (1994). The Plasma Membrane of *Arabidopsis thaliana* Contains a Mercury-Insensitive Aquaporin That Is a Homolog of the Tonoplast Water Channel Protein TIP. *Plant Physiol.* **106**: 1325-1333.

Gorin, M.B., et al. (1984). The major intrinsic protein (MIP) of the bovine lens fiber membrane: Characterization and structure based on cDNA cloning. *Cell* **39**: 49-59.

Guerrero, F.D., Jones, J.T.& Mullet, J.E. (1990). Turgor-responsive gene transcription and RNA levels increase rapidly when pea shoots are wilted. Sequence and expression of three inducible genes. *Plant Mol. Biol.* **15**: 11-26.

Henzler, T. & Steudle, E. (1995). Reversible closing of water channels in Chara internodes provides evidence for a composite transport model of the plasma membrane. *J. Exp. Botany* **46**: 199-209.

Inoue, K., et al. (1995). Characterization of two integral membrane proteins located int eh protein bodies of pumpkin seeds. *Plant Mol. Biol.* **28**: 1089-1101.

Ishibashi, K., et al. (1994). Molecular cloning and expression of a member of the aquaporin family with permeability to glycerol and urea in addition to water expressed at the basolateral membrane of kidney collecting duct cells. *Proc. Natl. Acad. Sci. USA* **91**: 6269-6273.

Jap, B.K. & Li, H. (1995). Structure of the Osmo-regulated H_2O-channel, AQP-CHIP, in Projection at 3.5 Å Resolution. *J. Molec. Biol.* **251**: 413-420.

Johansson, I., et al. (1996). The major integral proteins of spinach leaf plasma membranes are putative aquaporins and are phosphorylated in response to Ca^{2+} and apoplastic water potential. *Plant Cell* **8**: 1181-1191.

Jung, J.S., et al. (1994). Molecular Structure of the Water Channel Through Aquaporin CHIP The Hourglass Model. *J. Biol. Chem.* **269**: 14648-14654.
Kaldenhof, R., Kölling, A. & Richter, G. (1993). A novel blue light- and abscisic acid-inducible gene of *Arabidopsis thaliana* encoding an intrinsic membrane protein. *Plant Mol. Biol.* **23**: 1187-1198.
Kaldenhoff, R., et al. (1995). The blue light-responsive *AthH2* gene of *Arabidopsis thaliana* is primarily expressed in expanding as well as in differentiating cells and encodes a putative channel protein of the plasmalemma. *Plant J.* **7**: 87-95.
Kaldenhoff, R., Kolling, A. & Richter, G. (1996). Regulation of the Arabidopsis thaliana aquaporin gene AthH2 (PIP1b). *J. Photochem. Photobiol. B.* **36**: 351-354.
Kashiwagi, S., Kanamaru, K. & Mizuno, T. (1995). A *Synechoccus* gene encoding a putative pore-forming intrinsic membrane protein. *Biochim. Biophys. Acta* **1237**: 189-192.
Katsura, T., Ausiello, D.A. & Brown, D. (1996). Direct demonstration of aquaporin-2 water channel recycling in stably transfected LLC-PK1 epithelial cells. *Am. J. Physiology-renal fluid and electrolyte physiol.* **39**: 548-553.
Koefoed-Johnsen, V. & Ussing, H.H. (1953). The contributions of diffusion and flow to the passage of D_2O through living membranes. Effect of Neurohypophyseal hormone on isolated anuran skin. *Acta Physiol. Scand.* **28**: 60-76.
Kuwahara, M., et al. (1995). cAMP-dependent Phosphorylation Stimulates Water Permeability of Aquaporin-collecting Duct Water Channel Protein Expressed in *Xenopus* oocytes. *J. Biol. Chem.* **270**: 10384-10387.
Lande, M.V., et al. (1996). Phosphorylation of Aquaporin-2 Does Not Alter the Membrane Water Permeability of Rat Papillary Water Channel-containing Vesicles. *J. Biol. Chem.* **271**: 5552-5557.
Li, H., Lee, S.& Jap, B.K. (1997). Molecular design of aquaporin-1 water channel as revealed by electrol crystallography. *Nature Struc. Biol.* **4**: 263-265.
Loo, D.D.F., et al. (1996). Cotransport of water by the Na+/glucose cotransporter. *Proc. Natl. Acad. Sci. USA* **93**: 13367-13370.
Ludevid, D., et al. (1992). The Expression Pattern of the Tonoplast Intrinsic Protein γ-TIP in *Arabidopsis thaliana* Is Correlated with Cell Enlargement. *Plant Physiol.* **100**: 1633-1639.
Maggio, A. & Joly, R.J. (1995). Effects of Mercuric Chloride on the Hydraulic Conductivity of Tomato Root Systems. *Plant Physiol.* **109**: 331-335.
Maurel, C. (1997). Aquaporins and Water Permeability In Plant Membranes. *Annu. Rev. Plant Physiol. Plant Mol. Biol.* **48**: *in press*
Maurel, C., et al. (1993). The vacuolar membrane protein γ-TIP creates water specific channels in *Xenopus* oocytes. *EMBO J.* **12**: 2241-2247.
Maurel, C., et al. (1995). Phosphorylation regulates the water channel activity fo the seed-specific aquaporin a-TIP. *EMBO J.* **14**: 3028-3035.
Maurel, C., et al. (1997). Purified vesicles of tobacco cell vacuolar and plasma membranes exhibit dramatically different water permeability and water channel activity. *Proc. Natl. Acad. Sci. USA*: *in press*
Mitra, A.K., et al. (1995). The CHIP28 water channel visualized in ice by electron crystallography. *Nature Struc. Biol.* **2**: 726-729.
Mulders, S., et al. (1996). Physiology and pathophysiology of aquaporins. *Eur. J. Clin. Invest.* **26**: 1041-1050.

Mulders, S., et al. (1997). New mutations in the AQP2 gene in nephrogenic diabetes insipidus resulting in functional but misrouted water channels. *J. Am. Soc. Nephrology* **8**: 242-248.

Park, J.H. & Saier, M.H.J. (1996). Phylogenetic Characterization of the MIP Family of Transmembrane Channel Proteins. *J. Memb. Biol.* **153**: 171-180.

Phillips, A.L. Huttly, A.K. (1994). Cloning of two gibberellin-regulated cDNAs from *Arabidopsis thaliana* by subtractive hybridization: expression of the tonoplast water channel,γ-TIP, is increased by GA_3. *Plant Mol. Biol.* **24**: 603-615.

Preston, G.M., et al. (1992). Appearance of Water Channels in *Xenopus* Oocytes Expressing Red Cell CHIP28 Protein. *Science* **256**: 385-387.

Preston, G.M., et al. (1994). Mutations in Aquaporin-1 in phenotypically normal humans without functional CHIP water channels. *Science* **265**: 1585-1587.

Sabolic, I., et al. (1995). The AQP2 Water Channel: Effect of a Vasopressin Treatment, Microtubule Disruption and Distribution in Neonatal Rats. *J. Membrane Biol.* **143**: 165-175.

Smith, B.L. & Agre, P. (1991). Erythrocyte Mr 28,000 transmembrane protein exists as a multisubunit oligomer similar to channel proteins. *J. Biol. Chem.* **266**: 6407-6415.

Solomon, A.K. (1989). Water Channels across the Red Blood Cell and Other Biological Membranes. *Methods. Enzymol.* **173**: 192-222.

Steudle, E. & Boyer, J.S. (1985). Hydraulic resistance to radial water flow in growing hypocotyl of soybean measured by a new pressure-perfusion technique. *Planta* **164**: 189-200.

van Hoek, A.N., et al. (1991). Functional unit of 30 kDa for proximal tubule water channels as revealed by radiation inactivation. *J. Biol. Chem.* **266**: 16633-16635.

Ward, J.M., Pei, Z.-M. & Schroeder, J.I. (1995). Roles of Ion Channels in Initiation of Signal Transduction in Higher Plants. *Plant Cell* **7**: 833-844.

Weig, A.R., Deswarte, C. & Chrispeels, M.J. (1997). Analysis of the Aquaporin Gene Family in *Arabidopsis Thaliana. Plant Physiol. in press*

Willmer, C. & Fricker, M. (1996). Stomata. New York, Chapman & Hall.

Yamaguchi-Shinozaki, K., et al. (1992). Molecular Cloning and Characterization of 9 cDNAs for Genes That Are Responsive to Desiccation in *Arabidopsis thaliana*: Sequence Analysis of One cDNA Clone That Encodes a Putative Transmembrane Channel Protein. *Plant Cell Physiol.* **33**: 217-224.

Yamamoto, Y.T., et al. (1991). Characterization of *cis*-Acting Sequences regulating Root-Specific Gene Expression in Tobacco. *Plant Cell* **3**: 371-382.

Yool, A.J., Stamer, D.& Regan, J.W. (1996). Forskolin Stimulation of Water and Cation Permeability in Aquaporin 1 Water Channels. *Science* **273**: 1216-1218.

Zeiger, E., Farquhar, G.D. & Cowan, I.R., Ed. (1987). Stomatal Function. Stanford, CA, Stanford University Press.

Transport of Cytoplasmically Synthesized Proteins into Chloroplasts

Kenneth Keegstra, Mitsuru Akita, Jennifer Davila-Aponte, John Froehlich, Erik Nielsen and Sigrun Reumann

MSU-Department of Energy Plant Research Laboratory, Michigan State University, East Lansing, MI 48824, USA

Keywords: chloroplast, precursor, envelope membranes, import, binding, translocation

1. Introduction

Chloroplasts are metabolically complex organelles that participate in a diverse array of biochemical processes in addition to their well known role in photosynthesis. Consistent with their functional complexity, chloroplasts are structurally complex organelles. They possess three different lipid bilayer membranes enclosing three different aqueous compartments (Figure 1). Each membrane and each aqueous compartment has a unique set of proteins and enzyme activities that reflect their respective functions. Because the plastid genome has a limited coding capacity, most chloroplastic proteins are encoded by nuclear genes and synthesized in the cytoplasm as higher molecular weight precursors (Figure 1). The extra region of peptide is present at the amino terminus of a precursor, and is called a transit peptide because of its importance in directing the precursor into chloroplasts. Understanding how these precursor proteins are targeted from the cytoplasm to their proper location within chloroplasts is a major challenge and recent progress in this effort is briefly reviewed in this paper. This subject has been the topic of several reviews in recent years and readers should consult them for further details (de Boer and Weisbeek, 1991; Cline and Henry, 1996; Fuks and Schnell, 1997; Gray and Row, 1995; Kouranov and Schnell, 1996; Lübeck et al., 1997; Schnell, 1995; Theg and Scott, 1993).

Most of our current knowledge regarding protein import has derived from studies where protein import was reconstituted in vitro using isolated intact chloroplasts and radiolabeled precursor proteins. The results of these studies have demonstrated that the import pathway can be divided into several processes, some with multiple steps. Process 1 (Figure 1) is the transport of precursor proteins across the two envelope membranes. This process, which will be the major focus of this chapter, can be further subdivided into at least two discrete steps. The first, is the binding of precursor proteins to the surface of chloroplasts (step 1_B in Figure 1). This process requires low levels of ATP, presumably in the intermembrane space (Olsen and Keegstra, 1992). When ATP levels are adequate (generally greater than 1mM), precursors are translocated across both envelope membranes and into the stromal space (step 1_T in Figure 1).

NATO ASI Series, Vol. H 104
Cellular Integration of Signalling Pathways in Plant Development
Edited by F. Lo Schiavo, R. L. Last, G. Morelli, and N. V. Raikhel

During or immediately after envelope translocation, the transit peptide is removed by a stromal protease (vanderVere et al., 1995). Most proteins require further processing, either assembly into multisubunit complexes (step 2_A in Figure 1), insertion into the thylakoid membrane (step 2_I in Figure 1) or translocation across the thylakoid membrane to the thylakoid lumen (step 2_T in Figure 1). Although these are interesting and important processes, they will not be considered further here. Readers should consult the excellent recent review by Cline and Henry (1996) for additional details. Finally, some chloroplastic proteins are targeted to the outer envelope membrane via a different mechanism (process 3 in Figure 1). These proteins lack cleavable transit peptides and are inserted into the outer membrane via a process that does not require ATP (Chen et al., 1997; Fischer et al., 1994; Li et al., 1991; Li and Chen, 1996; Salomon et al., 1990). This is also an interesting and important process, however, it will not be considered further here. Before considering the envelope transport machinery in detail, the features of precursors that interact with this import apparatus will be briefly reviewed.

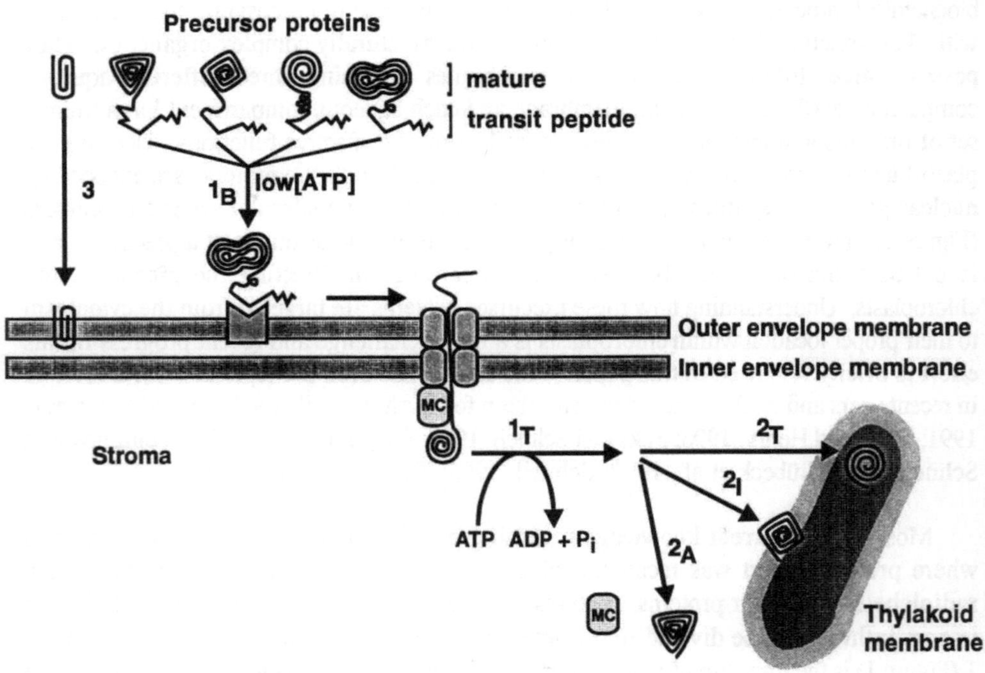

Figure 1. A schematic representation of the various processes involved in targeting cytoplasmically-synthesized precursor proteins into chloroplasts. Transport across the envelope membrane can be divided into two steps, binding of precursors to the chloroplastic surface (step 1_B) and translocation of precursors across the two envelope membranes (step1_T). Translocation is postulated to be driven by the action of some molecular chaperone (MC) that utilizes ATP in the stromal space as a source of energy. Many imported proteins need to be assembled into multisubunit complexes (step 2_A), inserted into the thylakoid membrane (step 2_I) or transported across the envelope membrane to the thylakoid lumen (step 2_T). Many proteins destined for the outer envelope membrane follow a different pathway, designated as process 3. Additional details on the steps process 1 are provided in the text.

2. Precursor Structure

Experiments from several laboratories have demonstrated that transit peptides are both necessary and sufficient for the proper targeting of precursors (Keegstra et al., 1989). Deletion of the transit peptide produces a protein that is not capable of being targeted into chloroplasts. Even more important is the observation that it is possible to redirect foreign proteins into chloroplasts by the addition of a transit peptide.

Efforts to define essential regions within the primary structure of transit peptides have not been successful. Transit peptides have a distinctive amino acid composition, being rich in hydroxylated, basic, and small hydrophobic amino acids. However, sequence comparisons have failed to identify conserved motifs present in all transit peptides (Keegstra et al., 1989; von Heijne et al., 1989). Given this variability among the sequences of transit peptides, it is unlikely that the essential features reside within the primary structure. Rather, it seems likely that the essential features lie within the secondary structure, as has been postulated for mitochondrial presequences (von Heijne et al., 1989). However, to date, the essential secondary or higher order features that determine the functionality of transit peptides have not been determined (von Heijne and Nishikawa, 1991)

One important conclusion regarding transit peptides is that they are responsible for organelle specificity. Because plant cells contain both mitochondria and chloroplasts, the question of how organelle specificity is determined is important. Boutry et al. (1987) examined this question with a series of *in vivo* experiments. They compared the targeting of chimeric precursors containing a chloroplastic transit peptide or a mitochondrial presequence fused to the same passenger protein, bacterial chloramphenicol acetyltransferase (CAT). These chimeric constructs were introduced into plant cells by Ti-mediated transformation. Upon expression of the chimeric genes, the chloroplastic transit sequence directed CAT into chloroplasts, but not mitochondria. On the other hand, the mitochondrial presequence directed CAT to mitochondria, but not chloroplasts. These results provide convincing evidence that organelle specificity resides in the respective targeting sequences.

3. The Import Apparatus and the Mechanism of Protein Translocation

All proteins that enter chloroplasts follow a similar pathway (process 1 in Figure 1). Several lines of evidence support the hypothesis that a single translocation apparatus mediates the transport of all internal proteins. The most compelling evidence comes from competition studies where excess unlabeled precursors block the binding and translocation of different radiolabeled precursors (Cline et al., 1993; Oblong and Lamppa, 1992; Perry et al., 1991). A second system for transport across the envelope membranes may exist, and some evidence points in this direction, but further work will be needed to define and characterize it.

As noted earlier, transport across the envelope membranes can be divided into two discrete stages and each can be studied separately. The first is the binding of precursors to the chloroplastic surface (step 1_B in Figure 1; see also Figure 2). Binding is mediated, at least in part, by a receptor protein, but it is likely that lipid-protein interactions have an important role (Figure 2) (van't Hof et al., 1991; van't Hof et al., 1993). Specific binding can be observed only if translocation is prevented; otherwise specifically bound precursors are transported across the envelope. Inhibition of translocation can be accomplished by limiting the levels of ATP because binding requires approximately 0.1mM ATP whereas translocation requires greater than 1mM ATP. The ATP needed for binding is utilized in the intermembrane space (Olsen and Keegstra, 1992)(Figure 2) whereas the ATP needed for translocation is utilized in the stroma (Theg et al., 1989) (Figure 2). While we postulate that some early forms of bound precursors interact in a reversible manner (see Figure 2), precursors bound in the presence of ATP are irreversibly attached to the surface of chloroplasts and cannot be removed by washing, even in the absence of ATP (Olsen and Keegstra, 1992). This observation, and other arguments, suggest that precursors attached to chloroplasts in the presence of low levels of ATP should be considered as an early translocation intermediate, rather than a bound form. The precise topology of this intermediate is not clear, but one possibility is shown in Figure 2.

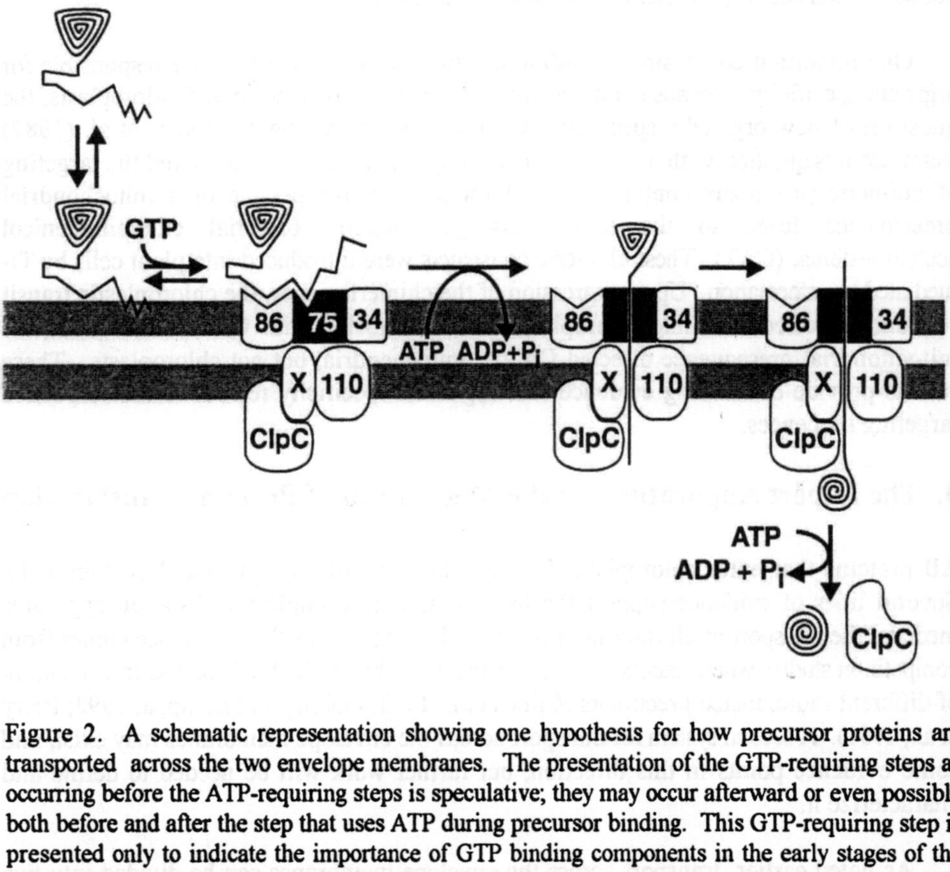

Figure 2. A schematic representation showing one hypothesis for how precursor proteins are transported across the two envelope membranes. The presentation of the GTP-requiring steps as occurring before the ATP-requiring steps is speculative; they may occur afterward or even possibly both before and after the step that uses ATP during precursor binding. This GTP-requiring step is presented only to indicate the importance of GTP binding components in the early stages of the transport process. Additional details about the individual components and the various steps in envelope transport are presented in the text.

Translocation across the envelope membranes is thought to involve unfolding of precursor proteins so that they can pass through a putative translocation channel (Figure 2). It is generally thought that translocation occurs at contact sites where the two envelope membranes are present in close physical proximity (Schnell and Blobel, 1993). Each envelope membrane is thought to have its own translocation apparatus. At present it is uncertain whether they can act independently or whether they must act in concert to accomplish the transport of precursors from the cytoplasm to the stroma. The observation that ATP for translocation is needed inside the organelle is generally interpreted in terms of the action of a molecular chaperone (Figures 1 and 2) which uses the hydrolysis of ATP to undergo repeated cycles of binding and release from the protein being translocated. Evidence that ClpC, a protein in the hsp100 family of molecular chaperones, is part of translocation complexes will be discussed below. However, it remains to be established that this protein functions as the molecular motor driving protein transport into chloroplasts.

Several research groups have recently made progress in identifying the polypeptides that mediate the binding and translocation of precursor proteins across the envelope membranes (see reviews by Gray and Row, 1995; Fuks and Schnell, 1997; Kouranov and Schnell, 1996; Lübeck et al., 1997; Schnell, 1995 for further details). The major conclusions from these studies are summarized schematically in Figure 2. Although additional polypeptides have been identified as putative transport components, the discussion here, and the summary presented in Figure 2, are limited to those components for which cDNA clones have been isolated. Finally, before discussing each component, a brief description of the nomenclature for transport components is needed. At least three different nomenclature systems were being used by the research groups studying this problem. However, all the groups involved have recently agreed upon a uniform nomenclature system (Schnell et al., 1997). Components from the outer envelope membrane are designated Toc followed by a number that indicates the molecular mass of the protein (Toc= Translocon of the outer envelope membrane of chloroplasts). Similarly, components from the inner envelope membrane are designated Tic followed by a number that indicates the molecular mass of the protein (Tic=Translocon of the inner envelope membrane of chloroplasts).

As shown in Figure 2, the outer envelope membrane contains three polypeptides that are thought to be involved in protein translocation: Toc86, Toc75 and Toc34. All three are integral membrane proteins that have no sequence similarity to any of the components in the mitochondrial import apparatus. Toc86 and Toc34 are GTP-binding proteins (Kessler et al., 1994; Seedorf et al., 1995). Toc86 is postulated to be a receptor that interacts initially with precursor proteins (Hirsch et al., 1994) whereas Toc75 is postulated to be a channel protein that allows the translocation of precursors across the outer envelope membrane(Tranel et al., 1995). The function of Toc34 is not known, but it is attractive to speculate that it functions to regulate the interaction of the other outer membrane components. Tic110 is the only inner membrane component for which a cDNA clone has been isolated (Kessler and Blobel, 1996; Lübeck et al., 1996). Again, it is an integral membrane protein with no sequence similarity to any of the components of the mitochondrial import apparatus. Its function in the chloroplastic import apparatus in unknown. Finally, ClpC, a stromal protein in the hsp100 family of molecular chaperones has been identified as a component of translocation complexes. It is postulated to serve as a molecular chaperone, providing the driving force for pulling precursor proteins into chloroplasts. However, it should be emphasized that all of the hypotheses regarding the functions of putative transport components are tentative and more validation is needed before they can be accepted as proven.

Two general strategies have been used to identify and characterize the transport components described above. Different groups have used slightly different variations of each strategy and consequently have obtained slightly different results. The first general strategy is chemical cross-linking; two variations of this strategy have been used. First, Perry and Keegstra (1994) used a radiolabeled, heterobifunctional, cleavable cross-linker to identify Toc86 and Toc75 as polypeptides that are in close physical proximity to bound precursors. Ma et al. (1996) used the same cross-linking reagent and slightly different conditions to confirm these results and to identify proteins in the inner envelope membrane that are in close physical proximity to bound precursors. In both cases, precursors were not cross-linked to either Toc34 or Tic110, indicating that these components may not interact directly with precursor proteins. A second cross-linking strategy, used by Wu et al. (1994) and Akita et al. (1997), is shown schematically in Figure 3. In this case, chloroplasts containing bound precursors are reacted with a reversible homobifunctional cross-linker that reacts both with translocation components and precursors thereby cross-linking them into a complex. This complex can be solubilized with strong detergents and purified. As shown in Figure 3 the complex can be analyzed directly or the cross-linkers cleaved and the individual components analyzed. Analysis of intact complexes can be accomplished by monitoring radioactivity, if radiolabeled precursors are used, or by monitoring the presence of individual components by immunoassays, when antibodies against individual

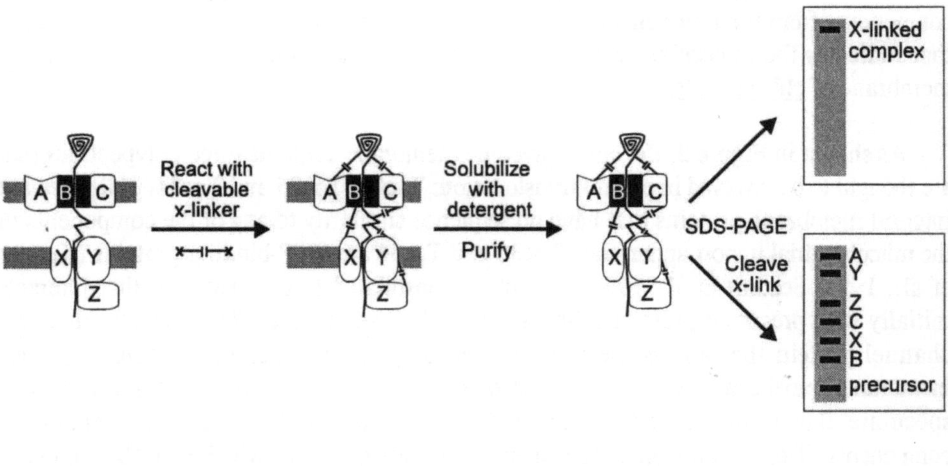

Figure 3. Schematic representation of a cross-linking strategy used to identify translocation complexes and the components contained within them. Additional details of the strategy and the results obtained using this strategy are presented in the text.

components are available. Alternatively, individual components released by cleavage of the cross-linkers can be monitored by immunoassays, when antibodies against these components are available. Using these strategies, Lübeck et al. (1996) and Akita et al. (1997) demonstrated that the translocation complexes contained Toc86, Toc75, Toc34, Tic110 and ClpC. Efforts are underway to purify sufficient quantities of the complexes so that general protein detection methods can be used to detect additional components. Using this strategy and antibodies against a 44 kD envelope membrane protein, Wu et al. (1994) identified Toc44/Tic44. A cDNA clone encoding this component was later isolated by Ko et al. (1995). This component is not shown in Figure 2 because of the ambiguity of whether it resides in the outer or the inner envelope membrane.

The second general strategy used by several groups is to use mild detergents to solubilize translocation complexes. Several variations of this strategy have been used with great success. For example, Schnell et al. (1994) used a chimeric precursor containing protein A and then used immobilized IgG to purify complexes containing several transport components complexed with the chimeric precursor. Nielsen et al. (1997) used the three methods shown in Figure 4 to characterize translocation complexes isolated from

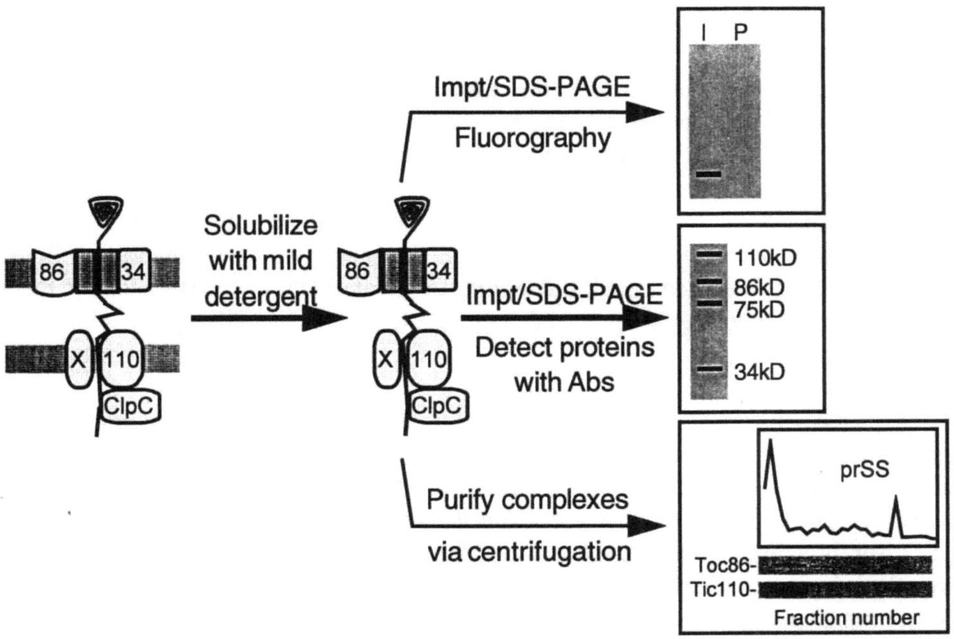

Figure 4. Schematic representation of a strategy used to identify and characterize translocation complexes and the components contained within them. Complexes solubilized using mild detergents have been analyzed by Nielsen et al. (1997) using the three different methods presented here. Additional details of the strategy and the results obtained using this strategy are presented in the text. Abbreviations used in the figure are: Impt=immunoprecipitation; Abs=antibodies

chloroplasts containing precursors bound under conditions of low ATP. First, complexes solubilized with decylmaltoside were subjected to immunoprecipitation with antibodies directed against Toc86, Toc75, Toc34, Tic110 or ClpC. In all cases, antibodies against these putative translocation components caused coimmunoprecipitation of radiolabeled precursors, as depicted schematically in the top panel of Figure 4. Second, complexes immunoprecipitated with antibodies to Toc75 or ClpC were subjected to gel electrophoresis and the presence of other translocation components was measured by immunoblotting. For example, as shown schematically in the middle panel of Figure 4, complexes immunoprecipitated by antibodies to ClpC contained Tic110, Toc86, Toc75 and Toc34. Finally, complexes were analyzed by sucrose gradient centrifugation as shown in the bottom panel of Figure 4. Most of the radioactive precursor and a substantial portion of each putative translocation component remained near the top of the gradient. These presumably represent translocation complexes that dissociated during the solubilization step. However, about 15 to 20% of both the radiolabled precursor and each of the putative translocation components migrated as a large complex near the bottom of the gradient. In addition to radiolabled precursor, the complexes contained Toc86, Toc75, Toc34, Tic110 and ClpC, but did not contain other membrane and soluble proteins not present in translocation complexes.

4. Summary and Prospects

Great progress has been made during the past three years in the identification and partial characterization of polypeptides that comprise the envelope-based translocation machinery. This allows the formulation of detailed working models for the translocation of proteins across the envelope membranes, such as the one shown in Figure 2. However, further progress is needed in several areas. First, it is clear that the translocation machinery contains additional polypeptides, especially in the inner envelope membrane. Some candidates have been identified and efforts to isolate cDNA clones encoding these components are underway in several laboratories. It is likely that other polypeptides still need to be identified.

Even more important is the need to determine the functions of individual polypeptides and to determine how they interact with each other and with the precursor proteins to accomplish the transport of precursors across two biological membranes. For example, while it is widely accepted that the Toc86 functions as a receptor during the recognition of precursor proteins, there is no direct evidence for this hypothesis. If the receptor hypothesis is correct, one prediction is that Toc86 should bind directly to transit peptides. Such measurements have not yet been reported. Another important area of investigation is the role of GTP during protein translocation. The observation that Toc86 and Toc34 are GTP-binding proteins (Hirsch et al., 1994; Kessler et al., 1994; Seedorf et al., 1995) fits well with the observation that nonhydrolyzable GTP analogs can inhibit precursor binding and translocation (Olsen and Keegstra, 1992; Kessler et al., 1994). However, the precise role of GTP during the protein transport remains unclear. Indeed, it is not certain that GTP is needed before ATP, as indicated in Figure 2. Another unproven postulate is that Toc75 forms a channel for the transport of polypeptides across the outer envelope membrane. Efforts to test this hypothesis are currently underway in our laboratory. Still another example is the hypothesis that ClpC functions as a molecular chaperone to pull precursor proteins into chloroplasts. Although ClpC has been identified in translocation complexes, it is not certain whether it provides the driving force for protein translocation. It is not even certain that ClpC interacts directly with precursor proteins, as would be required by this hypothesis. Finally, it is likely that a translocation complex is a dynamic structure, with some components binding to or dissociating from the complex at certain stages of precursor translocation. Our current models are static, with the same components present at all stages of translocation. These static models will certainly need to be revised as our understanding of the protein transport process increases.

5. References

Akita M, Nielsen E, Keegstra K (1997) Identification of protein transport complexes in the chloroplastic envelope membranes via chemical cross-linking. J Cell Biol 136: 983-994

Boutry M, Nagy F, Poulsen C, Aoyagi K, Chua NH (1987) Targeting of bacterial chloramphenicol acetyltransferase to mitochondria in transgenic plants. Nature 328: 340-342

Chen DD, Schnell DJ (1997) Insertion of the 34-kDa chloroplast protein import component, IAP34, into the chloroplast outer membrane is dependent on its intrinsic GTP-binding capacity. J Biol Chem 272: 6614-6620

Cline K, Henry R (1996) Import and routing of nucleus-encoded chloroplast proteins. Ann Rev Cell Biol 12: 1-26

Cline K, Henry R, Li C, Yuan J (1993) Multiple pathways for protein transport into or across the thylakoid membrane. EMBO J 12: 4105-4114

De Boer AD, Weisbeek PJ (1991) Chloroplast protein topogenesis: Import, sorting and assembly. Biochim Biophys Acta Rev Biomembr 1071: 221-253

Fischer K, Weber A, Arbinger B, Brink S, Eckerkorn C, Flügge U (1994) The 24 kDa outer envelope membrane protein from spinach chloroplasts: molecular cloning, in vivo expression and import pathway of a protein with unusual properties. Plant Mol Biol 25: 167-177

Fuks B, Schnell DJ (1997) Mechanism of protein transport across the chloroplast envelope. Plant Physiol 114:405-410

Gray JC, Row PE (1995) Protein translocation across chloroplast envelope membranes. Trends Cell Biol 5: 243-247

Hirsch S, Muckel E, Heemeyer F, von Heijne G, Soll J (1994) A receptor component of the chloroplast protein translocation machinery. Science 266: 1989-1992

Keegstra K, Olsen LJ, Theg SM (1989) Chloroplastic precursors and their transport across the envelope membranes. Ann Rev Plant Physiol Plant Mol Biol 40: 471-501

Kessler F, Blobel G, Patel HA, Schnell DJ (1994) Identification of two GTP-binding proteins in the chloroplast protein import machinery. Science 266: 1035-1039

Kessler F, Blobel G (1996) Interaction of the protein import and folding machineries in the chloroplast. Proc Natl Acad Sci USA 93:7684-7689

Ko K, Budd D, Wu CB, Seibert F, Kourtz L, Ko ZW (1995) Isolation and characterization of a cDNA clone encoding a member of the Com44/Cim44 envelope components of the chloroplast protein import apparatus. J Biol Chem 270: 28601-28608

Kouranov A, Schnell DJ (1996) Protein translocation at the envelope and thylakoid membranes of chloroplasts. J Biol Chem 271: 31009-31012

Li H-m, Moore T, Keegstra K (1991) Targeting of proteins to the outer envelope membrane uses a different pathway than transport into chloroplasts. Plant Cell 3: 709-717

Li H-m, Chen LJ (1996) Protein targeting and integration signal for the chloroplastic outer envelope membrane. Plant Cell 8: 2117-2126

Lübeck J, Soll J, Akita M, Nielsen E, Keegstra K (1996) Topology of IEP110, a component of the chloroplastic protein import machinery present in the inner envelope membrane. EMBO J 15: 4230-4238

Lübeck J, Heins L, Soll J (1997) Protein import into chloroplasts. Physiol Plant 100:53-64

Ma YK, Kouranov A, LaSala SE, Schnell DJ (1996) Two components of the chloroplast protein import apparatus, IAP86 and IAP75, interact with the transit sequence during the recognition and translocation of precursor proteins at the outer envelope. J Cell Biol 134: 315-327

Nielsen E, Akita M, Davila-Aponte J, Keegstra K (1997) Stable association of chloroplastic precursors with protein translocation complexes that contain proteins from both envelope membranes and a stromal Hsp 100 molecular chaperone. EMBO J 16: 935-946

Oblong JE, Lamppa GK (1992) Precursor for the light-harvesting chlorophyll *a/b*-binding protein synthesized in *Escherichia coli* blocks import of the small subunit of ribulose-1,5-bisphosphate carboxylase/oxygenase. J Biol Chem 267: 14328-14334

Olsen LJ, Keegstra K (1992) The binding of precursor proteins to chloroplasts requires nucleoside triphosphates in the intermembrane space. J Biol Chem 267: 433-439

Perry SE, Buvinger WE, Bennett J, Keegstra K (1991) Synthetic analogues of a transit peptide inhibit binding or translocation of chloroplastic precursor proteins. J Biol Chem 266: 11882-11889

Perry SE, Keegstra K (1994) Envelope membrane proteins that interact with chloroplastic precursor proteins. Plant Cell 6: 93-105

Salomon M, Fischer K, Flügge U-I, Soll J (1990) Sequence analysis and protein import studies of an outer chloroplast envelope polypeptide. Proc Natl Acad Sci USA 87: 5778-5782

Schnell DJ (1995) Shedding light on the chloroplast protein import machinery. Cell 83: 521-524

Schnell DJ, Blobel G (1993) Identification of intermediates in the pathway of protein import into chloroplasts and their localization to envelope contact sites. J Cell Biol 120: 103-115

Schnell DJ, Kessler F, Blobel G (1994) Isolation of components of the chloroplast protein import machinery. Science 266: 1007-1012

Schnell DJ, Blobel G, Keegstra K, Kessler F, Ko K, Soll J (1997) A nomenclature for the protein import components of the chloroplast envelope. Trends in Cell Biol 7: in press

Seedorf M, Waegemann K, Soll J (1995) A constituent of the chloroplast import complex represents a new type of GTP-binding protein. Plant J 7: 401-411

Theg SM, Bauerle C, Olsen LJ, Selman BR, Keegstra K (1989) Internal ATP is the only energy requirement for the translocation of precursor proteins across chloroplastic membranes. J Biol Chem 264: 6730-6736
Theg SM, Scott SV (1993) Protein import into chloroplasts. Trends in Cell Biol 3: 186-190
Tranel PJ, Froehlich J, Goyal A, Keegstra K (1995) A component of the chloroplastic protein import apparatus is targeted to the outer envelope membrane via a novel pathway. EMBO J 14: 2436-2446
Van't Hof R, Demel RA, Keegstra K, de Kruijff B (1991) Lipid-peptide interactions between fragments of the transit peptide of ribulose-1,5-bisphosphate carboxylase/oxygenase and chloroplast membrane lipids. FEBS Lett 291: 350-354
Van't Hof R, Van Klompenburg W, Pilon M, Kozubek A, De Korte-Kool G, Demel RA, Weisbeek PJ, de Kruijff B (1993) The transit sequence mediates the specific interaction of the precursor of ferredoxin with chloroplast envelope membrane lipids. J Biol Chem 268: 4037-4042
VanderVere PS, Bennett TM, Oblong JE, Lamppa GK (1995) A chloroplast processing enzyme involved in precursor maturation shares a zinc-binding motif with a recently recognized family of metalloendopeptidases. Proc Natl Acad Sci USA 92: 7177-7181
von Heijne G, Nishikawa K (1991) Chloroplast transit peptides: The perfect random coil. FEBS Lett 278: 1-3
von Heijne G, Steppuhn J, Herrmann R (1989) Domain structure of mitochondrial and chloroplast targeting peptides. Eur J Bioch 180: 535-545
Wu C, Seibert FS, Ko K (1994) Identification of chloroplast envelope proteins in close physical proximity to a partially translocated chimeric precursor protein. J Biol Chem 269: 32264-32271

On the Structure and Function of Plant K⁺ Channels

Rainer Hedrich, Stefan Hoth, Dirk Becker, Ingo Dreyer and Petra Dietrich

Institut für Molekulare Pflanzenphysiologie und Biophysik, Mittlerer Dallenbergweg 64, 97082 Würzburg, Germany

Abstract. K^+ uptake is essential throughout the plant life cycle. Transport and accumulation of this macronutrient in large quantities have been recognized e.g. during germination, cell division and growth, differentiation, movement and reproduction. Since K^+ transport is electrogenic - accompanied by the movement of charges - electrophysiological techniques have been used to unravel the molecular mechanisms of K^+ transporters and ion channels, pumps and carriers in general (Hedrich, 1995). This review will concentrate on the biophysical and molecular biological approaches the results of which have shaped our current picture of the structure and function of plant ion channels. For related studies on carriers and pumps see (Hedrich and Schroeder, 1989; Lohse and Hedrich, 1992; Walker et al., 1996; for review see Chasan and Schroeder, 1992; Maathuis and Sanders, 1992).

Patch clamp studies have identified K^+ uptake and K^+ release channels, two major channel types which are inversely activated by the membrane voltage (Blatt, 1988, 1992; Schroeder, 1988, 1989; for review see Schroeder et al., 1994; Hedrich and Dietrich, 1996). Besides, voltage-independent cation channels with often very broad selectivity were described as well (Hedrich and Neher, 1987; Enz et al., 1993; Schulz-Lessdorf and Hedrich, 1995; Allen and Sanders, 1996). Since for K^+ uptake and K^+ release channels extensive *in vivo*-studies, molecular cloning and functional expression analyses are available this review will focus on these channel types, only.

Keywords: K^+ channel, plant physiology, heterologous expression, cation sensitivity

Plant K^+ Channels Group into Families

K^+ channels have been found to be present in archaebacteria such as *Methanococcus* (Bult et al., 1996), eubacteria like *Streptomyces* (Schrempf et al., 1995), yeasts (Ketchum et al., 1995; Zhou et al., 1995; Lesage et al., 1996a; Reid et al., 1996), plants (Anderson et al., 1992; Sentenac et al., 1992; Müller-Röber et al., 1995; Ketchum and Slayman, 1996; Czempinski et al., 1997) and animals (for review see Warmke and Ganetzky, 1994; Jan and Jan, 1997). Genome sequencing projects have revealed evidence for the existence of novel K^+ channel genes thereby discovering eight K^+ channel families in the nematode *Caenorhabditis elegans* (Wei et al., 1996). From the existence of four different K^+ channel structural classes characterized by either 2, 4, 6 or 8 putative transmembrane domains in ancient bacteria to man one might conclude that survival of

NATO ASI Series, Vol. H 104
Cellular Integration of Signalling Pathways in Plant Development
Edited by F. Lo Schiavo, R. L. Last, G. Morelli, and N. V. Raikhel

organisms required the control of K^+ fluxes and membrane potential (Fig. 1). From the basic structural elements of 2,4 and 6 putative transmembrane domains in yeast and human and possibly in plants too, 8 transmembrane segments seem to evolve from assembly of a K^+ channel with 6 transmembrane domains and one with just two (Fig. 1, see Lesage et al., 1996a; Zhou et al., 1995).

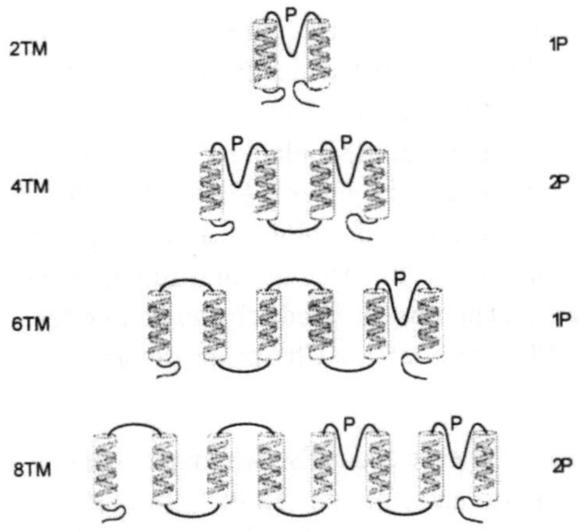

Fig. 1 K^+ channel structures. K^+ channel classes with 2, 4, and 8 transmembrane domains (TM) have been identified. They are characterized by either one or two pore domains (P).

So far the *2-transmembrane-channels* seem to represent voltage-independent K^+ uptake channels known as the 'classical' inward rectifiers (for review see Hille, 1992). Inward rectification - asymmetrical K^+ current in response to hyperpolarization - in these channels is a function of cytoplasmic gating particles such as divalent cations or polyamines (Ficker et al., 1994; Lu and MacKinnon, 1994; for review see Jan and Jan, 1997).The *4-transmembrane-channels* structurally look like two 2-transmembrane channels in tandem. In contrast to the latter the *4-transmembrane channels* are either K^+ release or uptake channels (Lesage et al., 1996b; Fink et al., 1996; Czempinski et al., 1997). ORK1 which is expressed in *Drosophila* neuromuscular tissue, however, appears to be ungated (Goldstein et al., 1996). The *6-transmembrane-channels* found in excitable animal cells comprise the 'classical' outward rectifiers (Hodgkin and Huxley, 1952) first identified in the *Shaker* mutant of *Drosophila melanogaster* (e.g. Papazian et al., 1987; Tempel et al., 1987). Upon depolarization these channels release K^+ and thus repolarize the membrane potential following an action potential. In plants, however, channels with related structure seem to mediate K^+ uptake upon hyperpolarization, only (for review see Hedrich and Becker, 1994; Hedrich and Dietrich, 1996).

From current sequences available the plant K^+ channels group into the two families of inward (K^+_{in}) and outward (K^+_{out}) rectifyiers (Fig. 2). While only recently the first member of the K^+ outward rectifiers has been cloned (Czempinski et al., 1997) the identification and sequence analyses of several K^+ uptake channel genes groups this family into at least four subfamilies (Fig. 2).

After initial patch-clamp studies on guard cells in 1984 (Schroeder et al., 1984) the K^+ uptake channel in these plant motor cells has been cloned from potato in 1995 (KST1 for <u>K</u>$^+$ channel <u>*S*</u>*olanum* <u>*t*</u>*uberosum* 1, Müller-Röber et al., 1995). The molecular identification of kst1 in a guard cell library was based on a homology screen taking advantage of *Shaker*-like sequences found in *Arabidopsis thaliana*(kat1: Anderson et al., 1992; akt1: Sentenac et al., 1992). With these members of two different K^+ channel families (Fig. 2) the door to structure-function, antisense and knock out analyses was thrown open.

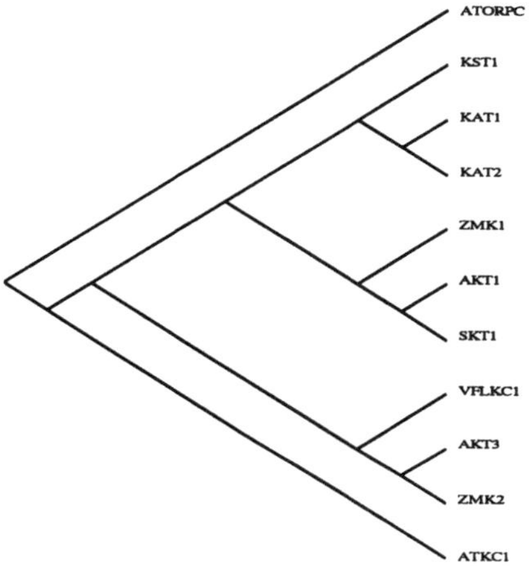

Fig. 2 Evolutionary tree of K^+ channels. The alignment of plant potassium channel sequences was performed using ClustalW (Thompson et al., 1994) and the tree was drawn with the program TreeView (R.D.M. Page, division of environmental and evolutionary biology, University of Glasgow).

K^+ Channel Properties *in vivo* and *in vitro*

Cloning of K^+ channel genes from cell types of specialized physiology such as guard cells (movement), coleoptile cells (growth), phloem cells (sugar loading and unloading) or pollen (polarity and tip growth- fertilization) allowed to relate the K^+ channel properties identified in patch-clamp studies on the 'mother' cells *in vivo* to those of the gene product after heterologous expression *in vitro* (Müller-Röber et al., 1995; Hoth et al., 1997a). This comparison further enabled to distinguish between channel-intrinsic properties and those resulting from plant- or cell-specific regulators present in the 'mother' cells, only.

K^+ uptake channels (Fig. 2) functionally express either in *Xenopus* oocytes (Schachtman et al., 1992; Véry et al., 1994, 1995; Hedrich et al., 1995; Müller-Röber et al., 1995), insect (Sf9, Sf21, Gaymard et al., 1996; Marten et al., 1996; Ehrhardt et al., 1997) or mammalian cell lines (CHO, own unpublished results) or yeast (Bertl et al., 1995). Selecting the expression system depends on three basic requirements:

1. Functional expression of plant channels in general: AKT1 and SKT1 α-subunit homomers so far do not form active channels in *Xenopus* oocytes (for functional coexpression of AKT1 and SKT1, see Dreyer et al., 1997).

2. Low background of endogenous channels (for oocyte-intrinsic channels see Yang and Sachs, 1989; Krause et al., 1996).

3. Sufficient membrane stability to withstand the voltage protocols required to analyse the voltage-dependence of the plant channels properly.

While activation by protons and block by calcium ions already differs between guard cell K^+ channels from different species *in vivo* (Dietrich et al., 1997), comparative studies to the *in vitro* properties revealed that the selectivity and voltage-dependence is conserved between K^+ channels in different cell types, species and expression systems (Hedrich et al., 1995; Müller-Röber et al., 1995; Véry et al., 1995; Becker et al., 1996; Dietrich et al., 1997). The lack of NH_4^+ permeability found with Sf9 cells and yeast (Marten et al., 1996; Bertl et al., 1995) compared to K^+ channels *in vivo* or after heterologous expression in oocytes (Müller-Röber et al., 1995; Becker et al., 1996; Dietrich et al., 1997) might result from posttranslational modifications such as phosphorylation and glycosylation.

Functional Domains

Since the voltage sensor of the K^+ channel which responds to membrane hyperpolarization in K^+ uptake channels should locate within a transmembrane domain the primary sequence was screened for regions with charged residues within the hydrophobic core. In analogy to the *Shaker* channels (Papazian et al., 1991; Perozo et al., 1994; Yusa et al., 1996) mutations in the fourth putative transmembrane segment (S4, Fig. 3) of KST1 and KAT1 strongly affected the voltage-dependence of the corresponding channel (Dreyer et al., 1997; Hoth et al., 1997b). This result is in line with S4 as a dipole moved by the electrical field which thereby changes the size of the permeation pore (Durell and Guy 1992, 1996; Mannuzzu et al., 1996). Although having S4-segments in common it is however still unclear why the 'red' *Shaker* activates upon depolarization and the 'green' *Shaker* (KAT1, KST1) upon hyperpolarization. It is therefore tempting to speculate that besides S4, C- and N-terminal regions make the difference (Marten and Hoshi, 1997).

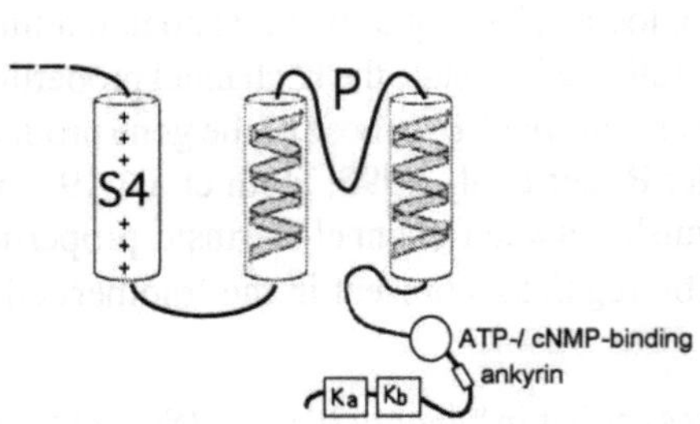

Fig. 3 Functional domains of plant *6-transmembrane K^+ channels*. The transmem-brane domains S4, S5 and S6 are shown with the charged (+) voltage sensor in S4 and the pore region (P). ATP-/cNMP-binding: putative (cyclic) nucleotide-binding domain; ankyrin: ankyrin-binding domain; Kb: basic consensus sequence; Ka: acid consensus sequence.

The selectivity filter is the site of interaction between the potassium ion and blockers with residues lining the permeation pathway, the pore (P). In order to allow the movement of the hydrophilic potassium ions through an hydrophobic membrane protein the pore should be formed by amphiphilic residues. Based on the stuctural model for the *Shaker* channel exhibiting one pore domain (P) between S5 and S6 (Durell and Guy, 1992, 1996) point mutations in the amphiphilic linker between S5 and S6 in KAT1 identified this domain as a major part of the plant K^+ channel pore as well (Fig. 3 and Becker et al., 1996; Uozumi et al., 1995; Nakamura et al., 1997). Mutations in the inner and outer mouth of the pore resulted in a change in selectivity as well as an alteration of the susceptibility towards the ion channel blockers Cs^+ and Ca^{2+} (Fig. 4). When expressed in yeast they showed a strong phenotype with respect to Cs^+- and Ca^{2+}-dependent growth. The K^+ channel mutant carrying a negatively charged residue in the inner pore developed a Ca^{2+} block absent in the KAT1 wild type. In all cases tested the yeast phenotype could be predicted from the electrophysiology and *vice versa* (Becker et al., 1996; Uozumi et al., 1995).

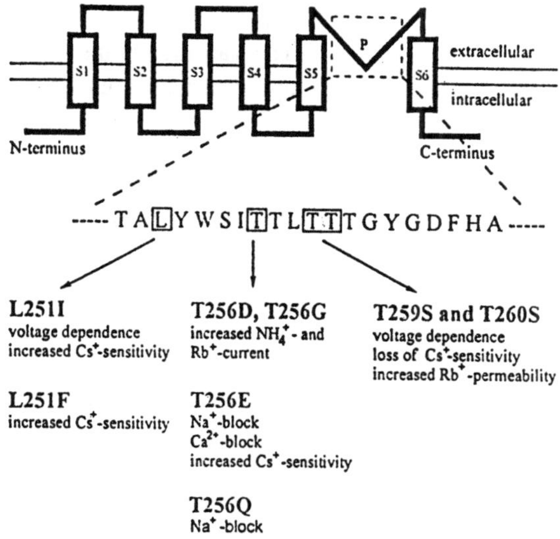

Fig. 4 Structure-function analysis of KAT1 residues. Point mutations in the pore alter cation selectivity and suceptibility to ionic blockers (Hoth et al., 1997b).

pH-sensor: During stomatal movement, coleoptile and pollen growth as well as phloem loading K^+ uptake is always accompanied by changes in external pH in a manner that K^+ uptake is dependent or at least strongly stimulated upon acidification (Weisenseel and Jaffe, 1976; Satter et al., 1988; Cosgrove, 1993). Since in KST1 the pH-sensitivity was most pronounced at pH-changes around 6.2 extracellular histidines were examined in detail (Hoth et al., 1997a; Dietrich et al., 1997). In all plant K^+ uptake channels but not K^+ release channels the motif HXXN was apparent (Fig. 5) pointing to a possible signature sequence for pH-dependent K^+ uptake. Indeed, when this histidine in the outer pore (H271 in KST1, Fig. 5) in addition to the only other histidine in the linker between S2-S3 (H160) was replaced by alanines (A) the activity of this *S. t.* guard cell K^+ channel mutant KST1-H160A/ H271A was pH-insensitive. This fact together with an inversion of the pH-dependence when the histidine was replaced by an arginine (alcaline- instead of acid-activation) provided good evidence for a major role of this residue in the pH-sensing.

Nucleotide-sensor: In the C-terminus of the KAT- and AKT-related channel sequences sites for interaction with ATP or cyclic nucleotides were found (Fig. 3). On the other hand KAT1- and KST1-activity strongly depend on cytoplasmic ATP (Hoshi, 1995; Müller-Röber et al., 1995). The role of this ATP-binding site in KST1 as well as a cNMP-binding domain in AKT-related channels for nucleotide activation now awaits detailed mutational analysis.

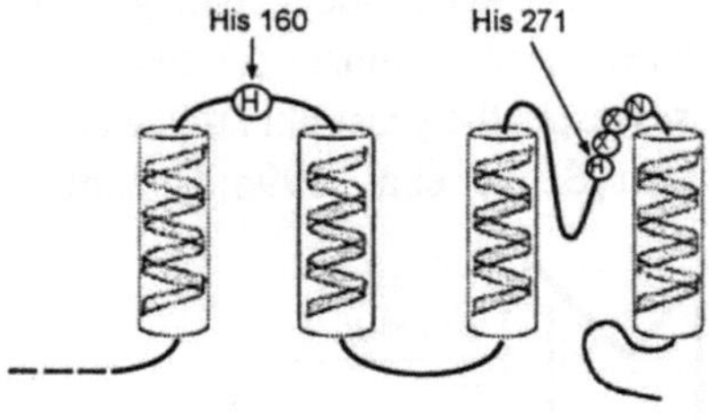

Fig. 5 Putative pH-sensing domains in KST1. The transmembrane segments S3 to S6 and the apparent consensus sequence HXXN are shown.

Heterologous Assembly - a Basis for Functional Diversity

From the biophysical analyses and in analogy to the *Shaker* channels one could propose that the plant K^+ uptake channels functionally assemble as tetramers (Dreyer et al., 1997). Recently, Daram et al. (1997) applying the two-hybrid system to AKT1 found that the cytoplasmic C-terminus alone formed tetrameric structures when expressed in Sf9 cells. For this interaction within AKT1 α-subunits two sides were required, the cNMP-binding domain specific to the AKT-related families (see Fig. 2) and an acidic and basic domain common to all plant K^+ uptake channels (Fig. 4; Daram et al., 1997; Ehrhardt et al., 1997). Furthermore, Dreyer et al. (1997) were able to demonstrate that plant K^+ channel α-subunits from different subfamilies assemble indiscriminately when coinjected into *Xenopus* oocytes. Using α-subunits with different strong phenotypes or creating them by site-directed mutagenesis they could show that in contrast to animal K^+ channels plant α-subunits formed functional heteromers even when originated from different species (Fig. 6 and Dreyer et al., 1997).

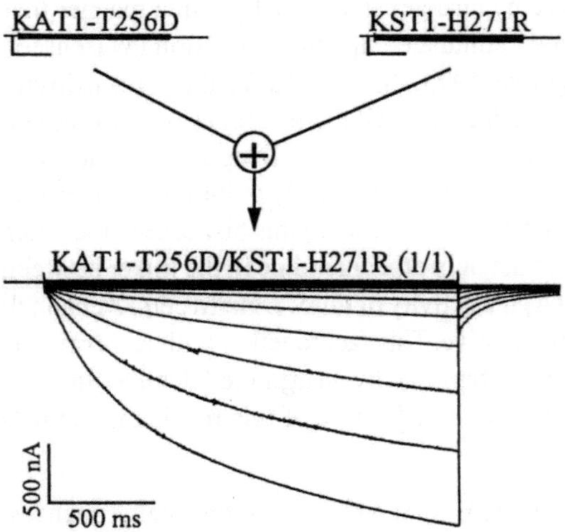

Fig.6 Rescue of the K^+ channel phenotype by coexpression of two silent subunits. The KAT1 mutant KAT1-T256D does not mediate potassium currents (Uozumi et al., 1995) whereas the mutant KST1-H271R is unable to activate at moderate voltages under acidic conditions (Hoth et al., 1997a). Representative current families from oocytes expressing KAT1-T256D, KST1-H271R, and KAT1-T256D/KST1-H271R (1:1 mixture), respectively. Currents were elicited by 2-s voltage steps from +20 mV to -170 mV (10-mV steps) followed by a voltage step to -70 mV from a holding potential of -20 mV.Experiments were performed in 30 mM KCl, pH 5.6.

Outlook

Besides studies on the stucture-function analysis of sites common to K^+ channels and those specific for the 'green' ones cellular expression patterns of plant K^+ channels and their relation to developmental processes are under current investigation. Knowing the cellular localization and developmental pattern the relation between channel distribituion and cell polarity will come to a central question. In epithelia for example cell polarity is directly linked to membrane protein distribution between apical and basolateral sites as a result of immobilization by the cytoskeleton. Within the family of AKT-related channels the ankyrin binding domain might point to a potential site for interaction with the cytoskeleton (see Fig. 3). Despite the absence of the ankyrin binding motif in KAT1, inhibitiors of cytoskeleton dynamics affect K^+ channel activity (Marten and Hoshi, 1997). The molecular mechanism of this interaction has to await further characterization using methods suited to resolve protein-protein interactions like the yeast two-hybrid system or phage display.

Many examples have shown that the electrical properties of a cell membrane depend on the existence of a single predominant channel type or just a few channel types. Thus, ongoing studies aimed on the identification of channel 'knock-out' mutants in transgenic plants very likely will enable to relate phenotypes in for example K^+ uptake efficiency, growth rate, or even plant architecture to the presence or absence of an ion channel.

References

Allen G.J. & Sanders D. 1996. Control of ionic currents in guard cell vacuoles by cytosolic and luminal calcium. *Plant J.* 10, 1055-1069.

Anderson, J. A., Huprikar, S. S., Kochian, L. V., Lucas, W. J. & Gaber, R. F. 1992. Functional expression of a probable *Arabidopsis thaliana* potassium channel in *Saccharomyces cerevisiae*. *Proc. Natl. Acad. Sci. USA* 89, 3736-3740.

Becker, D., Dreyer, I., Hoth, S., Reid, J.D., Busch, H., Lehnen, M., Palme, K. & Hedrich, R. 1996. Changes in voltage-activation, Cs^+ sensitivity, and ion permeability in H5 mutants of a plant K^+ channel, KAT1. *Proc. Natl. Acad. Sci. USA*, 93, 8123-8128.

Bertl, A., Anderson, J. A., Slayman, C. L. & Gaber, R. F. 1995. Use of *Saccharomyces cerevisiae* for patch-clamp analysis of heterologous membrane proteins: Characterization of Kat1, an inward-rectifying K^+ channel from *Arabidopsis thaliana*, and comparison with endogeneous yeast channels and carriers. *Proc. Natl. Acad. Sci. USA* 92, 2701-2705.

Blatt, M. R. 1988. Potassium-dependent, bipolar gating of K^+ channels in guard cells. *J. Membr. Biol.* 102, 235-246.

Blatt, M. R. 1992. K^+ channels of stomatal guard cells / Characteristics of the inward rectifier and its control by pH. *J. Gen. Physiol.* 99, 615-644.

Bult C.J., White O, Olsen G.J., Zhou L., Fleischmann R.D., Sutton G.G., Blake J.A., FitzGerald L.M., Clayton R.A., Gocayne J.D., KerlavageA.R., Dougherty B.A., Tomb J.F., Adams M.D., Reich C.I., Overbeek R., Kirkness E.F, Weinstock K.G., Merrick J.M., Glodek A., Scott J.L., Geoghagen N.S.M., Weidman J.F., Fuhrmann J.L., Venter J.C., et al. 1996. Complete genome sequence of the methanogenic archaeon, *Methanococcus jannaschii*. *Science* 273, 1058-1073.

Chasan, R. and Schroeder, J.I. 1992. Excitation in plant membrane biology. *The Plant Cell*, 1180-1188.

Cosgrove, J. 1993. How do plant cell walls extend? *Plant Physiol.* 102, 1-6.

Czempinski, K., Zimmermann, S., Ehrhardt, T. & Müller-Röber, B. 1997. New structure and function in plant K^+ channels: KCO1, an outward rectifier with a steep Ca^{2+} dependency. *EMBO J.* 16, 2565-2575.

Daram, P., Urbach, S., Gaymard, F., Sentenac, H. & Charel, I. 1997. Tetramerization of the AKT1 plant potassium channel involves its C-terminal cytoplasmic domain. *EMBO J.* 16, 3455-3463.

Dietrich, P., Dreyer, I., & Hedrich. 1997. Cation-sensitivity and kinetics of guard cell potassium channels differ among species-dependent. *Plant J.*, submitted.

Dreyer, I., Antunes, S., Hoshi, T., Müller-Röber, B., Palme, K., Pongs, O., Reintanz, B. & Hedrich, R. 1997. Plant K^+ channel α-subunits assemble indiscriminately. *Biophysical Journal* 72, 2143-2150.

Durell, S. R. & Guy, H. R. 1992. Atomic scale structure and functional models of voltage-gated potassium channels. *Biophys. J.* 62, 238-250.

Durell, S. R. & Guy, H. R. 1996. Structural model of the outer vestibule and selectivity filter of the *Shaker* voltage-gated K^+ channel. *Neuropharmacology* 35, 761-773.

Ehrhardt, T., Zimmermann, S. & Müller-Röber, B. 1997. Association of plant K^+_{in} channels is mediated by conserved C-termini and does not affect subunit assembly. *FEBS Lett.* 409, 166-170.

Enz, C., Steinkamp, T. & Wagner, R. 1993. Ion channels in the thylakoid membrane. *Biochim. Biophys. Acta* 1143, 67-76.

Ficker, E., Taglialatela, M., Wible, B.A., Henley, C.M. & Brown, A.M. 1994. Spermine and spermidine as gating molecules for inward rectifier K^+ channels. *Science* 266, 1068-1072.

Fink, M., Duprat, F., Lesage, F., Reyes, R., Romey, G., Heurteaux, C. & Lazdunski, M. 1996. Cloning, functional expression and brain localization of a novel unconventional outward rectifier K^+ channel. *EMBO J.* 15, 6854-6862.

Gaymard, F., Cerutti, N., Horeau, C., Lemaillet, G., Urbach, S., Ravallec, M., Devauchelle, G., Sentenac, H. & Thibaud, J.B. 1996. The bacculovirus/insect cell system is an alternative to *Xenopus* oocytes. First characterization of the AKT1 K^+ channel from *Arabidopsis thaliana. J. Biol. Chem.* 271, 22863-22870.

Goldstein, S.A., Price, L.A., Rosenthal, D.N. & Pausch, M.H. 1996. ORK1, a potassium-selective leak channel with two pore domains cloned from *Drosophila melanogaster* by expression in *Saccharomyces cerevisiae*. *Proc. Natl. Acad. Sci. USA* 93, 13256-13261.

Hedrich, R. 1995. Technical approaches to studying specific properties of ion channels in plants. In: *Single Channel Recording*, 2nd edition, Sakmann, B. & Neher, E., eds., Ch. 12, 277-305.

Hedrich, R. & Becker, D. 1994. Green circuits - The potential of plant specific ion channels. In: *Plant Molecular Biology*, 26. Ausg. Palme, K. Hrsg., Kluwer Academic Publishers, Dordrecht, Belgium, 1637-1650.

Hedrich, R. & Dietrich, P. 1996. Plant K^+ channels: Similarity and diversity. *Bot. Acta* 109, 94-101.

Hedrich, R. & Neher, E. 1987. Cytoplasmic calcium regulates voltage-dependent ion channels in plant vacuoles. *Nature* 329, 833-836.

Hedrich, R. & Schroeder, J. I. 1989. The physiology of ion channels and electrogenic pumps in higher plants. *Ann. Rev. Plant Physiol.* 40, 539-569.

Hedrich, R., Moran, O., Conti, F., Busch, H., Becker, D., Gambale, F., Dreyer, I., Küch, A., Neuwinger, K. & Palme, K. 1995. Inward rectifier potassium channels in plants differ from their animal counterparts in response to voltage and channel modulators. *Eur. Biophys. J.* 24, 107-115.

Hille, B. 1992. *Ionic channels of excitable membranes*, Sinauer Associates Inc., Sunderland, Massachusetts.

Hodgkin, A. L. & Huxley, A. F. 1952. A quantitative description of membrane current and its application to conduction and excitation in nerve. *J. Physiol.* 117, 500-544.

Hoshi, T. 1995. Regulation of voltage dependence of the KAT1 channel by intracellular factors. *J. Gen. Physiol.* 105, 309-328.

Hoth, S., Dreyer, I., Dietrich, P., Becker, D., Müller-Röber, B. & Hedrich, R. 1997a. Molecular basis of plant-specific acid activation of K^+ uptake channels. *Proc. Natl. Acad. Sci. USA* 94, 4806-4810.

Hoth, S., Dreyer, I. & Hedrich, R. 1997b. Mutational analysis of functional domains within plant K^+ uptake channels. *J. Exp. Bot.* 48 Special Issue, 415-420.

Jan, L.Y. & Jan, Y.N. 1997. Cloned potassium channels from eukaryotes and prokaryotes. *Annu. Rev. Neurosci.* 20, 91-123.

Ketchum, K. A. & Slayman, C. W. 1996. Isolation of an ion channel gene from *Arabidopsis thaliana* using the H5 signature sequence from voltage-dependent K^+ channels. *FEBS Lett.* 378, 19-26.

Ketchum, K. A., Joiner, W. J., Sellers, A. J. Kaczmarek, L. K. & Goldstein, S.A. 1995. A new family of outwardly rectifying potassium channel proteins with two pore domains in tandem. *Nature* 376, 690-695.

Krause, J.D., Foster, C.D. & Reinhart, P. H. 1996. *Xenopus laevis* oocytes contain endogenous large conductance Ca^{2+}-activated K^+ channels. *Neuropharmacology.* 35, 1017-1022.

Lesage, F., Guillemare, E., Fink, M., Duprat, F., Lazdunski, M. Romey, G. & Barhanin, J. 1996a. A pH-sensitive yeast outward rectifier K^+ channel with two pore domains and novel gating properties. *J. Biol. Chem.* 271, 4183-4187.

Lesage, F., Guillemare, E., Fink, M., Duprat, F., Lazdunski, M., Romey, G. & Barhanin, J. 1996b. TWIK-1, a ubiquitous human weakly inward rectifying K^+ channel with a novel structure. *EMBO J.* 15, 1004-10011.

Lohse, G. & Hedrich, R. 1992. Characterization of the plasmamembrane H^+ATPase from *Vicia faba* guard cells. *Planta* 188, 206-214.

Lu, Z. & MacKinnon, R. 1994. Electrostatic tuning of Mg^{2+} affinity in an inward-rectifier K^+ channel. *Nature* 371, 243-246.

Maathuis, F.J.M. & Sanders, D. 1992. Plant membrane transport. *Current Opinion in Cell Biology* 4, 661-669.

Mannuzzu, L. M., Moronne M. M., & Isacoff E. Y, 1996. Direct physical measure of conformational rearrangement underlying potassium channel gating. *Science* 271, 213-216.

Marten, I. & Hoshi, T. 1997. Voltage-dependent gating characteristics of the K^+ channel KAT1 depend on the N and C termini. *Proc. Natl. Acad. Sci.* USA 94, 3448-3453.

Marten, I., Gaymard, F., Lemaillet, G., Thibaud, J.-B., Sentenac, H., & Hedrich, R. 1996. Cytoplasmic Ca^{2+} and nucleotides do not affect the voltage-dependent activity of KAT1, *FEBS Lett.* 380, 229-232.

Müller-Röber, B., Ellenberg, J., Provart, N., Willmitzer, L., Busch, H., Becker, D., Dietrich, P., Hoth, S. & Hedrich, R. 1995. Cloning and electrophysiological analysis of KST1, an inward-rectifying K^+ channel expressed in potato guard cells. *EMBO J.* 14, 2409-2416.

Nakamura, R.L., Anderson, J.A. & Gaber R.F. 1997. Determination of key structural requirements of a K^+ channel pore. *J. Biol. Chem.* 272, 1011-1018.

Papazian, D. M., Schwarz, T. L., Tempel, B. L., Jan, Y. N. & Jan, L. Y. 1987. Cloning of genomic and complementary DNA from *Shaker*, a putative potassium channel gene from *Drosophila. Science* 237, 749-753.

Papazian, D. M., Timpe, L. C., Jan, Y. N. & Jan, L. Y. 1991. Alteration of voltage-dependence of *Shaker* potassium channel by mutations in the S4 sequence. *Nature* 349, 305-310.

Perozo, E., Santacruz-Toloza, L., Stafani, E., Bezanilla, F. & Papazian, D. M. 1994. S4 mutations alter gating currents of *Shaker* K^+ channels. *Biophys. J.* 66, 345-354.

Reid, J. D., Lukas, W., Shafaatian, R., Bertl, A., Scheurmann-Kettner, C., Guy, H. R. & North, R. A. 1996. The *S. cerevisiae* outwardly-rectifying potassium channel (DUK1) identifies a new family of channels with duplicated pore domains. *Recept. Channels* 4, 51-62.

Satter, R.L., Morse, M.J., Lee, Y., Crain, R.C. Coté, G.G. & Moran, N. 1988. Light- and clock-controlled leaflet movements in *Samanea saman*: A physiological, biophysical and biochemical analysis. *Bot. Acta* 101, 205-213.

Schachtman, D.P., Schroeder, J.I., Lucas, W.J., Anderson, J.A. & Gaber R.F. 1992. Expression of an inward-rectifying potassium channel by the *Arabidopsis* KAT1 cDNA. *Science* 258, 1645-1658.

Schrempf, H., Schmidt, O., Kümmerlen, R., Hinnah, S., Müller, D., Betzler, M., Steinkamp, T. & Wagner, R. 1995. A procaryotic potassium ion channel with two predicted transmembrane segments from *Streptomyces lividans. EMBO J.* 14, 5170-5178.

Schroeder, J. I. 1988. K^+ transport properties of K^+ channels in the plasma membrane of *Vicia faba* guard cells. *J. Gen. Physiol.* 92, 667-683.

Schroeder, J. I. 1989. Qantitative analysis of outward rectifying K^+ channel currents in guard cell protoplasts from *Vicia faba. J. Membr. Biol.* 107, 229-235.

Schroeder, J.I., Hedrich, R. & Fernandez, J.M. 1984. Potassium-selective single channels in guard cell protoplasts of *Vicia faba. Nature* 312, 361-362.

Schroeder, J. I., Ward, J. M. & Gassmann, W. 1994. Perspectives on the physiology and structure of inward-rectifying K^+ channels in higher plants: Biophysical implications for K^+ uptake. *Annu. Rev. Biophys. Biomol. Struct.* 23, 441-4471.

Schulz-Lessdorf, B. & Hedrich, R. 1995. Protons and calcium modulate SV-type channels in the vacuolar-lysosomal compartment - channel interaction with calmodulin inhibitors. *Planta* 197, 655-671.

Sentenac, H., Bonneaud, N., Minet, M., Lacroute, F., Salmon, J. -M., Gaymard, F. & Grignon, C. 1992. Cloning and expression in yeast of a plant potassium ion transport system. *Science* 256, 663-665.

Tempel, B., Papazian, D. M., Schwarz, T. L., Jan, Y. N. & Jan, L. Y. 1987. Sequence of a probable potassium channel component encoded at the *Shaker* locus of *Drosophila*. *Science* 237, 770-775.

Thompson, J.D., Higgins, D.G. & Gibson, T.J. 1994. CLUSTAL W: improving the sensitivity of progressive multiple sequence alignment through sequence weighting, position-specific gap penalties and weight matrix choice. *Nucleic Acids Res.* 22, 4673-4680.

Uozumi, N., Gassmann, W., Cao, Y.& Schroeder, J.I. Identification of strong modifications in cation selectivity in an Arabidopsis inward rectifying potassium channel by mutant selection in yeast. *Biol. Chem.* 270, 24276-24281.

Véry, A.-A., Bosseux, C., Gaymard, F., Sentenac, H. & Thibaud, J. -B. 1994. Level of expression in *Xenopus* oocytes affects some characteristics of a plant inward-rectifying voltage-gated K^+ channel. *Pflügers Arch.* 428, 422-424.

Véry, A.-A., Gaymard, F., Bosseux, C., Sentenac, H. & Thibaud, J. -B. 1995. Expression of a cloned plant K^+ channel in *Xenopus* oocytes: analysis of macroscopic currents. *Plant J.* 7, 321-332.

Walker N.A., Sanders D., & Maathuis F.J. 1996. High-affinity potassium uptake in plants. *Science* 273, 977-979.

Warmke, J. W. & Ganetzky, B. 1994. A family of potassium channel genes related to *eag* in *Drosophila* and mammals. *Proc. Natl. Acad. Sci. USA* 91, 3438-3442.

Wei, A., Jegla, T. & Salkoff, L. 1996. Eight potassium channel families revealed by the C. elegans genome project. *Neuropharmacology* 35, 805-829.

Weisenseel, M.H. & Jaffe, L.F. 1976. Natural H^+ currents traverse growing roots and root hairs of barley (*Hordeum vulgare* L.). *Planta* 133, 1-7.

Yang, X.-C. & Sachs, F. 1989. Block of stretch-activated ion channels in *Xenopus laevis* oocytes by gadolinium and calcium ions. *Science* 243, 1068-1071.

Yusa, S.P., Wray, D.& Sivaprasadarao, A. 1996. Measurement of the movement of the S4 segment during the activation of a voltage-gated potassium channel. *Pflügers Arch.* 433, 91-97.

Zhou, X. L., Vaillant, B., Loukin, S. H., Kung C. & Saimi Y. 1995. YKC1 encodes the depolarization-activated K^+ channel in the plasma membrane of yeast. *FEBS Lett.* 373, 170-176.

Integration of Ion Channel Activity in Calcium Signalling Pathways

Dale Sanders, Gethyn J Allen, Shelagh R Muir and Stephen K Roberts

The Plant Laboratory, Biology Department, University of York, PO Box 373, York YO1 5YW, UK

Abstract. Perception of a wide range of developmental and stress signals by plants results in rapid elevation of cytosolic free calcium ($[Ca^{2+}]_c$: Bush, 1995). The change in $[Ca^{2+}]_c$ is widely accepted to comprise an early step in signal transduction, with downstream targets of the Ca^{2+} signal including activation of calmodulin-dependent enzymes, calmodulin-domain protein kinases (CDPKs: Roberts & Harmon, 1992), or activation of ion channels (Schroeder & Hagiwara, 1989). Yet this simple notion gives rise to a number of equally elementary questions:

- From which compartment is Ca^{2+} mobilised?
- Which membrane pathways (ion channels) facilitate passive Ca^{2+} flow into the cytosol in response to the primary signal?
- How do these Ca^{2+}-permeable channels interact with other cellular response elements (including other ion channels, ligands and phosphorylation cascades) to evoke the Ca^{2+} signal?
- How is stimulus specificity encoded in the Ca^{2+} signal?

Answers to all four questions require as a first step an understanding of the properties of Ca^{2+}-permeable channels in plant cells: which membranes the channels are located in, what activates (gates) them, and how their activities might be integrated with other signalling pathways in the cell. The remainder of this chapter highlights areas of achievement and of ignorance in our attempts to address these questions.

Keywords: calcium channel; patch clamp; calcium-induced Ca^{2+} release; inositol 1,4,5-trisphosphate; cyclic ADP-ribose; phytochrome; anion channel

1 The Source of Ca^{2+}

Pioneering work by Schumaker & Sze (1987), Ranjeva et al. (1988) and Alexandre et al. (1990) demonstrated that Ca^{2+} release channels gated by inositol 1,4,5-trisphosphate ($InsP_3$) reside in the vacuolar membrane of plants. These studies led to a growing appreciation that the vacuole - as quantitatively the most significant intracellular Ca^{2+} store in most cell types - might play a role in Ca^{2+} mobilisation during signal transduction processes (Johannes et al., 1992). In principle, a role for the vacuole in intracellular Ca^{2+} mobilisation adds an extra dimension to plant cell signalling which is not embraced by the majority of animal cell models where the ER is thought to comprise the intracellular source of Ca^{2+} during signalling. Such arguments do not, of course, exclude a role for plant ER or for extracellular Ca^{2+} as the origin of cytosolic Ca^{2+} signals. Indeed recent studies on Ca^{2+} signalling in pollen

NATO ASI Series, Vol. H 104
Cellular Integration of Signalling Pathways in Plant Development
Edited by F. Lo Schiavo, R. L. Last, G. Morelli, and N. V. Raikhel

tubes have pointed to a non-vacuolar (possibly ER) $InsP_3$-labile pool (Franklin-Tong et al., 1996), and membrane fractionation of cauliflower florets has yielded a similar conclusion (Muir & Sanders, 1997).

In practice, dissecting the contributions of the variety of possible Ca^{2+} sources for a given Ca^{2+} signal is not a simple one. While the continuation of a Ca^{2+} signal in the absence of external Ca^{2+} (e.g. after complexation of Ca^{2+} with EGTA) is a clear indication that the signal must be reliant on intracellular stores, the converse is not necessarily the case. Thus, although abolition of a Ca^{2+} signal by EGTA implies a contribution of extracellular Ca^{2+}, it is possible, by analogy with some classes of signalling in animal cells (Gillot & Whitaker, 1993; Shen & Larter, 1995), that this Ca^{2+}-dependence is required only for initiating the process of Ca^{2+} release from intracellular stores.

An outstandingly imaginative approach to the problem of the origins of Ca^{2+} signals was developed by Marc Knight and colleagues (Knight et al., 1996). Using membrane-targeted aequorin to report on the free Ca^{2+} in the microdomain around specific membranes, the dynamics of the Ca^{2+} signal in that microdomain can be compared with that in the soluble cytosolic phase. Using this approach, the cytosolic Ca^{2+} transient after cold shock has been concluded to arise from a joint contribution of Ca^{2+} entry from outside the cell and release from the vacuole. Conversely, the touch-induced Ca^{2+} transient emanates from a separate source, possibly though mobilisation of ER stores (MR Knight, personal commune: see also Klüsener et al., 1995). An early and important conclusion from this work is that it is likely that a number of Ca^{2+} sources - both intracellular and extracellular - will be drawn upon during signalling, and that these sources might be signal-specific. This is supported by separate studies on the response of the *Fucus* rhizoid to osmotic stress. There an initial Ca^{2+} influx from outside the cell is likely to be augmented by Ca^{2+} release from internal stores.

2 Calcium-Permeable Ion Channels

Until the advent of single channel recording techniques, virtually nothing was known concerning the properties of plant Ca^{2+} channels. This absence of information can be understood in terms of the overall contribution of prospective Ca^{2+} channels to the electrical properties of vacuolar or plasma membranes: the contribution of Ca^{2+} channels is likely to be low, even when they are activated during Ca^{2+}-based signalling events. This point can be appreciated from two distinct quantitative perspectives: the macroscopic (whole cell/whole vacuole), and the microscopic (single channel).

First, at the macroscopic level, let us take the case of global elevation of $[Ca^{2+}]_c$ by 2 μM during signalling. This comprises an increase in total cytosolic Ca^{2+} of 2 mM, assuming a buffer capacity of 99.9% (a value intermediate between those derived for animal cells (Thomas, 1982) and for a fungus (Miller et al., 1990). For a 30 s rise time and a cytosolic thickness of 2 μm, the $[Ca^{2+}]_c$ change converts to a membrane flux of 13 $pmol \cdot cm^{-2} \cdot s^{-1}$, or a membrane current of 1.3 $\mu Amp \cdot cm^{-2}$. This current is very small indeed in comparison with the currents of several tens or even hundreds of $\mu Amp \cdot cm^{-2}$ which normally flow across the membranes of intact plant cells (e.g.

Blatt 1987; Ermolayeva et al., 1997). At a qualitative level, the limited magnitude of any Ca^{2+} current can be understood in the context of the role of Ca^{2+} as a signalling ion, whereas currents through K^+ channels (which dominate membrane electrical properties) must often provide for major sustained changes in intracellular K^+ concentration of the order of tens or hundreds of mM.

Second, at the microscopic level, and for the same $[Ca^{2+}]_c$ change as that considered above, we can ask about the likely density of channels required to generate the $[Ca^{2+}]_c$ signal. The turnover time for a typical channel is of the order 10^7 ions·s^{-1}, or $1.7 \cdot 10^{-17}$ mol·s^{-1}. This compares with the flux of Ca^{2+} estimated above as $1.3 \cdot 10^{-19}$ mol·µm^{-2}·s^{-1}, and therefore implies a channel density of only one Ca^{2+} channel every 100 µm^2 of membrane, even assuming that the channel resides in its open state 100% of the time. Since membrane patches used for patch clamp recordings typically have an area of between 1 and 5 µm^2, this calculation, if taken at face value, implies that attempts at patch clamp recordings of single Ca^{2+} channels would be futile.

However, recordings are possible, at both macroscopic and microscopic levels. At a macroscopic level, it is possible in patch clamp experiments to reduce the contribution of other channels to the overall membrane current by appropriate choice of ionic conditions and the use of inhibitors. At a microscopic level, several studies have demonstrated a significant K^+ permeability through plant Ca^{2+}-permeable channels (Johannes et al., 1992; Ward & Schroeder, 1994; White, 1994; Piñeros & Tester, 1995; Allen & Sanders, 1996). Detailed analysis of permeability properties suggests that the Ca^{2+} flux through such channels accounts for only a few per cent of the overall current in physiological conditions with mixture of K^+ and Ca^{2+} (Gradmann et al., 1997; G.J. Allen, D. Sanders & D. Gradmann, unpublished data). These findings, which to date relate only to vacuolar channels, reconcile the relatively higher channel density than would be predicted based on assumptions of pure Ca^{2+} selectivity because the Ca^{2+} currents through them are really very small indeed . A similar conclusion regarding the limited fractional contribution of Ca^{2+} to the overall ionic current through Ca^{2+} permeable channels has also been reached in animal cells through more direct measurements of Ca^{2+} fluxes (Mayer et al., 1987; Schneggenburger et al., 1993).

2.1 Calcium-permeable channels at the plasma membrane

Several studies, using both patch clamp and planar lipid bilayer approaches, have revealed the presence of channels with significant degrees of Ca^{2+} permeability in roots (e.g. White, 1994; Piñeros & Tester, 1995; Roberts & Tester, 1997), cultured cells (Thuleau et al., 1994), marine algal cells (Taylor et al., 1996) and epidermal cells (Ding & Pickard, 1993). [These studies are discussed in critical detail by Piñeros & Tester (1997) to whom the reader referred for a more comprehensive description of channel properties.] With the exception of epidermal cells (where the channels are stretch-activated), all channels appear to be gated by membrane depolarisation. This finding implies that activation of other, depolarising channels in the membrane must comprise an earlier stage in the response to the primary signal. A more detailed account of the integration of channel behaviour at the plasma membrane is presented below.

2.2 Calcium-permeable channels at the vacuolar membrane

At least four distinct classes of Ca^{2+}-permeable channel have been reported at the vacuolar membrane. These channels can be distinguished through consideration of their gating factors. Two classes of channel are gated by ligands: these are respectively $InsP_3$ (Alexandre et al., 1990; Allen & Sanders, 1994a) and cyclic ADP-ribose (cADPR: Allen et al., 1995; Muir & Sanders, 1996). In the latter case, gating is likely by analogy with animal channels to involve indirect association between the channel and the ligand. Membrane voltage gates the other two classes of channel: the vacuolar voltage-gated Ca^{2+} (VVCa) channel is activated by membrane hyperpolarisation (Johannes et al., 1992; Johannes & Sanders, 1995), while the slowly activating vacuolar (SV) channel is activated by membrane depolarisation (Ward & Schroeder, 1994; Allen & Sanders, 1995). More detailed properties of all four classes of channel are discussed by Allen & Sanders (1997). The manner in which the channels might interact during Ca^{2+} signalling is discussed in more detail below.

2.3 Calcium-permeable channels at the ER

The discovery of voltage-gated Ca^{2+}-permeable channels in the ER of tendrils from *Bryonica dioica* by Klüsener et al. (1995) has served to emphasize the probable role of the ER in Ca^{2+}-based signalling in plants. In many respects (apparent Ca^{2+} selectivity, single channel conductance, voltage-dependence and pharmacology), this channel resembles the VVCa channel of broad bean guard cell vacuoles (Allen & Sanders, 1994b). However, until molecular details emerge for these channels, further speculation on their identity is not possible.

3 Interaction of Calcium-permeable Channels with other Cellular Response Elements: Two Case Studies

3.1 Vacuolar Ca^{2+} release through the SV channel

In almost all respects, the $InsP_3$-gated Ca^{2+} channel of plant vacuoles strongly resembles its animal counterpart at the ER: specificity and affinity for $InsP_3$, high affinity block by low M_r heparin, inhibition by TMB-8, and binding of ATP (Schumaker & Sze, 1987; Ranjeva et al., 1988; Brosnan & Sanders, 1990; 1993; Muir et al., 1997). Likewise, the cADPR-activated Ca^{2+} release pathway in the same membrane displays many of the properties of animal ryanodine receptors, including similar EC_{50} and specificity for cADPR and pharmacological properties (Allen et al., 1995; Muir & Sanders, 1996). Despite these points of commonality, one highly significant difference has emerged between the plant and animal channels: while the animal channels are activated by low (sub-μM) concentrations of Ca^{2+} on the cytosolic side, neither of the plant channel-types is similarly-activated. The significance of this activation by $[Ca^{2+}]_c$ is profound, since the resultant Ca^{2+}-induced Ca^{2+} release (CICR) provides not only amplification but also highly characteristic dynamics to Ca^{2+} signals in animal systems. The question therefore arises of whether plants cells display CICR, and, if so, how this might occur.

Although the SV channel had long been known to be Ca^{2+}-activated (Hedrich & Neher, 1987), its significance as a Ca^{2+} release channel has only relatively recently become apparent once a Ca^{2+} permeability had been established. Thus, Ward & Schroeder (1994), working with guard cell vacuoles, proposed that CICR might be executed by the SV channel after dual activation mechanisms by $[Ca^{2+}]_c$. In a "chemical" mechanism, Ca^{2+} in the low or sub-μM range activates, probably via calmodulin (Bethke & Jones, 1994) . In an "electrical" mechanism, Ward and Schroeder (1994) proposed that Ca^{2+}-activation of a highly selective K^+ channel (the VK channel) in the same membrane could serve to depolarise the membrane and bring the voltage-gated SV channel into a responsive range.

This is an attractive hypothesis, but there are nevertheless some issues which remain to be resolved. First, the SV channel activates at cytosol-positive potentials, and whether the equilibrium potential for K^+ is sufficiently positive to facilitate opening has not yet been established. Second, while the SV channel is ubiquitous among plant vacuoles (Hedrich et al., 1988), the VK channel has not been observed in systems other than guard cell vacuoles. A recent attempt to observe VK channels in the storage root of red beet met with failure (T. Jelitto & D. Sanders, unpublished data). Thus, in systems other than guard cell vacuoles, it is not easy to envisage how a positive swing in membrane potential would be generated away from the slightly negative values which are thought to pertain in the steady-state as a result of activity of the electrogenic H^+ pumps.

For cases in which Ca^{2+} is released from the vacuole, a further interesting difference between intracellular Ca^{2+} mobilisation in plants and animals must relate to the vast reservoir of Ca^{2+} which is stored in the vacuolar lumen. Free Ca^{2+} can attain mM levels there, and taking into account also the physical dominance of the vacuole in mature cells, it is inconceivable that vacuolar Ca^{2+} stores could become significantly depleted during signalling. This consideration has several implications. First, mechanisms for regulation of Ca^{2+} mobilisation which invoke control by luminal Ca^{2+} levels (Taylor & Marshall, 1992) cannot apply to plant vacuolar Ca^{2+} release channels. Second, for cases in which vacuolar Ca^{2+} release comprises the dominant means for elevation of $[Ca^{2+}]_c$, there is no requirement to invoke a role for so-called store-operated channels at the plasma membrane (which have been proposed in animal cells to respond to the size of the intracellular pool and to be involved in refilling it). Third, Ca^{2+} release across the vacuolar membrane must be acutely controlled, since simple emptying of the vacuolar Ca^{2+} pool would have severe (lethal) consequences for the cell.

The requirement for control is particularly acute if CICR operates at the vacuolar membrane, and one might anticipate some kind of negative feedback regulation on the SV channel, perhaps sensing $[Ca^{2+}]_c$. It seems possible that these regulatory requirements are met, at least in part, by protein phosphatase 2B (PP2B). This phosphatase is Ca^{2+}-activated and, in the presence of calmodulin, exerts strong downregulation of SV channel activity (Allen & Sanders, 1995). Of course, in order to obtain CICR via the SV channel, it must be proposed that the Ca^{2+} sensitivity of PP2B is lower than that of the channel itself, or that there is a delay after the start of the Ca^{2+} signal before PP2B activity becomes sufficient to have an effect on the channel.

3.2 Plasma membrane Ca^{2+} entry, cation and anion channels in phytochrome signalling

The early stages in many signalling processes involve plasma membrane depolarisation. Ward et al. (1995) have suggested that membrane depolarisation might play a fundamental role in activating voltage-gated Ca^{2+} channels which have been documented by several studies at plasma membrane (see discussion above on channel types). It seems possible that the initial depolarisation is provided by activation of anion channels (Ward et al., 1995).

Testing hypotheses on interactions of plant ion channels in the early stages of signalling is not easy. In patch clamp experiments, it is important that the voltage transient in response to the primary signal is preserved, and this is rarely the case, either because the ionic conditions normally chosen to maximise ion channel currents reflect poorly the conditions pertaining *in vivo*, or because essential intracellular control factors have been diluted out of the system by equilibration of pipet contents. Ideally, then, it would be advantageous to record ionic currents in intact cells using conventional impalement techniques. There are few cell types with permit this approach either, because intercellular connections via plasmadesmata provide pathways for current flow spread throughout a tissue in an ill-defined and non-quantifiable way. Notable exceptions to this statement are guard cells (which do not possess stomata) and root hairs (in which current flow along the projecting hair can be directly monitored and passage of current to adjoining cells is minimal (Lew 1991; Meharg et al., 1994). Another further system which we have recently developed for impalement studies is the moss *Physcomitrella patens*. The caulonemal cells have a defined, linear geometry in which the pattern of current flow both along a cell and between cells can be defined and corrected for.

In a phytochrome-mediated response, caulonemal cells flashed with red light develop side branch initials (Cove & Knight, 1993). Phytochrome signalling pathways are well-documented to involve elevation of $[Ca^{2+}]_c$ (Shacklock et al., 1992; Bowler et al., 1994). Red light also induces a dramatic but transient membrane depolarisation in the caulonemal cells of *P. patens* of around 90 mV (Ermolayeva et al., 1996). The electrical response occurs within seconds of the light flash and appears to be central to the signal transduction pathway. Thus, if the red light treatment is performed in the presence of channel blockers which inhibit the depolarisation, side branch initial formation is similarly inhibited (Ermolayeva et al., 1997).

An involvement of Ca^{2+}-permeable channels in the response is suggested by the observation that the depolarisation is blocked in Ca^{2+}-free media (Ermolayeva et al., 1996). Current-voltage analysis has confirmed and extended this suggestion by demonstrating that at the peak of the depolarisation the membrane exhibits an increased permeability not only to Ca^{2+} but also to K^+ and Cl^- (Ermolayeva et al., 1997). Direct measurements of net ionic fluxes shows red light-induced movement of these ions in accord with the prevailing electrochemical potential differences across the plasma membrane: Ca^{2+} moves into the cells, while K^+ and Cl^- leave.

By what sequence of events does ion channel opening facilitate the depolarisation? The Ward et al. model - at least at its simplest level - predicts that the depolarisation should still occur in Ca^{2+}-free media because the inward current should be provided by opening of anion channels. However the block of the depolarisation in Ca^{2+}-free solutions militates against that model. An alternative might be that Ca^{2+} entry provides the stimulus for opening of anion channels. Calcium-activated anion channels are ubiquitous in plant plasma membranes (Tyerman, 1992), and their opening provides the principal depolarising swing in the action potential of Charophytes (Beilby, 1984). The effective blockage of the depolarisation by the anion channel blocker niflumic acid is consistent with this notion.

The depolarisation is also blocked by the classical K^+ channel blocker tetraethylammonium (TEA). This finding is at first sight surprising, since E_K lies well negative of the depolarising peak, and K^+ channels would normally be thought of as contributing to membrane *re*polarisation. However, the observation of TEA inhibition can be reconciled with the scheme in which Ca^{2+} entry provides the stimulus for anion channel opening if it is proposed that K^+ channels have sufficient Ca^{2+} permeability to provide the principal pathway for initial Ca^{2+} uptake. It is therefore of considerable interest that patch clamp experiments have revealed the presence of a Ca^{2+} -permeable K^+ outward rectifying channel in the plasma membrane of *P. patens* (Johannes et al., 1997).

Even so, as already stated, the depolarisation itself seems to be central to the stimulus-transduction pathway, and it therefore seems likely that although a small amount of Ca^{2+} entry provides the stimulus for anion channel opening, it is the resultant depolarisation which in turn provides the trigger for the bulk of the Ca^{2+} entry. The anion channels therefore play a central role in amplifying the initial Ca^{2+} signal in a process akin to CICR at internal membranes.

4 Encoding Stimulus-Specificity

The basic problem of how each of the wide array of signals, which appear to be transduced via Ca^{2+}-dependent pathways, can encode its own specific response remains unsolved. However, some of the properties of Ca^{2+} which have emerged over the past few years and which are discussed above suggest solutions to this problem.

The expectation is that the variety of Ca^{2+}-permeable channels which appear to reside in most cell-types are placed to encode specificity by virtue of (1) their location and (2) their dynamic properties. Spatio-temporal aspects of Ca^{2+} signalling in plants is only just beginning to be addressed, but even some of the earliest non-imaging studies have revealed large differences in the amplitude and duration of Ca^{2+} signals in response to different stimuli (e.g. Knight et al., 1991). The expectation is that by drawing on different pools and by using different Ca^{2+} channels which access those pools, plant cells are endowed with an inherent flexibility in the Ca^{2+} signal which retains specificity on response.

Of course, Ca^{2+} signalling pathways do not operate in isolation, but as part of wider signalling networks involving other soluble messengers and protein phosphorylation. There are emerging hints of how such pathways interact and the level of Ca^{2+} and cyclic nucleotides (Bowler et al., 1994), and Ca^{2+} and phosphorylation (Allen & Sanders, 1995). Expanding upon and understanding the significance of these links will be a major research task for the years ahead.

References

Alexandre J, Lassalles JP & Kado RT (1990) Opening of Ca^{2+} channels in isolated red beet root vacuole membrane by inositol 1,4,5-trisphosphate. Nature 343: 567-570

Allen GJ, Muir SR & Sanders D (1995) Release of Ca^{2+} from individual plant vacuoles by both insp(3) and cyclic ADP-ribose. Science 268: 735-737

Allen GJ & Sanders D (1994a) Osmotic-stress enhances the competence of beta-vulgaris vacuoles to respond to inositol 1,4,5-trisphosphate. Plant J. 6: 687-695

Allen GJ & Sanders D (1994b) Voltage-gated, calcium-release channels coreside in the vacuolar membrane of broad bean guard-cells. Plant Cell 6: 685-694

Allen GJ & Sanders D (1995) Calcineurin, a type 2B protein phosphatase, modulates the Ca^{2+} permeable slow vacuolar ion-channel of stomatal guard-cells. Plant Cell 7: 1473-1483

Allen GJ & Sanders D (1996) Control of ionic currents in guard cell vacuoles by cytosolic and luminal calcium. Plant J 10: 1055-1067

Allen GJ & Sanders D (1997) Vacuolar ion channels of higher plants. Adv. Bot. Res. 25: 217-252

Beilby MJ (1984) Calcium and plant action-potentials. Plant Cell Environ. 7: 415-421

Bethke PC & Jones RL (1994) Ca^{2+}-calmodulin modulates ion-channel activity in storage protein vacuoles of barley aleurone cells. Plant Cell 6: 277-285

Blatt MR (1987) Electrical characteristics of stomatal guard cells: the contribution of ATP-dependent, "electrogenic" transport revealed by current-voltage and different-current-voltage analysis. J. Membrane Biol. 102: 235-246

Bowler C, Neuhaus G, Yamagata H & Chua N-H (1994) Cyclic-GMP and calcium mediate phytochrome phototransduction. Cell 77: 73-81

Brosnan JM & Sanders D (1990) Inositol trisphosphate-mediated calcium release in beet microsomes is inhibited by heparin. FEBS Letts. 260: 70-72

Brosnan JM & Sanders D (1993) Identification and characterization of high-affinity binding sites for inositol trisphosphate in red beet. Plant Cell 5: 931-940

Bush DS (1995) Calcium regulation in plant cells and its role in signalling. Annu. Rev. Plant Physiol. Plant Mol. Biol. 46: 95-122

Cove DJ & Knight CD (1993) The moss physcomitrella-patens, a model system with potential for the study of plant reproduction. Plant Cell 5: 1483-1488

Ding JP & Pickard BG (1993) Mechanosensory calcium-selective cation channels in epidermal cells. Plant J. 3: 83-110

Ermolayeva E, Hohmeyer H, Johannes E & Sanders D (1996) Calcium-dependent membrane depolarization activated by phytochrome in the moss *Physcomitrella patens*. Planta 199: 352-358

Ermolayeva E, Sanders D & Johannes E (1997) Ionic mechanism and role of phytochrome-mediated membrane depolarisation in caulonemal side branch initial formation in the moss *Physcomitrella patens*. Planta 201:109-118

Franklin-Tong VE, Drøbak BK, Allan AC, Watkins PAC & Trewavas AJ (1996) Growth of pollen tubes of *Papaver rhoeas* is regulated by a slow-moving calcium wave propagated by inositol 1,4,5-trisphosphate. Plant Cell 8: 1305-1321

Gillot I & Whitaker M (1993) Imaging calcium waves in eggs and embryos. J. Exp. Biol. 184: 213-219

Gradmann D, Johannes E & Hansen U-P (1997) Kinetic analysis of Ca^{2+}/ K^{+} selectivity of an ion channel by single-binding-site models. J. Membrane Biol. in press

Hedrich R, Barbier-Brygoo H, Felle H, Flügge UI, Lüttge U, Maathuis FJM, Marx S, Prins HBA, Raschke K, Schnabl H, Schroeder JI, Struve I, Taiz L & Ziegler P (1988) General mechanisms for solute transport across the tonoplast of plant vacuoles: a patch-clamp survey of ion channels and proton pumps. Botanica Acta 101: 7-13

Hedrich R & Neher E (1987) Cytoplasmic calcium regulates voltage dependent ion channels in plant vacuoles. Nature 329: 833-836

Johannes E, Brosnan JM & Sanders D (1992) Parallel pathways for intracellular Ca^{2+} release from the vacuole of higher plants. Plant J. 2: 97-102

Johannes E, Ermolayeva E & Sanders D (1997) Red light-induced membrane potential transients in the moss *Physcomitrella patens*: ion channel interaction in phytochrome signalling. J. Exp. Bot. (special issue) 48: 599-608

Johannes E & Sanders D (1995) Lumenal calcium modulates unitary conductance and gating of an endomembrane calcium release channel. J. Membr. Biol. 146: 211-224

Klüsener B, Boheim G, Liss H, Engelberth J & Weiler EW (1995) Gadolinium-sensitive, voltage-dependent calcium-release channels in the endoplasmic-reticulum of a higher plant mechanoreceptor organ. EMBO J 14: 2708-2714

Knight H, Trewavas AJ & Knight MR (1996) Cold calcium signalling in *Arabidopsis* involves 2 cellular pools and a change in calcium signature after acclimation. Plant Cell 8: 489-503

Knight MR, Campbell AK, Smith SM & Trewavas AJ (1991) Transgenic plant aequorin reports the effects of touch and cold-shock and elicitors on cytoplasmic calcium. Nature 352: 524-526

Lew RR (1991) Electrogenic transport-properties of growing *Arabidopsis* root hairs - the plasma-membrane proton pump and potassium channels. Plant Physiol. 97: 1527-1534

Mayer ML, MacDermott AB, Westbrook GL, Smith SJ & Barker JL (1987) Agonist-gated and voltage-gated calcium entry in cultured mouse spinal-cord neurons under voltage clamp measured using arsenazo-III. J. Neurosci. 7: 3230-3244

Meharg AA, Maurousset & Blatt MR (1994) Cable correction of membrane currents recorded from root hairs of *Arabidopsis thaliana* L. J. Exp. Bot. 45: 1-6

Miller AJ, Vogg G & Sanders D (1990) Cytosolic calcium homeostasis in fungi - roles of plasma membrane transport and intracellular sequestration of calcium. Proc. Natl. Acad. Sci. USA 87: 9348-9352

Muir SR, Bewell M, Sanders D & Allen GJ (1997) Ligand-gated Ca^{2+} channels and Ca^{2+} signalling in higher plants. J. Exp. Bot. (special issue) 48: 589-597

Muir SR & Sanders D (1996) Pharmacology of Ca^{2+} release from red beet plants. FEBS Letts. 395: 39-42

Muir SR & Sanders D (1997) Inositol 1,4,5-trisphosphate-sensitive Ca^{2+} release across nonvacuolar membranes in cauliflower. Plant Physiol. in press

Piñeros M & Tester M (1997) Calcium channels in higher plant cells: selectivity, regulation and pharmacology. J. Exp. Bot. (special issue) 48: 551-577

Piñeros M & Tester M (1995) Characterization of a voltage-dependent Ca^{2+} -selective channel from wheat roots. Planta 195: 478-488

Ranjeva R, Carrasco A & Boudet AM (1988) Inositol trisphosphate stimulates the release of calcium from intact vacuoles from *Acer* cells. FEBS Letts. 230: 137-141

Roberts DM & Harmon AC (1992) Calcium-modulated proteins: Targets of intracellular calcium signals in higher plants. Annu. Rev. Plant Physiol. Plant Mol. Biol. 43: 375-414

Roberts SK & Tester M (1997) Permeation of Ca^{2+} and monovalent cations through an outwardly rectifying channel in maize root stelar cells. J. Exp. Bot. 48: 839-846

Schneggenburger R, Zhou Z, Konnerth A & Neher E (1993) Fractional contribution of calcium to the cation current through glutamate-receptor channels. Neuron 11: 133-143

Schroeder JI & Hagiwara S (1989) Cytosolic Ca^{2+} regulates ion channels in the plasma membrane of *Vicia faba* guard cells. Nature 338: 427-430

Schumaker KS & Sze H (1987) Inositol 1,4,5-trisphosphate releases Ca^{2+} from vacuolar membrane vesicles of oat roots. J. Biol. Chem. 262: 3944-3946

Shacklock PS, Read ND & Trewavas AJ (1992) Cytosolic free calcium mediates red light-induced photomorphogenesis. Nature 358: 753-755

Shen P & Larter R (1995) Chaos in intracellular Ca^{2+} oscillations in a new model for non-excitable cells. Cell Calcium 17: 225-232

Taylor AR, Manison NFH, Fernandez C, Wood J & Brownlee C (1996) Spatial organization of calcium signalling involved in cell volume control in the *Fucus* rhizoid. Plant Cell 8: 2015-2031

Taylor CW & Marshall ICB (1992) Calcium and inositol 1,4,5-trisphosphate receptors: a complex relationship. TIBS 17: 403-407

Thomas MV (1982) Techniques in calcium research. London, Academic Press

Thuleau P, Ward JM, Ranjeva R & Schroeder JI (1994) Recruitment of plasma membrane voltage-dependent calcium-permeable channels in carrot cells. EMBO J. 13: 2970-2975

Tyerman SD (1992) Anion channels in plants. Annu. Rev. Plant Physiol. Plant Mol. Biol. 43: 351-373

Ward JM, Pei Z-M & Schroeder JI (1995) Roles of ion channels in initiation of signal transduction in higher plants. Plant Cell 7: 833-844

Ward JM & Schroeder JI (1994) Calcium-activated K^+ channels and calcium-induced calcium release by slow vacuolar channels in guard cell vacuoles implicated in the control of stomatal closure. Plant Cell 6: 669-683

White PJ (1994) Characterization of a voltage-dependent cation-channel from the plasma membrane of rye (*Secale cereale* L.) roots in planar lipid bilayers. Planta 193: 186-193

Protein Kinase and Phosphatase Regulation During Abscisic Acid Signaling and Ion Channel Regulation in Guard Cells

Julian I. Schroeder, Martin Schwarz and Zhen-Ming Pei

Department of Biology and Center for Molecular Genetics, University of California at San Diego, La Jolla, CA 92093-0116, USA

Abstract. The regulation of gas exchange between leaves and the atmosphere is controlled by stomatal pore apertures. Water vapor diffusion through stomatal pores is responsible for transpirational water loss of plants and therefore is detrimental to plants during drought stress (Mansfield et al., 1990). Plants adapt to changes in environmental conditions by adjusting the aperture of stomatal pores in the leaf epidermis. Guard cells, which in pairs surround stomatal pores, determine the stomatal pore aperture. Stomatal closing is mediated by potassium and anion efflux from guard cells and parallel malate metabolism (Raschke, 1979; MacRobbie, 1981). Environmental factors such as CO_2 concentrations, humidity, temperature, light quality and intensity and the plant hormone abscisic acid affect stomatal movements. Abscisic acid (ABA), is synthesized in response to drought, and triggers the signaling cascade in guard cells leading to stomatal closing.

Guard cells provide a well-suited system for characterizing the physiological functions of ion channels in higher plants and for unraveling very early signal transduction events. Several major classes of ion channels were identified by patch clamp studies of the plasma membrane of *Vicia faba* guard cells including outward- and inward-conducting K^+ channels and anion channels (for reviews; Schroeder and Hedrich, 1989; Hetherington and Quatrano, 1991; Assmann, 1993; Ward et al., 1995). In addition, non-selective ABA-activated Ca^{2+}-permeable ion channels (Schroeder, 1990) and several differing types of single stretch-activated channels (Cosgrove and Hedrich, 1991) were found.

In a proposed model for roles of guard cell ion channels, stomatal opening and stomatal closing are accompanied by regulation of several plasma membrane ion channels and proton pumps in parallel (Schroeder and Hedrich, 1989). Stomatal opening requires K^+ uptake (Raschke, 1979). Based on biophysical, pharmacological and cell physiological studies inward-rectifying K^+ channels may provide a major pathway for enabling K^+ uptake during stomatal opening (Schroeder et al., 1987). The driving force for channel-mediated K^+ uptake can be generated by the plasma membrane proton pump, which is activated by light signals and inactivated by ABA (Assmann et al., 1985; Shimazaki et al., 1986; Goh et al., 1996). Calcium is known to reduce stomatal opening (DeSilva et al., 1985). In correlation to this physiological response, elevation in the cytosolic Ca^{2+} concentration inhibits both plasma membrane proton pump activities (Kinoshita et al., 1995) and inward-rectifying K^+ channels (Schroeder and Hagiwara, 1989). This review will focus mainly on new findings pertaining to ion channel regulation during ABA-induced stomatal closing and to the proposed underlying signal transduction mechanisms.

Key Words: ABA, anion channel, stomata, Ca^{2+} channel, *abi*

NATO ASI Series, Vol. H 104
Cellular Integration of Signalling Pathways in Plant Development
Edited by F. Lo Schiavo, R. L. Last, G. Morelli, and N. V. Raikhel

1 S-type Anion Channels and Stomatal Movements

Stomatal closing requires release of solutes from guard cells. Studies have indicated that anion channels in the plasma membrane of guard cells may play a key role in the chain of events which produces stomatal closure (Schroeder and Hagiwara, 1989; Hedrich et al., 1990). One type of anion channel in guard cells shows slow and sustained activation properties (Schroeder and Hagiwara, 1989; Linder and Raschke, 1992; Schroeder, and Keller, 1992). When these "slow" or "S-type" anion channels are activated, the resulting sustained efflux of anions is proposed to result in long-term depolarization of guard cells. Abscisic acid has been shown to cause long-term depolarizations in guard cells (Kusamo, 1981; Ishikawa et al., 1983; Thiel et al., 1992). In addition another type of anion channel has been identified, which shows rapid and transient activation (named R-Type) (Hedrich et al., 1990) . These rapid anion channels can also produce depolarization, mediated by anion efflux. Depolarization by S-type and/or R-type anion channels in turn can activate outward-rectifying potassium channels, proposed to mediate K^+ - efflux required for stomatal closing (Schroeder et al., 1987). The resulting simultaneous efflux of potassium and anions would significantly lower the turgor and volume of guard cells, which leads to closure of stomatal pores (Raschke, 1979; MacRobbie, 1981; Schroeder and Hedrich, 1989).

Compelling experimental support for the proposed central function of slow anion channels during stomatal closing has been derived from the sustained activation properties allowing sustained ion release, cytosolic regulation by Ca^{2+} and phosphorylation events, pharmacological studies, genetic studies of ABA-insensitive mutants, ABA activation of S-type currents and from stomatal movement analyses (e.g. Schroeder and Hagiwara, 1989; Linder and Raschke, 1992; Schroeder and Keller, 1992; Schwartz et al., 1995; Pei et al., 1997). Interestingly, slow anion channels also appear to constitute an important control mechanism for stomatal opening. Down-regulation of slow anion channels enhances light-induced stomatal opening and partially reverses ABA inhibition of stomatal opening (Schroeder et al., 1993; Schmidt et al., 1995; Schwartz et al., 1995). These results can be explained by the depolarization that is caused by anion channels which would impede stomatal opening. Taken together these data point to a model in which activation of slow anion channels in guard cells is crucial for stomatal closing and down-regulation of anion channels contributes to stomatal opening. In the following we discuss new insights into the regulation of guard cell ion channels and stomatal movements by phosphorylation/dephosphorylation events.

2 S-type Channel Regulation by Phosphorylation Events

The weak voltage dependence of S-type anion channels (Schroeder and Hagiwara, 1989) suggests the need for a large degree in regulation of their activity during stomatal movements. Accordingly in *Vicia faba*, both elevation in cytosolic Ca^{2+} and intracellular conditions which can induce protein phosphorylation have been shown to strongly activate S-type anion channel activity; while dephosphorylating conditions strongly down-regulate channel activity (Schroeder and Hagiwara, 1989;

Schmidt et al., 1995). In *Vicia faba* guard cells, depletion of the intracellular ATP pool leads to inactivation of S-type anion currents, which cannot be restored by GTP and other non-hydrolyzable nucleotides including AMP-PNP or GTP-γ-S (Schmidt et al., 1995). These data showed that slow anion currents require hydrolyzable ATP for activation suggesting phosphorylation events in their activation. This hypothesis was supported by data showing that the protein kinase inhibitors K252a and H7 abolished slow anion channel activity in the presence of excess cytoplasmic ATP. Furthermore, ABA-induced stomatal closing was abolished by the same protein kinase inhibitors (Schmidt et al., 1995; Esser et al., 1997), providing evidence for an important role of active protein kinases in ABA signaling. Biochemical studies have recently demonstrated a rapid enhancement of a 48 kDa protein kinase activity by ABA in *Vicia faba* guard cells (Li and Assmann, 1996; Mori and Muto, 1997), supporting such models. Whether the described protein kinase activities are involved in S-type channel regulation during ABA signaling remains to be determined.

A recent study has shown that, the R-type anion channels are not regulated by phosphorylation events (Schultz-Lessdorf et al., 1996). Also R-type activation occurs at acidic cytosolic pH (Schultz-Lessdorf et al., 1996), which contrasts ABA-induced cytosolic alkalization (Irving et al., 1992; Blatt and Armstrong, 1993). In spite of these findings indicating insensitivity of R-type channels to phosphorylation events, it has been discussed that the properties of R-type anion channels still point to a possible activation during ABA-induced stomatal closing and that these ion channels could provide a partial physiological contribution to ABA-induced stomatal closing, in parallel to S-type channels (Schroeder and Keller, 1992; Schroeder et al., 1993). Complete block of R-type anion channels by DIDS, did not affect ABA-induced stomatal closing, indicating a rate-limiting requirement for S-type anion channels in ABA-induced stomatal closing, but logistically not excluding a contribution of R-type channels (Schroeder et al., 1993). In particular the enhancement of R-type currents by extracellular $CaCl_2$ and by depolarization indicates that these anion channels could be activated during ABA signaling (Hedrich et al., 1990; Schroeder and Keller, 1992). Furthermore, it was initially demonstrated in studies on tobacco protoplasts, that R-type anion channels may be converted to S-type anion channels (Zimmermann et al., 1994). The question whether ABA also regulates R-type channels and whether cytosolic Ca^{2+}, depolarization or other unknown signals play a role in their activation will require further investigation. At present, the close correlations between ABA signal transduction in guard cells and diverse S-type anion channel properties, including ABA activation (Pei et al., 1997; Schwarz and Schroeder, Submitted), together demonstrate that S-type anion channel activation is a rate-limiting step in the ABA signaling cascade.

3 Multiple Protein Phosphatases Affect ABA Signaling

The finding that the *Arabidopsis* gene *ABI1* (*ABA-insensitive* locus) encodes for a protein phosphatase type 2C (PP2C) provided first evidence that protein dephosphorylation is crucial for ABA signaling (Meyer et al., 1994; Leung et al., 1994). Further evidence for dephosphorylation events showed that the protein phosphatase inhibitor okadaic acid (OA) maintained slow anion current activity even in the absence of intracellular ATP (Schmidt et al., 1995) and that ABA-induced stomatal closing is enhanced by the protein phosphatase inhibitor okadaic acid in *Vicia faba* and *Commelina communis* (Schmidt et al., 1995; Esser et al., 1997).

Interestingly, okadaic acid enhances stomatal closing only when ABA is also present, while okadaic acid alone showed no strong effects when applied to pre-opened stomata (Schmidt et al., 1995). Similar findings were made for the partial stomatal closing response by extracellular malate, which was also enhanced by co-application of okadaic acid, but okadaic acid alone did not close stomata (Esser et al., 1997).

These data together suggest that ABA signaling and S-type anion channel regulation in Vicia faba are mediated by protein kinase activation and may also include inhibition of okadaic acid-sensitive protein phosphatases. The question whether ABA inhibits protein phosphatases was further addressed in a recent study. The inactivation of S-type channels by removal of cytosolic ATP was found to be reversible, when ATP was reintroduced into the cytosol of these guard cells as expected for a kinase mediated activation (Schwarz and Schroeder, Submitted). In a previous study, inactivation of S-type channels by ATP removal was suggested to be mediated by protein phosphatases, because this response can be inhibited by okadaic acid (Schmidt et al., 1995). In further experiments it was found that also ABA can inhibit S-type anion channel inactivation in Vicia faba guard cells with zero cytosolic ATP (Schwarz and Schroeder, Submitted). Because in the absence of cytosolic ATP, kinase activation could not account for the ABA response, these data lead to a model in which ABA inhibits an okadaic acid-sensitive protein phosphatase. These data indicate that ABA simultaneously activates protein kinases and inhibits protein phosphatases during signal transduction in Vicia guard cells.

The identification of the *ABI1* and *ABI2* genes as PP2Cs suggests that protein phosphatases are involved in ABA signaling (Meyer et al., 1994; Leung et al., 1994; Leung et al., 1997). Note that based on both the primary structure of PP2Cs and biochemical studies with the ABI1 protein, *ABI1* is not sensitive to okadaic acid (Bertauche et al., 1996) (M. Leube and E. Grill, personal communication). Therefore, these data indicate that at least two distinct protein phosphatases function in ABA signaling, okadaic insensitive PP2Cs (Leung et al., 1994; Meyer et al., 1994; Leung et al., 1997) and okadaic acid sensitive protein phosphatases (Schmidt et al, 1995; Schwarz and Schroeder, Submitted). The recent findings that *ABI2* also encodes a protein phosphatase 2C (Leung et al., 1997) (E. Grill personal communication) suggests that multiple protein phosphatases affect ABA signaling. The question whether ABA activates, inhibits or does not modulate the ABI1 and ABI2 PP2C phosphatases requires further analysis.

Similarly it is possible that multiple protein kinases are activated by ABA. Both Li and Assmann (1996) and Mori and Muto (1997) showed ABA and Ca^{2+}-independent activation of a 48 kDa protein kinase activity. In addition, data suggested that upstream of the 48 kDa kinase, ABA may activate a Ca^{2+}-dependent protein kinase, leading to the suggestion that a kinase cascade is activated in response to ABA in *Vicia faba* guard cells (Mori and Muto, 1997).

4 Regulation of K^+ Channels by Phosphorylation Events

The substrates of the recently revealed ABA-dependent protein kinase and phosphatase events remain to be determined. It is possible the S-type anion channels represent a direct target of protein kinases and phosphatases, although this has not yet been analyzed. Furthermore, transgenic expression of the *abi1-1* gene in tobacco also affects K^+ channel regulation, in a manner consistent with the stomatal phenotype (Armstrong et al., 1995). Therefore protein phosphatases and kinases regulate multiple ion channels in parallel in guard cells, giving rise to the parallel activation (and inactivation) of multiple transporters leading to stomatal closing.

In addition, the protein phosphatase inhibitor okadaic acid also inhibits inward-rectifying K^+ channels in *Vicia faba* guard cells (Li et al., 1994; Thiel and Blatt, 1994). Whether this okadaic acid effect is related to ABA signaling in *Vicia faba* guard cells has not yet been reported. Nevertheless, the inactivation of K^+_{in} channel currents by okadaic acid (Li et al., 1994; Thiel and Blatt, 1994), appears to correlate with the enhancement of ABA-induced stomatal closing by okadaic acid (Schmidt et al., 1995). While K^+_{in} channels have not been directly implicated in mediating stomatal closing, reduction in K^+_{in} channel currents would reduce stomatal opening, which is also an effect of ABA. [Note that inhibition of outward-rectifying K^+ channels by okadaic acid in *Vicia faba* guard cells has also been reported (Thiel and Blatt 1994) (but see Li et al., 1994). Inhibition of K^+_{out} channels by okadaic acid does not correlate to the enhancement of ABA-induced stomatal closing by okadaic acid and might therefore represent a different pathway in K^+_{out} channel regulation.]

It is possible that other protein phosphatases, than okadaic acid sensitive phosphatases, activate K^+_{in} channels. Immuno-suppressants, which act as inhibitors of calcineurin-type phosphatases (PP2Bs) have the opposite effect of okadaic acid and enhance K^+_{in} currents (Luan et al., 1993), leading to a model in which calcineurin phosphatases would counter-act the okadaic acid-sensitive phosphatases in regulating K^+_{in} channels (Luan et al., 1993; Li et al., 1994; Thiel and Blatt, 1994). Because calcineurins are okadaic acid-insensitive and because calcineurin effects would counteract those of okadaic acid discussed above, the proposed calcineurins would need to be distinct from the ABA-down-regulated (okadaic acid-sensitive) protein phosphatases characterized in S-type channel studies (Schmidt et al., 1995; Schwarz and Schroeder, Submitted). Calcineurin phosphatases and/or the targets of immuno-suppressants have not yet been identified in plants, which should further insight into this putative mode of K^+_{in} channel regulation.

5 Species Specific Differences in Guard Cell Ion Channel Regulation

Recent studies have shown that differences in the regulation of guard cell ion channels by phosphorylation events exist when comparing plant species. The underlying reasons for this variation remain unknown and could be based on species-specific differences in the signal transduction cascades or emphasis on different elements of identical cascades in different species (Pei et al., 1997). The most striking species-dependent differences in guard cell ion channel regulation found to date are related to regulation of K^+ and anion channels by phosphorylation events: For example, in *Vicia faba* guard cells the inward-rectifying K^+ channels, are down-regulated by the protein phosphatase inhibitor okadaic acid (Li et al., 1994; Thiel and Blatt, 1994), as discussed above. On the other hand, in tobacco guard cells K^+_{in} channel activities are enhanced by the phosphatase inhibitor okadaic acid (Armstrong et al., 1995).

A recently characterized example of differences in guard cell signaling, that is consistent with the above differences in K^+_{in} channel regulation, is the regulation of S-type anion channels (Schmidt et al., 1995; Pei et al., 1997). In *Vicia faba* guard cells, S-type anion channels are strongly activated by phosphorylation events and require hydrolysable ATP for activation as discussed above (Schmidt et al., 1995). S-type channel activity was inhibited by protein kinase inhibitors and maintained without ATP in the presence of protein phosphatase inhibitors or ABA in *Vicia faba* (Schmidt et al, 1995; Schwarz and Schroeder, Submitted).

On the other hand, S-type anion channels in *Arabidopsis* guard cells are activated by dephosphorylation events and protein phosphatase inhibitors inhibit ABA activation of these anion channels in *Arabidopsis* (Pei et al., 1997). Further evidence for differences in signaling specificity between *Vicia faba* and *Arabidopsis* guard cells is demonstrated by the fact that ABA regulation of stomatal movements *in vivo* correlates closely to anion channel regulation in the respective species showing opposite regulation by phosphorylation modulators and dephosphorylation modulators in the two species (Schmidt et al, 1995; Esser et al., 1997; Pei et al., 1997). Thus okadaic acid partially inhibits ABA-induced stomatal closing in *Arabidopsis* (Pei et al., 1997). These data demonstrate that the revealed differences are not dependent on electrophysiological experiments, and correlate closely to the *in vivo* responses. Further detailed analysis of *Arabidopsis* guard cell signaling has led to the suggestion that protein kinases act as a major negative regulator of ABA signaling in *Arabidopsis* (Pei et al., 1997). For example, the kinase inhibitor K252a restores ABA-induced stomatal closing and ABA activation of S-type anion channels in guard cells of the ABA insensitive mutant *abi1-1* (Pei et al., 1997).

Two possible models to explain the substantial differences in ABA signaling in *Vicia faba* and *Arabidopsis* guard cells have been proposed by Pei et al. (1997): (1) Distinct differences in the ABA signaling cascades may exist when comparing *Arabidopsis* and *Vicia faba*, or (2) the same signaling cascades could emphasize different rate-limiting components in the two species. The second hypothesis would necessitate a network of multiple kinases and phosphatases in ABA signaling as implicated above.

During stomatal closing ABA enhances the activity of K^+_{out} channels (Blatt and Armstrong, 1993). Together with the depolarization provided by anion channels, solute efflux from guard cells is enabled, which is required for stomatal closing. In tobacco guard cells, kinase and phosphatase modulators have been shown to modulate these outward K^+ channels (Armstrong et al., 1995). In tobacco guard cells kinase inhibitors stimulate K^+_{out} channels and enhance stomatal closing (Armstrong et al., 1995), which cannot be compared to the ABA activation of protein kinases in guard cells and inhibition of ABA-induced stomatal closing by kinase inhibitors found in other species (*Vicia faba* and *Commelina*). [Note in this context that a model (Li and Assmann, 1996) suggesting that the demonstrated enhancement of outward-rectifying K^+ (K^+_{out}) channel activities by protein kinase inhibitors in tobacco (Armstrong et al., 1995) can be explained by the recent evidence for ABA-activated protein kinase activities in *Vicia faba* (Li and Assmann, 1996), would not fit into physiological models of K^+_{out} channel function. ABA-induced stomatal closing in tobacco could not proceed if ABA-activated kinases inhibit K^+_{out}

channels in tobacco, based on present models of K^+_{out} channel function]. Significant differences among species again may account for this difference between tobacco vs. *Vicia faba* and *Commelina* guard cells as discussed above.

This dicotomy in S-type anion channel and outward and inward K^+ channel regulation, when comparing regulation in different species, correlates closely to the dicotomy in ABA signaling among the discussed species, suggesting that the observed regulatory mechanisms also determine the physiological response *in vivo* (Schmidt et al., 1995; Pei et al., 1997; Armstrong et al., 1995). This dicotomy extends to four species to date, namely to ABA signaling in *Vicia faba, Commelina communis,* tobacco and *Arabidopsis* stomata: In both *Vicia faba* and *Commelina communis,* studies of ABA-induced stomatal closing and the large degree of regulation of S-type anion channels suggest that ABA signals are transduced by activation of protein kinases and in parallel by down-regulation of okadaic acid sensitive protein phosphatases (Schmidt et al., 1995; Li and Assmann, 1996; Esser et al., 1997; Mori and Muto, 1997; Schwarz and Schroeder, Submitted). On the other hand, in both *Arabidopsis* and tobacco guard cells, ABA induced stomatal movements appear to involve up-regulated protein phosphatases and rate-limiting negatively regulating protein kinases (Armstrong et al., 1995; Pei et al., 1997). These above comparative studies, suggest that by analyzing guard cell ion channels in different species, new regulation mechanisms or differences among species may be revealed and that studies involving multiple species or transplanting genes or proteins (e.g. by microinjection) among species will require careful analysis.

6 *abi1-1* and *abi2-1* Mutants Disrupt Early ABA Signaling Events

Electrophysiological studies of transgenic tobacco guard cells expressing the mutant *abi1-1* gene from *Arabidopsis* provided initial evidence that ABI1 affects early signaling events (Armstrong et al., 1995). These findings are consistent with results in *Arabidopsis* guard cells showing that ABA activation of S-type anion channels is suppressed in the *abi1-1* mutant (Pei et al., 1997). The finding that 100 μM ABA does not even cause a small degree of stomatal closing in *abi1-1* guard cells (Roelfsema and Prins, 1995; Pei et al., 1997), confirms that S-type anion channel activation is also impaired *in vivo* in *abi1-1*. If S-type anion channels were activated to a similar degree by ABA in wildtype and *abi1-1* guard cells, at least a partial ABA-induced stomatal closing response would be found in *abi1-1 in vivo.* Furthermore, kinase inhibitors partially suppressed both the stomatal closing phenotype and the ABA insensitivity of S-type anion channels (Pei et al., 1997). These data indicate that *abi1* and protein kinases are involved in negative regulation of the ABA response (Pei et al., 1997).

Furthermore, patch clamping of *Arabidopsis* guard cells provided initial evidence that the *abi2-1* mutant also affects very early signal transduction in guard cells. Patch clamping of *abi2* guard cells, showed that the *abi2* mutants also disrupts ABA activation of S-type anion channels (Pei et al., 1997). Interestingly, the *abi2-1* phenotype could not be suppressed by protein kinase inhibitors, showing that patch clamp studies provide quantitative resolution of differences in effects of the *abi1-1* and *abi2-1* mutants (Pei et al., 1997).

These findings are particularly interesting in light of the recent isolation of the *ABI2* gene, showing that it also encodes a protein phosphatase 2C (Leung et al., 1997). Furthermore, the mutation in the *abi2-1* allele (Leung et al., 1997) mimics the mutation found in the *abi1-1* allele (Meyer et al., 1994; Leung et al., 1994). In addition to the above patch clamp studies, several other studies have demonstrated that the *abi1-1* and the *abi2-1* mutants show clearly distinct phenotypes in ABA signaling (Söderman et al., 1996; Gilmour and Thomashaw, 1991; Gosti et al., 1995; Vartanian et al., 1994). Strong differences in signaling among these 2 mutant loci have been revealed for ABA-induced regulation of the *ATHB-7* gene (Söderman et al., 1996), drought-induced regulation of the *AtD* and *ATHB-7* genes (Söderman et al., 1996; Gosti et al., 1995), drought-dependent rhizogenes (Vartanian et al., 1994) and cold tolerance (Gilmour and Thomashaw, 1991), as well as in S-type anion channel regulation in guard cells (Pei et al., 1997). These comparative studies on various ABA responses in *abi1-1* and *abi2-1* clearly demonstrate that these two mutations can be distinguished physiologically and in terms of signal transduction.

The fact that both mutant loci encode for PP2Cs with comparable mutations will render further analysis of the mutant specificity interesting for dissecting early signal transduction cascades. It is possible that both PP2Cs share certain overlapping targets (Leung et al., 1997). But specific targets for each PP2C are likely to also exist, based on the above summarized findings. In this respect it is interesting that the N-terminal sequences of ABI1 and ABI2 are divergent (Leung et al., 1997). As the N-termini of protein phosphatases interact with the targets of protein phosphatases, these may provide an approach for identifying different interactors in signaling. Although some interactors may overlap, the differences in signaling and in N-terminal sequences suggest certain distinct interactors and certain distinct functions in signaling. On the other hand strong overexpression of either one of the mutant ABI1-1 or ABI2-1 proteins in transgenic tissue could lead to overlapping functions, as high levels of phosphorylating enzymes (e.g kinases or phosphatases) are known to cross react with multiple non-specific targets when concentrations are enhanced above physiological levels (or localization is modified). This concern points to possible caveats of future transgenic plant studies, which could be circumvented for example by comparing strong and weak expressing independent transgenic lines.

The question whether ABA activates, inhibits or does not directly regulate one or the other of the ABI PP2Cs will require further analysis. The question whether either one of the *abi* mutants can be attributed to neomorphic effects of the mutation will also be of interest in this regard. The finding that both *abi1* and *abi2* RNA expression levels are regulated by ABA has led to the suggestion that neomorphic effects of the mutants may not be occurring (Leung et al., 1997). The combined use of high-resolution approaches to study early signal transduction, including *Arabidopsis* molecular genetics and *Arabidopsis* guard cell patch clamping, as well as biochemical assays and microinjection approaches should lead to a more detailed understanding of the network of early signal transduction mechanisms during guard cell ion channel regulation and ABA signal transduction.

Acknowledgments

Research in the authors' laboratory was supported by NSF grant MCB9506191.

References

Armstrong F, Leung J, Grabov A, Brearley J, Giraudat J and Blatt MR (1995) Sensitivity to abscisic acid of guard cell K^+ channels is suppressed by ABI1-1, a mutant *Arabidopsis* gene encoding a putative protein phosphatase. Proc. Natl. Acad. Sci. USA **92:** 9520-9524.

Assmann SM (1993) Signal transduction in guard cells. Ann. Rev. Cell Biol. **9:** 345-375.

Assmann SM, Simoncini L and Schroeder JI (1985) Blue light activates electrogenic ion pumping in guard cell protoplasts of *Vicia faba*. Nature **318:** 285-287.

Bertauche N, Leung J and Giraudat J (1996) Protein phosphatase activity of abscisic acid insensitive 1 (ABI1) protein from *Arabidopsis thaliana*. Eur. J. Biochem. **241:** 193-200.

Blatt MR and Armstrong F (1993) K^+ channels of stomatal guard cells: Abscisic-acid-evoked control of the outward-rectifier mediated by cytoplasmic pH. Planta **191:** 330-341.

Cosgrove DJ and Hedrich R (1991) Stretch-activated chloride, potassium, and calcium channels co-existing in the plasma membranes of guard cells of *Vicia faba* L. Planta **186:** 143-153.

DeSilva DLR, Cox RC, Hetherington AM and Mansfield TA (1985) Synergism between calcium ions and abscisic acid in preventing stomatal opening. New Phytol. **101:** 555-563.

Esser JE, Liao YJ and Schroeder JI (1997) Characterization of ion channel modulator effects on ABA- and malate-induced stomatal movements: Strong regulation by kinase and phosphatase inhibitors, and relative insensitivity to mastoparans. J. Exp. Bot. **48:** 539-550.

Gilmour SJ and Thomashow MF (1991) Cold acclimation and cold-regulated gene expression in ABA mutants of *Arabidopsis thaliana*. Pl. Mol. Biol. **17:** 1233-1240.

Goh CH, Kinoshita T, Oku T and Shimazaki KI (1996) Inhibition of blue light-dependent H^+ pumping by abscisic acid in *Vicia* guard-cell protoplasts. Plant Physiol. **111:** 433-440.

Gosti F, Bertauche N, Vartanian N and Giraudat J (1995) Abscisic acid-dependent and - independent regulation of gene expression by progressive drought in *Arabidopsis thaliana*. Mol. Gen. Genet. **246:** 10-18.

Hedrich R, Busch H and Raschke K (1990) Ca^{2+} and nucleotide dependent regulation of voltage dependent anion channels in the plasma membrane of guard cells. EMBO J. **9:** 3889-3892.

Hetherington AM, and Quatrano RS (1991) Mechanisms of action of abscisic acid at the cellular level. New Phytol. **119:** 9-32.

Irving HR, Gehring CA and Parish RW (1992) Changes in cytosolic pH and calcium of guard cells precede stomatal closure. Proc. Natl. Acad. Sci. USA **89:** 1790-1794.

Ishikawa H., Aizawa H, Kishira H, Ogawa T and Sakata M (1983) Light-induced changes of membrane potential in guard cells of *Vicia faba*. Plant Cell Physiol. **24:** 769-772.

Kinoshita T, Nishimura M and Shimazaki KI (1995) Cytosolic concentration of Ca^{2+} regulates the plasma membrane H^+-ATPase in guard cells of fava bean. Plant Cell **7:** 1333-1342.

Kusamo K (1981) Effect of abscisic acid on the K^+ efflux and membrane potential of *Nicotiana tabacum* leaf cells. Plant Cell Physiol. **22:** 1257-1267.

Leung J, Bouvier-Durand M, Morris P-C, Guerrier D, Chefdor F and Giraudat J (1994) *Arabidopsis* ABA response gene *ABI1* - Features of a calcium-modulated protein phosphatase. Science **264:** 1448-1452.

Leung J, Merlot S and Giraudat J (1997) The *Arabidopsis* abscisic acid-insensitive (ABI2) and ABI1 genes encode homologous protein phosphatases 2C involved in abscisic acid signal transduction. Plant Cell **9:** 759-771.

Li J and Assmann SM (1996) An abscisic acid-activated and calcium-independent protein kinase from guard cells of fava bean. Plant Cell **8:** 2359-2368.

Li W, Luan S, Schreiber SL and Assmann SM (1994) Evidence for protein phosphatase and 2A regulation of K^+ channels in two types of leaf cells. Plant Physiol. **106:** 963-970.

Linder B and Raschke K (1992) A slow anion channel in guard cells, activation at large hyperpolarization, may be principal for stomatal closing. FEBS Lett. **131:** 27-30.

Luan S, Li W, Rusnak F, Assmann SM and Schreiber SL (1993) Immunosuppressants implicate protein phosphatase regulation of K^+ channels in guard cells. Proc. Natl. Acad. Sci. USA **90:** 2202-2206.

MacRobbie EAC (1981) Effects of ABA on "isolated" guard cells of *Commelina communis* L. J. Exp. Bot. **32:** 563-572.

MacRobbie EAC (1983) Effects of light/dark on cation fluxes in guard cells of *Commelina communis* L. J. Exp. Bot. **34:** 1695-1710.

Mansfield TA, Hetherington AM and Atkinson CJ (1990) Some current aspects of stomatal physiology. Ann. Rev. Pl. Physiol. & Pl. Mol. Biol. **41:** 55-75.

Meyer K, Leube MP and Grill E (1994) A protein phosphatase 2C involved in ABA signal transduction in *Arabidopsis thaliana*. Science **264:** 1452-1455.

Mori IC, and Muto S (1997) Abscisic acid activates a 48-kilodalton protein kinase in guard cell protoplasts. Plant Physiol. **113:** 833-840.

Pei Z-M, Kuchitsu K, Ward JM, Schwarz M and Schroeder JI (1997) Differential abscisic acid regulation of guard cell slow anion channels in *Arabidopsis* wild-type and *abi1* and *abi2* mutants. Plant Cell **9:** 409-423.

Raschke K (1979) In: *Encyclopedia of Plant Physiology*, Feinleib, W.H.. (ed.), Berlin: Springer-Verlag. p. 384-441.

Roelfsema MRG and Prins HBA (1995) Effect of abscisic acid on stomatal opening in isolated epidermal strips of *abi* mutants of *Arabidopsis thaliana*. Physiol. Plant **95:** 373-378.

Schmidt C, Schelle I, Liao YJ and Schroeder JI (1995) Strong regulation of slow anion channels and abscisic acid signaling in guard cells by phosphorylation and dephosphorylation events. Proc. Natl. Acad. Sci. USA **92:** 9535-9539.

Schroeder JI and Hagiwara S (1989) Cytosolic calcium regulates ion channels in the plasma membrane of *Vicia faba* guard cells. Nature **338:** 427-430.

Schroeder JI and Hagiwara S (1990) Repetitive increases in cytoslic Ca^{2+} of guard cells by abscisic acid activation of non-selective Ca^{2+}-permeable channels. Proc. Natl. Acad. Sci. USA **87:** 9305-9309.

Schroeder JI and Keller BU (1992) Two types of anion channel currents in guard cells with distinct voltage regulation. Proc. Natl. Acad. Sci. USA **89:** 5025-5029.

Schroeder JI, Raschke K and Neher E (1987) Voltage dependence of K^+ channels in guard cell protoplasts. Proc. Natl. Acad. Sci. USA **84:** 4108-4112.

Schroeder JI, Schmidt R and Sheaffer J (1993) Identification of high-affinity slow anion channel blockers and evidence for stomatal regulation by slow anion channels in guard cells. Plant Cell **5:** 1831-1841.

Schroeder JI. and Hedrich R (1989) Involvement of ion channels and active transport in osmoregulation and signaling of higher plant cells. Trends Biochem. Sci. **14:** 187-192.

Schultz-Lessdorf B, Lohse G and Hedrich R (1996) GCAC1 recognizes the pH gradient across the plasma membrane: A pH-sensitive and ATP-dependent anion channel links guard cell membrane potential to acid and energy metabolism. Plant Journal **10:** 993-1004.

Schwartz A, Ilan N, Schwarz M, Scheaffer J, Assmann SM and Schroeder JI (1995) Anion-channel blockers inhibit S-type anion channels and abscisic acid responses in guard cells. Plant Physiol. **109:** 651-658.

Schwarz M and Schroeder JI (Submitted) Abscisic acid regulation of slow anion channels in *Vicia faba* guard cells implicates ABA-induced down-regulation of okadaic acid-sensitive protein phosphatases.

Shimazaki K, Iino M and Zeiger E (1986) Blue light-dependent proton extrusion by guard-cell protoplasts of *Vicia faba*. Nature **319:** 324-326.

Söderman E, Mattsson J and Engstrom P (1996) The *Arabidopsis* homeobox gene ATHB-7 is induced by water deficit and by abscisic acid. The Plant Journal **10:** 375-381.

Thiel G and Blatt MR (1994) Phosphatase antagonist okadaic acid inhibits steady-state K^+ currents in guard cells of *Vicia faba*. Plant Journal **5:** 727-733.

Thiel G, MacRobbie EAC and Blatt MR (1992) Membrane transport in stomatal guard cells: The importance of voltage control. J. Memb. Biol. **126:** 1-18.

Vartanian N, Marcotte L and Giraudat J (1994) Drought rhizogenesis in *Arabidopsis thaliana*. Pl. Physiol. **104:** 761-767.

Ward JM, Pei Z-M and Schroeder JI (1995) Roles of ion channels in initiation of signal transduction in higher plants. The Plant Cell **7:** 833-844.

Zimmermann S, Thomine S, Guern J and Barbier-Brygoo H (1994) An anion current at the plasma membrane of tobacco protoplasts shows ATP-dependent voltage regulation and is modulated by auxin. Plant J. **6:** 707-716.

UV and blue light signal transduction in the regulation of flavonoid biosynthesis gene expression in *Arabidopsis*

Gareth I. Jenkins

Plant Molecular Science Group, Division of Biochemistry and Molecular Biology, Institute of Biomedical and Life Sciences, Bower Building, University of Glasgow, Glasgow G12 8QQ, UK

Abstract. Biochemical and genetic approaches in *Arabidopsis* have produced new insights into the photoreceptors and signal transduction processes involved in the regulation of gene expression by UV and blue light. Several distinct photoreceptors mediate the effects of UV-B, UV-A and blue light on the expression of the gene encoding the flavonoid biosynthesis enzyme chalcone synthase (CHS). Information on the signal transduction components involved in the induction of *CHS* gene expression has been obtained in experiments with an *Arabidopsis* cell culture. Experiments with intact plants indicate that interactions between phototransduction pathways maximise *CHS* expression. Mutants altered in negative regulators of *CHS* transcription in response to UV and blue light have been isolated.

Keywords. *Arabidopsis*, blue light, chalcone synthase, flavonoids, mutants, photomorphogenesis, signal transduction, UV-A, UV-B

1 Introduction

Light is of vital importance to the growth and development of plants. In addition to its pivotal role in driving photosynthesis, light is a key factor in the regulation of numerous developmental processes. These effects of light are mediated by several distinct photoreceptors coupled to signal transduction networks. Much of the focus of present research is on understanding the roles of the different photoreceptors in controlling development and the mechanisms of signal transduction. The most powerful way to investigate these complex processes is to employ a combination of biochemical, molecular, genetic and cell physiological approaches. *Arabidopsis thaliana* is an excellent subject for this research because of its suitability, in particular, for molecular and genetic studies. In fact, the isolation and characterisation of *Arabidopsis* mutants altered in the light regulation of development has already resulted in novel insights into photoreception and signal transduction.

Several different classes of photoreceptors mediate the effects of light on plant growth and development. The best characterised photoreceptors are the phytochromes, which effect responses principally to red and far-red light. The study of *Arabidopsis* mutants deficient in specific phytochromes has yielded important information on the roles of the different members of this photoreceptor family in controlling development (Whitelam and Harberd 1994). It is evident, however, that many aspects of plant

NATO ASI Series, Vol. H 104
Cellular Integration of Signalling Pathways in Plant Development
Edited by F. Lo Schiavo, R. L. Last, G. Morelli, and N. V. Raikhel

development are controlled, entirely or in part, by light in the UV-A (320-390 nm) and blue (390-500 nm) regions of the spectrum. Examples of responses regulated by UV/blue light include: phototropism, stem extension, leaf expansion, chloroplast development, stomatal opening and the expression of various genes. Although the phytochromes absorb UV/blue as well as red and far-red light, most of the responses of plants to UV/blue wavelengths are mediated by separate photoreceptors (Kaufman 1993; Liscum and Hangarter 1994; Short and Briggs 1994; Jenkins *et al.* 1995; Ahmad and Cashmore 1996). It is evident that several different UV/blue photoreceptors are present in plants, but they are much less well characterised than the phytochromes.

UV-B (280-320 nm) wavelengths also have a profound influence on the growth and development of plants (Tevini and Teramura 1989; Stapleton 1992; Jenkins *et al.* 1995). Although UV-B may cause damage to macromolecules such as DNA, at moderate levels it has morphogenetic effects (Tevini and Teramura 1989). Furthermore, as discussed below, UV-B induces the expression of genes encoding enzymes of the phenylpropanoid and flavonoid biosynthesis pathways, which produce UV-protective pigments in the epidermal layers. Thus, the effects of UV-B are not confined to macromolecular damage, and the physiological responses to UV-B are likely to involve a further distinct photoreception system.

Although responses to UV-B, UV-A and blue wavelengths are evidently of key importance, our knowledge of the photoreceptors and signal transduction processes involved is very limited (Kaufman 1993; Short and Briggs 1994; Jenkins *et al.* 1995). This paper discusses recent progress in research on UV and blue light signal transduction in *Arabidopsis*, focusing on the regulation of flavonoid biosynthesis genes.

2 Light-Regulation of *Arabidopsis* Flavonoid Biosynthesis Genes

The flavonoids are a large group of compounds derived from a branch of the general phenylpropanoid biosynthetic pathway (Dixon and Paiva 1995). Several genes and products of flavonoid biosynthesis, in particular the red-purple anthocyanins, have been the focus of numerous studies of the light-regulation of gene expression. Chalcone synthase (CHS) is the first committed step in flavonoid biosynthesis and it is well established that *CHS* gene expression may be controlled by UV/blue photoreceptors and by phytochrome, depending on the species and the stage of development. In mature leaves of several species, including white mustard and parsley (Batschauer *et al.* 1991; Frohnmeyer *et al.* 1992), *CHS* expression is regulated principally by UV and blue light, whereas in young or dark-grown seedlings, phytochrome is effective. This is also the case in *Arabidopsis*, in that phytochrome regulation is confined to dark-grown seedlings less than 6 days old (Kaiser *et al.* 1995) and UV/blue light regulation predominates (Feinbaum *et al.* 1991; Jackson *et al.* 1995; Fuglevand *et al.* 1996). Kubasek *et al.* (1992) have shown that other genes of the phenylpropanoid and flavonoid biosynthetic pathways are regulated similarly to *CHS* in *Arabidopsis*. Phytochrome-deficient *Arabidopsis* mutants retain the UV/blue light-induction of *CHS* expression, indicating that the response to UV/blue light is independent of phytochrome (Batschauer *et al.* 1996). However, anthocyanin accumulation in response to blue light is much reduced in phytochrome deficient mutants (Ahmad and Cashmore 1997), demonstrating that the blue light regulation of a later step in the pathway is phytochrome dependent.

Since the regulation of *CHS* in mature *Arabidopsis* leaves is specifically regulated by UV/blue light, it is an excellent system for studying the UV/blue photoreceptors and signal transduction processes involved in the regulation of transcription.

3 UV/Blue Photoreception and Signal Transduction in *Arabidopsis*

A major step forward in the understanding of UV/blue light responses in plants was the cloning of the first UV/blue photoreceptor. Ahmad and Cashmore (1993) reported the cloning of the *Arabidopsis* CRY1 (cryptochrome) photoreceptor encoded by the *HY4* gene. Mutants in the *HY4* gene are impaired in the suppression of hypocotyl extension in UV-A, blue and green light (Koornneef *et al.* 1980; Ahmad *et al.* 1995) and in several other extension growth responses, such as petiole extension and leaf expansion (Jackson and Jenkins 1995). In addition, as discussed in detail below, *hy4* mutants have reduced induction of flavonoid biosynthesis gene expression and anthocyanin accumulation in UV-A/blue light (Jackson and Jenkins 1995; Ahmad *et al.* 1995). However, *hy4* mutants are not altered in responses such as phototropism and stomatal opening, which are therefore mediated by separate UV/blue photoreceptors.

There is evidence that the photoreceptor for phototropism is encoded by the *NPH1* gene of *Arabidopsis* (Liscum and Briggs 1995). Mutants at this locus are defective in phototropism and in phosphorylation of a plasma membrane protein which appears to be an integral component of the photoreceptor. Information on the DNA sequence of the *NPH1* gene is presently being obtained and hence the extent of the similarity of this photoreceptor to CRY1 will soon be evident.

Little detailed information is yet available on the signal transduction processes initiated by CRY1. It is likely that branching signal transduction pathways are responsible for the different responses mediated by CRY1, such as extension growth and gene regulation, but the initial steps in phototransduction may be common. The CRY1 protein expressed in *E.coli* and insect cells binds flavin and pterin chromophores (Lin *et al.* 1995; Malhotra *et al.* 1995). Similar chromophores are bound by the microbial DNA photolyases, which are related in sequence to CRY1, and catalyse DNA repair by a mechanism involving electron transfer. It is therefore possible that electron transport is an initial event in CRY1 signal transduction. Redox activity has been implicated in several blue light responses (Jenkins *et al.* 1995).

There is evidence that CRY1 is a soluble protein rather than an integral membrane protein (Lin *et al.* 1996). Nevertheless CRY1 could interact with membranes to initiate signal transduction. The C-terminal domain is unrelated to DNA photolyases and has sequence similarity to rat smooth muscle tropomyosin A. This domain may be important in initiating signal transduction, perhaps by interacting with a membrane-associated component. There is evidence that membrane processes may be early events in blue light signal transduction (Jenkins *et al.* 1995). Membrane potential changes and ion fluxes are associated with blue light-induced extension growth responses (Spalding and Cosgrove 1989; 1992; Cho and Spalding 1996). Moreover, blue light-induced membrane potential changes and H^+ fluxes are associated with stomatal opening and pulvinal movements in several species (Assmann *et al.*

1985; Shimazaki *et al.* 1986; Nishizaki 1994). Recent studies, discussed in detail below, indicate that calcium fluxes are involved the signalling events which couple CRY1 to the regulation of transcription (Christie and Jenkins 1996). Additional research is required to elucidate how photoreception by CRY1 initiates these membrane associated events and which membranes are involved.

A further putative UV/blue photoreceptor, related to CRY1 in the sequence of the N-terminal domain, has recently been identified by Cashmore and coworkers (Ahmad and Cashmore 1996). This protein, designated CRY2, has a different C-terminal domain to CRY1, which may be significant with regard to the initiation of signal transduction. Research is in progress to investigate the function of CRY2.

4 Distinct UV-B and UV-A/Blue Light Signalling Pathways Induce *CHS* Expression in *Arabidopsis*

Experiments with a *hy4* mutant allele, *hy42.23N*, indicate that there are at least two distinct UV/blue photoreception systems responsible for the induction of *CHS* transcription in *Arabidopsis*. The *hy42.23N* mutant is a null allele, since CRY1 protein is not detected among *hy42.23N* leaf proteins by an antibody raised against CRY1 (Lin *et al.* 1996) and, moreover, *hy42.23N* shows the same alteration in hypocotyl extension in blue light as other null alleles (Ahmad and Cashmore 1993; Ahmad *et al.* 1995). The induction of *CHS* transcripts by blue light in *hy42.23N* is much reduced compared to the wild type and induction by UV-A is nearly undetectable (Jackson and Jenkins 1995; Fuglevand *et al.* 1996). Reduced *CHS* expression and anthocyanin accumulation in a range of *hy4* alleles has been reported by Ahmad *et al.* (1995). These observations demonstrate that the CRY1 photoreceptor mediates the UV-A/blue light-induction of *CHS*. However, the residual induction by blue light cannot be accounted for by CRY1 photoreception and is therefore mediated by another, as yet undefined photoreceptor. In addition, Fuglevand *et al.* (1996) reported that the *hy42.23N* mutant retains normal induction of *CHS* transcripts in response to UV-B exposure, indicating that a separate light detection system mediates this response. The putative UV-B photoreceptor has yet to be identified.

5 Components of the UV-B and UV-A/Blue Light Signal Transduction Pathways Inducing *CHS* Expression

Christie and Jenkins (1996) used a pharmacological approach with an *Arabidopsis* cell suspension culture to identify signal transduction components involved in the induction of *CHS* expression by UV-B and UV-A/blue light. They demonstrated that *CHS* transcripts accumulated in the cell culture following exposure to UV-B and UV-A/blue light, but not red or far-red light. Thus the response was equivalent to that in mature *Arabidopsis* leaf tissue. *CHS* transcripts accumulated in the cells within four hours of illumination, which was sufficiently rapid to enable the effects of various inhibitors and agonists of signal transduction components on *CHS* expression to be studied. The cells were preincubated with the compounds for one hour under non-inductive conditions before transfer to UV-B or UV-A/blue light.

Using this approach, Christie and Jenkins (1996) obtained evidence that cellular calcium was involved in both the UV-B and UV-A/blue light signal transduction pathways inducing *CHS* expression. The induction of *CHS* transcripts by UV-B and

UV-A/blue light was impaired by both nifedipine, which inhibits voltage-gated calcium channels, and ruthenium red, which inhibits calcium channels in vacuolar, and probably other, membranes (Allen *et al.* 1995). It contrast, it was found that verapamil, a different calcium channel blocker, had no effect, suggesting that specific types of calcium channels are involved. Nifedipine and ruthenium red did not have general inhibitory effects on the cells because they did not affect a control response, the induction of *PAL* transcripts by the protein phosphatase inhibitor cantharidin.

Since the effects of nifedipine and ruthenium red indicated an involvement of calcium in *CHS* induction, Christie and Jenkins (1996) investigated the effect of lanthanum on the response. Lanthanum ions compete externally with calcium for uptake into cells and therefore inhibit responses requiring an influx of calcium through plasma membrane channels as opposed to release from intracellular pools. It was found that lanthanum had no effect on the induction of *CHS* expression by either UV-B or UV-A/blue light at concentrations which inhibit other gene expression responses in the same cells (J. Christie and G.I. Jenkins, unpublished data) and in other systems. The lanthanum and ruthenium red data together suggest that the UV-B and UV-A/blue light-induction of *CHS* each require calcium flux across an internal membrane and a specific pool of intracellular calcium may therefore be involved. Interestingly, it was not possible to induce *CHS* expression in the *Arabidopsis* cells by artificially elevating cytosolic calcium with the ionophore A23187 and 10 mM external calcium. However, the ionophore treatment did stimulate the accumulation of transcripts of a control gene, the *Arabidopsis TCH3* (*TOUCH3*) gene, demonstrating that the ionophore treatment had raised the cytosolic calcium concentration.

If the cytosolic calcium concentration is raised from intracellular stores, the failure to induce expression by artificially elevating cytosolic calcium with the ionophore treatment could indicate that UV-B and UV-A/blue illumination produce an additional essential signal: that is, the increase in calcium is necessary but not sufficient for the response. The initiation of redox reactions, for instance, may also be necessary. A further possibility is that UV-B and UV-A/blue light promote a localised increase in calcium, for example in a microdomain associated with a particular intracellular membrane, or in a specific organelle. Hence there may not be a general increase in the cytosolic calcium concentration. Moreover, a localised calcium pool may not be elevated by the ionophore treatment, accounting for the failure to induce *CHS* transcripts. Consistent with the latter model, UV-B and UV-A/blue light do not induce a large, rapid increase in cytosolic calcium in *Arabidopsis*, as detected by the calcium-sensing protein aequorin in transgenic seedlings (J. Christie, J. Long, M.R. Knight and G.I. Jenkins, unpublished data). Direct measurements are now required to establish the location and magnitude of the increases in calcium ions in response to illumination predicted by the pharmacological experiments and to determine whether UV-B and UV-A/blue light affect the same calcium pools. It is also important to investigate whether other processes, such as redox reactions, are initiated following illumination which are essential for the gene expression response to occur.

Further experiments with the cell culture showed that the UV-B and UV-A/blue light signalling pathways differed with respect to the involvement of calmodulin (Christie and Jenkins 1996). The calmodulin antagonist, W-7, strongly inhibited the induction of *CHS* expression by UV-B, but not by UV-A/blue light. This finding is consistent with the genetic evidence described above which shows that the UV-B and UV-A/blue light responses are mediated by distinct photoreception systems.

Furthermore, it is evident that the UV-B and UV-A/blue phototransduction pathways regulating *CHS* expression in *Arabidopsis* are different to the phytochrome signalling pathway regulating *CHS* expression that has been identified in other species (Neuhaus *et al.* 1993; Bowler *et al.* 1994a). The phytochrome signal transduction pathway involves cGMP rather than calcium (Bowler *et al.* 1994a; 1994b). However, dibutyryl-cGMP does not induce *CHS* expression in the *Arabidopsis* cells (J.M. Christie and G.I. Jenkins, unpublished data). A further feature of the phytochrome signalling pathway regulating *CHS* in soybean cells is that it is inhibited by the protein histidine/tyrosine kinase inhibitor genistein (Bowler *et al.* 1994b). Christie and Jenkins (1996) did not find any inhibition of the UV-B and UV-A/blue light-induction of *CHS* in *Arabidopsis* cells by genistein, although expression was greatly reduced by inhibitors of protein serine/threonine kinase activity and protein phosphatase activity. The identity of the specific protein kinases and phosphatases involved in the responses requires further investigation.

In summary, pharmacological experiments with the *Arabidopsis* cell culture have generated new insights into the signal transduction processes through which UV-B and UV-A/blue light signals are coupled to *CHS* gene expression. Moreover, it is likey that experiments with further specific inhibitors will generate more detailed information. However, the pharmacological approach clearly has its limitations and it is therefore essential to combine it with cell physiological and biochemical approaches which identify directly the signal transduction processes initiated by UV-B and UV-A/blue light.

6 Effectors of the Transcriptional Response

The molecular targets of the UV-B and UV-A/blue light signalling pathways are ultimately the specific transcription factors that interact with the promoter of the *CHS* gene and probably other genes concerned with flavonoid biosynthesis. Hence, in order to understand fully the light response it is important to identify the relevant transcription factor proteins and to establish how they effect transcription. In *Arabidopsis* there is a single *CHS* gene, so investigation of the molecular basis of *CHS* transcription is not complicated by questions of differential expression within a gene family.

Information about the *cis*-acting DNA sequence elements and transcription factors that effect the light-regulation of *CHS* genes has been obtained in several species, in particular parsley (Schulze-Lefert *et al.* 1989; Weisshaar *et al.* 1991; Feldbrugge *et al.* 1997). The parsley *CHS* promoter contains two Light Regulatory Units (LRU's), each of which has two DNA sequence elements, boxes I and II in LRU1 and boxes III and IV in LRU2, that bind protein following illumination of parsley cells with UV-containing white light (Schulze-Lefert *et al.* 1989). Transient expression experiments with protoplasts derived from cultured parsley cells demonstrated that LRU1 is sufficient for the UV-induction of transcription. Box II in LRU1 interacts with the basic leucine zipper (bZIP) class of transcription factors (Schulze-Lefert *et al.* 1989; Weisshaar *et al.* 1991) whereas box I has recently been shown to be a recognition element for MYB transcription factors (Feldbrugge *et al.* 1997).

The *Arabidopsis CHS* promoter (Feinbaum and Ausubel 1988) contains a 41 bp sequence with considerable homology to LRU1 in the parsley *CHS* promoter. There is no obvious equivalent of LRU2. Transient expression assays with protoplasts from cultured *Arabidopsis* cells show that the *Arabidopsis CHS* LRU is sufficient to confer strong UV-B and UV-A/blue light-induction of transcription when fused to the β-glucuronidase (GUS) coding sequence (J.M. Christie, U. Hartmann, W.J. Valentine,

B. Weisshaar and G.I. Jenkins, unpublished data). Mutation of either the putative bZIP or MYB binding sites in this LRU greatly reduces transcriptional activity.

Both the bZIP and MYB classes of transcription factors in *Arabidopsis* are encoded by gene families and it is not yet known which specific members of the families mediate the UV-B and UV-A/blue light responses. In addition to identifying the specific transcription factors involved, it is important to establish how light regulates their activity. Harter *et al.* (1994) reported that the bZIP transcription factors that associate with the parsley LRU1 are activated by phosphorylation and that the proteins move between the cytosolic and nuclear compartments. In addition, there is evidence from experiments with cycloheximide that protein synthesis is required for the light induction of *CHS* expression in parsley (Feldbrugge *et al.* 1996) and for both the UV-B and UV-A/blue light induction of *CHS* in *Arabidopsis* (Christie and Jenkins, 1996). It therefore appears that one or more effectors of *CHS* transcription, such as a transcription factor, or possibly a kinase, needs to be synthesised rapidly in response to the inductive stimulus. Identification of the component(s) synthesised is therefore of key importance in understanding how *CHS* is regulated.

7 Interactions Between Light Signal Transduction Pathways Regulating *CHS*

The inductive UV-B and UV-A/blue light responses discussed above are insufficient to account for the level of stimulation of *CHS* expression observed in mature *Arabidopsis* leaves in natural daylight. The reason for this is that the response to UV-B is enhanced by synergistic interactions with UV-A and blue light signalling pathways that maximise the level of *CHS* expression. Evidence for these interactions was provided by Fuglevand *et al.* (1996) in experiments with transgenic *Arabidopsis* expressing a *CHS-GUS* fusion. They reported that GUS activity was four to eight-fold greater in plants exposed to UV-B and blue light at the same time than in plants given either light quality alone. A similar result was obtained with UV-B and UV-A light. Since the combined effects of the different light qualities was much more than additive, it was evident that the responses were the result of a synergistic interaction between separate phototransduction pathways. In contrast, UV-A and blue light together gave an additive effect. Evidence for a synergistic interaction between UV-B and blue light was reported previously for *CHS* expression in parsley cell cultures (Ohl *et al.* 1989).

Further experiments with *Arabidopsis* showed that blue light given before UV-B resulted in a synergistic increase in *CHS-GUS* expression whereas UV-B given before blue light did not (Fuglevand *et al.* 1996). Blue light therefore appeared to generate a signal that enhanced the subsequent response to UV-B. Dark periods of varying duration were introduced between the blue and UV-B light treatments to investigate the stability of the signal. It was observed that the blue pretreatment still enhanced the subsequent response to UV-B when 5 to 10 hours of darkness intervened, indicating that the signal derived from blue light was relatively stable. This effect was observed on *CHS* transcript levels as well as *CHS-GUS* expression. In contrast, Fuglevand *et al.* (1996) found that the signal generated by UV-A light was transient, not stable, because a synergistic interaction between UV-A and UV-B was observed only when the two treatments were given simultaneously. The different stabilities of the signals produced by UV-A and blue light indicate that they are generated by different phototransduction pathways. Fuglevand *et al.* (1996) went on to demonstrate that both synergistic interactions can operate together to enhance *CHS* expression. When plants were illuminated with blue light followed by UV-A and UV-B together, the level of *CHS-GUS* expression was approximately twice that observed with either synergistic combination alone (UV-B plus blue light or UV-B plus UV-A light). The resultant stimulation of expression over the basal level was approximately 150-fold, compared to approximately 10-fold with either of the single light qualities.

Further experiments indicated that neither the blue nor the UV-A synergistic interaction with UV-B was mediated by the CRY1 photoreceptor, because both responses were present in the *hy42.23N* mutant. It therefore appears that several distinct UV/blue light detection systems are involved in the regulation of *CHS* expression. In addition to the UV-A/blue (CRY1) and UV-B inductive phototransduction pathways there are UV-A and blue light signalling pathways responsible for the synergistic interactions. A further blue light phototransduction pathway is hypothesised to account for the residual blue light induction of *CHS* in the *hy4* mutant. This degree of complexity is unexpected and indicates how much remains to be learnt about the photoreceptors and signal transduction networks that mediate the UV/blue light regulation of *CHS* and other genes concerned with flavonoid biosynthesis.

8 A Genetic Approach to the UV/Blue Light Regulation of *CHS* Expression

The isolation and characterisation of mutants altered in *CHS* expression will provide a powerful means of identifying, and investigating the functions of, signalling components and effectors of the UV/blue light signalling pathways. In addition to providing information on the functions of particular gene products, the availability of mutants enables the functional interactions between genes to be studied by the analysis of double mutants. Furthermore, in *Arabidopsis*, genes identified by mutation can be isolated. This approach has already been successful in the investigation of several aspects of plant signal transduction (Bowler and Chua 1994).

The isolation of mutants altered specifically in gene expression requires the design and execution of transgene expression screens. In this approach, a transgenic line containing a specific promoter-reporter fusion is mutagenised and the M2 generation screened for altered reporter gene expression. Various reporter systems, or combinations thereof, can be used for mutant isolation. Jackson *et al.* (1995) developed a transgene expression screen to isolate mutants altered specifically in the light-regulation of *CHS* gene expression. They produced an isogenic population of transgenic *Arabidopsis* expressing a *CHS-GUS* fusion and screened M2 seedlings for altered GUS activity in the light. This screen resulted in the isolation of a novel mutant, *icx1* (increased chalcone synthase expression) which has very low *CHS* expression in darkness, like the wild-type, but an enhanced response to light in the induction of *CHS* expression (Jackson *et al.* 1995). Subsequent analysis reveals that the mutant has an elevated response to both UV-B and UV-A/blue light (H.K. Wade and G.I. Jenkins, unpublished data). In addition, transcript levels of other flavonoid biosynthesis genes are elevated in the light and there is a two- to three-fold increase in anthocyanin induction (Jackson *et al.* 1995). The *ICX1* gene product therefore appears to constrain the UV/blue light induction of flavonoid biosynthesis genes in the wild type, suggesting that it functions as a downstream negative regulator of the phototransduction pathways. A further feature of the *icx1* mutant is that it has a pleiotropic visible phenotype: it has altered leaf shape, fewer trichomes, altered leaf epidermal surface morphology and abnormal root development (Jackson *et al.* 1995; J.A. Jackson, B.A. Brown and G.I. Jenkins unpublished data). These data suggest that ICX1 is concerned with a range of functions in epidermal gene expression and development. *ICX1* is located on the lower half of chromosome 1 and experiments are in progress to isolate the gene.

Additional mutants with altered *CHS* expression have since been isolated using the above screen (G. Fuglevand and G.I. Jenkins, unpublished data). One, which is not allelic to *icx1*, shows greatly increased *CHS-GUS* expression, *CHS* transcript levels and anthocyanin induction in response to UV/blue light. There is no alteration to morphology in light or darkness. This mutant, designated *icx2*, therefore defines a further negative regulator of *CHS* expression.

The *icx* mutants are among several mutants known to affect the regulation of flavonoid biosynthesis gene expression. The various *det/cop/fus* mutants are fundamentally altered in photomorphogenesis and have elevated flavonoid biosynthesis gene expression in young seedlings both in the presence and absence of inductive signals (Mol *et al.* 1996). It appears that the corresponding gene products are involved in the transduction of a range of cellular signals. Another *Arabidopsis* mutant, *ttg* (*transparent testa glabra*), is altered in a similar range of aspects of epidermal gene expression and development to *icx1*, although the specific alterations are different. *TTG* is likely to encode a key regulator of transcription in the epidermis. As a further example, the tomato *hp* mutants (Peters *et al.* 1992) have elevated anthocyanin induction in response to light and may be altered in a signal transduction component that mediates the response to phytochrome. An important goal of future research will be to understand how the functions of these different gene products interact to regulate *CHS* and other flavonoid biosynthesis genes. In addition, it will be important to integrate the functions of the regulators identified genetically into the signal transduction pathways characterised by pharmacological experiments.

9 Conclusions and Perspectives

The application of genetic and biochemical approaches in *Arabidopsis* has generated new insights into the photoreceptors and signal transduction processes that mediate the effects of UV and blue light on gene expression and development. Nevertheless, our understanding of UV and blue light perception and signal transduction remains limited. Although likely components of UV/blue light signal transduction pathways have been identified, in many cases there is little direct evidence relating signalling events either to specific photoreceptors or to specific responses (Kaufman 1993; Short and Briggs 1994; Jenkins *et al.* 1995). It is therefore important to isolate further genes encoding UV-B, UV-A and blue light receptors and to identify and characterise the corresponding mutants. Additional mutants altered in the UV/blue light regulation of gene expression should be isolated as these may identify novel signalling or effector components. Further information on UV/blue light signal transduction will come from biochemical and cell physiological studies with the *Arabidopsis* cell culture and from integration of the biochemical and genetic approaches.

Acknowledgements

The author is indebted to members of his laboratory for discussions of UV and blue light signalling and to the BBSRC, the Gatsby Charitable Foundation and the British Council for their support of research in his laboratory.

References

Ahmad M. and Cashmore A.R. (1993) *HY4* gene of *A. thaliana* encodes a protein with the characteristics of a blue-light photoreceptor. *Nature* **366**, 162-166.

Ahmad M. and Cashmore A.R. (1996) Seeing blue: the discovery of cryptochrome. *Plant Mol. Biol.* **30**, 851-861.

Ahmad M. and Cashmore A.R. (1997) The blue-light receptor cryptochrome 1 shows functional dependence on phytochrome A or phytochrome B in *Arabidopsis thaliana. Plant J.* **11**, 421-427.

Ahmad M., Lin C. and Cashmore A.R. (1995) Mutations throughout an *Arabidopsis* blue-light photoreceptor impair blue-light-responsive anthocyanin accumulation and inhibition of hypocotyl extension. *Plant J.* **8**, 653-658.

Allen G.J., Muir S.R. and Sanders D. (1995) Release of Ca^{2+} from individual plant vacuoles by both $InsP_3$ and cyclic ADP-ribose. *Science* **268**, 735-737.

Assmann S.M., Simoncini L. and Schroeder J.I. (1985) Blue light activates electrogenic ion pumping in guard cell protoplasts of *Vicia faba. Nature* **318**, 285-287.

Batschauer A., Ehmann B. and Schäfer E. (1991) Cloning and characterisation of a chalcone synthase gene from mustard and its light-dependent expression. *Plant Mol. Biol.* **16**, 175-185.

Batschauer A., Rocholl M., Kaiser T., Nagatani A., Furuya M. and Schäfer E. (1996) Blue and UV-A light-regulated *CHS* expression in *Arabidopsis* independent of phytochrome A and phytochrome B. *Plant J.* **9**, 63-69.

Bowler C., Neuhaus G., Yamagata H. and Chua N-H. (1994a) Cyclic GMP and calcium mediate phytochrome phototransduction. *Cell* **77**, 73-81.

Bowler C., Yamagata H., Neuhaus G. and Chua N-H. (1994b) Phytochrome signal transduction pathways are regulated by reciprocal control mechanisms. *Genes & Development* **8**, 2188-2202.

Cho M.H. and Spalding E.P. (1996) An anion channel in *Arabidopsis* hypocotyls activated by blue light. *Proc. Natl. Acad. Sci. USA* **93**, 8134-8138.

Christie J.M. and Jenkins G.I. (1996) Distinct UV-B and UV-A/blue light signal transduction pathways induce chalcone synthase gene expression in Arabidopsis cells. *Plant Cell* **8**, 1555-1567.

Dixon R.A. and Paiva N.L. (1995) Stress-induced phenylpropanoid metabolism. *Plant Cell* **7**, 1085-1097.

Feinbaum R.L. and Ausubel F.M. (1988) Transcriptional regulation of the *Arabidopsis thaliana* chalcone synthase gene. *Mol. Cell Biol.* **8**, 1985-1992.

Feinbaum R.L., Storz G. and Ausubel F.M. (1991) High intensity and blue light regulated expression of chimeric chalcone synthase genes in transgenic *Arabidopsis thaliana* plants. *Mol. Gen. Genet.* **226**, 449-456.

Feldbrugge M., Hahlbrock K. and Weisshaar B. (1996) The transcriptional regulator CPRF1: expression analysis and gene structure. *Mol. Gen. Genet.* **251**, 619-627.

Feldbrugge M., Sprenger M., Hahlbrock K. and Weisshaar B. (1997) PcMYB1, a novel plant protein containing a DNA-binding domain with one MYB repeat, interacts *in vivo* with a light-regulatory promoter unit. *Plant J.* **11**, in press.
Frohnmeyer H., Ehmann B., Kretsch T., Rocholl M., Harter. K., Nagatani A., Furuya M., Batschauer A., Hahlbrock K. and Schäfer E. (1992) Differential usage of photoreceptors for chalcone synthase gene expression during plant development. *Plant J.* **2**, 899-906.
Fuglevand G., Jackson J.A. and Jenkins G.I. (1996) UV-B, UV-A and blue light signal transduction pathways interact synergistically to regulate chalcone synthase gene expression in Arabidopsis. *Plant Cell* **8**, 2347-2357.
Harter K., Kircher S., Frohnmeyer H., Krenz M., Nagy F. and Schäfer, E. (1994) Light-regulated modification and nuclear translocation of cytosolic G-box binding factors in parsley. *Plant Cell* **6**, 545-559.
Jackson J.A., Fuglevand G., Brown B.A., Shaw M.J. and Jenkins G.I. (1995) Isolation of *Arabidopsis* mutants altered in the light-regulation of chalcone synthase gene expression using a transgenic screening approach. *Plant J.* **8**, 369-380.
Jackson J.A. and Jenkins G.I. (1995) Extension growth responses and flavonoid biosynthesis gene expression in the *Arabidopsis hy4* mutant. *Planta* **197**, 233-239.
Jenkins G.I., Christie J.M., Fuglevand G., Long J.C. and Jackson J.A. (1995) Plant responses to UV and blue light: biochemical and genetic approaches. *Plant Sci.* **112**, 117-138.
Kaiser T., Emmler K., Kretsch T., Weisshaar B., Schäfer E. and Batschauer, A. (1995) Promoter elements of the mustard *CHS1* gene are sufficient for light-regulation in transgenic plants. *Plant Mol. Biol.* **28**, 219-229.
Kaufman L.S. (1993) Transduction of blue light signals. *Plant Physiol.* **102**, 333-337.
Koornneef M., Rolff E. and Spruit C.J.P. (1980) Genetic control of light-inhibited hypocotyl elongation in *Arabidopsis thaliana. Z. Pflanzenphysiol.* **100**, 147-160.
Kubasek W.L., Shirley B.W., McKillop A., Goodman H.M., Briggs W.R. and Ausubel F.M. (1992) Regulation of flavonoid biosynthetic genes in germinating Arabidopsis seedlings. *Plant Cell* **4**, 1229-1236.
Lin C., Ahmad M. and Cashmore A.R. (1996) *Arabidopsis* cryptochrome 1 is a soluble protein mediating blue-light-dependent regulation of plant growth and development. *Plant J.* **10**, 893-902.
Lin C., Robertson D.E., Ahmad M., Raibekas R.A., Jorns S., Dutton L. and Cashmore A.R. (1995) Association of flavin adenine dinucleotide with the *Arabidopsis* blue light receptor CRY1. *Science* **269**, 968-970.
Liscum E. and Briggs W.R. (1995) Mutations in the *NPH1* locus of Arabidopsis disrupt the perception of phototropic stimuli. *Plant Cell* **7**, 473-485.
Liscum E. and Hangarter R.P. (1994) Mutational analysis of blue-light sensing in *Arabidopsis. Plant, Cell & Env.* **17**, 639-648.
Malhotra K., Kim S-T., Batschauer A., Dawut L. and Sancar A. (1995) Putative blue-light photoreceptors from *Arabidopsis thaliana* and *Sinapis alba* with a high degree of sequence homology to DNA photolyase contain the two photolyase cofactors but lack DNA repair activity. *Biochemistry* **34**, 6892-6899.
Mol J., Jenkins G.I., Schäfer E. and Weiss D. (1996) Signal perception, transduction, and gene expression involved in anthocyanin biosynthesis. *Crit. Rev. Plant Sci.* **15**, 525-557.

Neuhaus G, Bowler C., Kern R. and Chua N-H. (1993) Calcium/calmodulin-dependent and -independent phytochrome signal transduction pathways. *Cell* **73**, 937-952.

Nishizaki Y. (1994) Vanadate and dicyclohexylcarbodiimide inhibit the blue light-induced depolarisation of the membrane in pulvinar motor cells of *Phaseolus. Plant & Cell Physiol.* **35**, 841-844.

Ohl S., Hahlbrock K. and Schäfer E. (1989) A stable blue-light-derived signal modulates ultraviolet-light-induced activation of the chalcone synthase gene in cultured parsley cells. *Planta* **177**, 228-236.

Peters J.L., Schreuder M.E.L., Verduin S.J.W. and Kendrick R.E. (1992) Physiological characterization of a high-pigment mutant of tomato. *Photochem. Photobiol.* **56**, 75-82.

Schulze-Lefert P., Becker-Andre M., Schulz W., Hahlbrock K. and Dangl J.L. (1989) Functional architecture of the light-responsive chalcone synthase promoter from parsley. *Plant Cell* **1**, 707-714.

Shimazaki K., Ino M. and Zeiger E. (1986) Blue light-dependent proton extrusion by guard-cell protoplasts of *Vicia faba. Nature* **319**, 324-326.

Short T.W. and Briggs W.R. (1994) The transduction of blue light signals in higher plants. *Ann. Rev. Plant Physiol. Plant Mol. Biol.* **45**, 143-171.

Spalding E.P. and Cosgrove D.J. (1989) Large plama-membrane depolarization precedes rapid blue-light-induced growth inhibition in cucumber. *Planta* **178**, 407-410.

Spalding E.P. and Cosgrove D.J. (1992) Mechanism of blue-light-induced plasma-membrane depolarisation in etiolated cucumber hypocotyls. *Planta* **188**, 199-205.

Stapleton A.E. (1992) Ultraviolet radiation and plants: burning questions. *Plant Cell* **4**, 1353-1358.

Tevini M. and Teramura A.H. (1989) UV-B effects on terrestrial plants. *Photochem. Photobiol.* **50**, 479-487.

Weisshaar B., Armstrong G.A., Block A., da Costa e Silva O. and Hahlbrock K. (1991) Light-inducible and constitutively expressed DNA-binding proteins recognizing a plant promoter element with functional relevance in light-responsiveness. *EMBO J.* **10**, 1777-1786.

Whitelam G.C. and Harberd, N.P. (1994) Action and function of phytochrome family members revealed through the study of mutant and transgenic plants. *Plant, Cell & Env.* **17**, 615-625.

Molecular Approaches to Biochemical Purification: The COP9 Complex Paradigm

Daniel A. Chamovitz[1] and Xing-Wang Deng[2]

[1]Department of Biology, Tel Aviv University, Tel Aviv 69978, Israel
[2]Department of Biology, Yale University, New Haven, CT 06520-8104, USA

Correspondence to Daniel A. Chamovitz, email: chamd@post.tau.ac.il

1. Abstract

Light signals perceived from the photoreceptors are transduced into the nucleus to regulate gene expression and development. The nuclear-localized COP9 protein complex has a central role in modulating light-regulated development, and may have a role in regulating development in general in all organisms. We have employed a novel molecular approach to traditional biochemical methodologies in order to purify the COP9 complex. As the *COP9* gene has been cloned without knowing the biochemical activity of its protein product, the COP9 complex was purified from cauliflower by assaying for the presence of COP9 by immuno-blot analysis. This approach should be applicable to studying other proteins which have been identified through genetic and molecular means, but whose activities are not known.

2. Introduction

For the greater part of the past century, a major focus of biology was the *in vitro* study of proteins. This biochemical approach provided most of our current understanding of a variety of biological phenomena from alcohol fermentation to DNA replication. This "classical" approach is based on being able to assay in vitro for a given biological activity, and then purifying this activity until it is free from contaminating proteins.

The study of the RNA polymerase II transcription factors provides a classic example of biochemical studies leading to an in depth understanding of molecular mechanisms (Wampler et al., 1990, Dynlacht et al., 1991). These studies were based on having an assayable biological activity - in vitro transcription from a specific promoter, a ready source of material for purifying the proteins - Drosophila embryos, and provided the basis for isolating the components of the TATA-binding protein containing TFIID complex from Drosophila and the cloning of their genes (reviewed in Chen et al., 1994). These studies provided the groundwork which subsequently allowed sophisticated and in depth mechanistic and structural studies employing recombinant proteins (Chen et al., 1994, Dickstein et al., 1996, Thut et al., 1995).

However, many biological phenomena present no obvious biochemical activity. Indeed, morphological phenomena are not readily adapted to a cell free system. These limitations have lead to the prevalence of genetic and molecular approaches over the

NATO ASI Series, Vol. H 104
Cellular Integration of Signalling Pathways in Plant Development
Edited by F. Lo Schiavo, R. L. Last, G. Morelli, and N. V. Raikhel

past decade. Genetic approaches have enabled the dissection of complex biological phenomena and have enabled the cloning of numerous genes, yet they have left the basic mechanistic questions unsolved.

2.1 Analysis of Light-Signaling Mutant Plants

The study of light signaling in plants has been greatly aided through the analysis of mutants defective in various aspects of light responses (reviewed in Chamovitz and Deng, 1996). Light is the most important developmental signal impinging plants, and as such, sophisticated photosensory mechanisms exist which enable a plant to modulate development in response to changes in light quantity, duration, quality and direction.

The effect of light on plant development is most evident in seedlings of dicotyledonous plants. Seedling development proceeds through one of two distinct developmental pathways dependent on the ambient light conditions. In darkness, seedlings go through skotomorphogenesis, developing an etiolated phentype of an elongated hypocotyl, apical hook and closed cotyledons. In contrast, light-grown seedlings go through photomorphogenesis, exhibiting short hypocotyls, open and expanded cotyledons, chloroplast development, and continuation of development to adult stages (reviewed in von Arnim and Deng, 1996, McNellis and Deng, 1995).

Genetic screens have been devised to identify mutants defective in the switch between skotomorphogenic to photomorphogenic growth. Mutant screens for plants which show photomorphogenic phenotypes in darkness have lead to the identification of more than a dozen Arabidopsis loci (Deng, 1994; Chory, 1993; Misera et al., 1994). Mutations in 10 loci, *COP1*, *COP8-10*, *DET1*, and *FUS4, 5, 6, 11, 12*, show the most pleiotropic phenotypes in darkness, essentially mimicking light-grown plants. Though grown in total darkness, these mutant seedlings have short hypocotyls, open, expanded cotyledons, which show extensive cellular differentiation and contain chloroplast-like plastids, and show high expression levels of light-inducible genes such as *Cab*, *Chs* and *PsbA*. As mutations in these loci are all recessive, it was proposed that the corresponding wild-type gene products act as repressors of photomorphogenesis in the dark (Deng, 1994). Mutations in these loci are epistatic to mutations in phytochrome and CRY1 blue-light receptor, consistent with their gene products being downstream of the photoreceptors (Chory, 1992; Wei et al., 1994a; Wei et al., 1994b; Deng and Quail, 1992). In addition to their role in repressing photomorphogenesis in the dark, these gene products are essential for plant development in the light, as severe (null) mutations are lethal after the seedling stage.

The genes for four of these loci, *COP1* (Deng et al., 1992), *COP9* (Wei et al., 1994a), *FUS6*: (Castle and Meinke, 1994), and *DET1* (Pepper et al., 1994), have been cloned. Each of these genes encode nuclear proteins (von Arnim and Deng, 1994; Chamovitz et al., 1996; Pepper et al., 1994; Staub et al., 1996). Extensive molecular and genetic analysis have indicated that these genes are not transcriptionally or translationally regulated by light. While COP9, DET1 and FUS6 are localized to the nucleus in both light and dark conditions, nucleo-cytoplasmic partitioning of COP1 is both light and tissue dependent (von Arnim and Deng, 1994).

While the genetic and molecular analysis of these mutants has greatly furthered our understanding of light signaling in seedling development, they have not led to a mechanistic description of these gene products. The primary sequence of COP9, DET1 and FUS6 did not provide clues as to the biochemical activity of the proteins. To further study the function of COP9, we have adapted molecular techniques for a biochemical study of COP9.

As outlined above, classical biochemistry is based on an assayable biochemical activity to purify an unknown protein (or protein complex). *COP9* encodes a small hydrophilic protein with a calculated molecular mass of 22.5 kD (Wei et al., 1994) and no known biochemical activity. However, the cloning of COP9 provided the means for further biochemical studies by providing a source of recombinant protein for the production of antibodies (Wei et al., 1994). *In our adaptation of classical biochemistry, in place of assaying for a known biochemical activity, we assay for the presence of our known protein by immuno-blot analysis.*

3. Results and Discussion

As mutations in the 10 pleiotropic *COP/DET/FUS* loci lead to essentially identical phenotypes, it was hypothesized that at least some of their gene products function in proximity to each other, possibly with some directly interacting, to control the primary switch between skotomorphogenesis and photomorphogenesis (Kwok et al., 1996; Wei et al., 1994b). The affinity purified anti-COP9 antibodies allowed us to directly test this hypothesis. Gel filtration studies over both Superdex 200 and Sephacryl 300 columns indicated that COP9 functions *in vivo* in a high molecular weight protein complex of approximately 560 kD (Wei et al., 1994). Gel filtration analysis of the complex in various mutant lines further indicated that COP8 and FUS6 are likely additional components of the complex (Chamovitz and Deng 1997).

In order to further understand the role of the COP9 complex in regulating plant development it is necessary to identify the proteins which make up the COP9 complex, and to have a source of complex and individual subunits to carry out biochemical analysis. While a variety of advanced molecular techniques have been developed that enable the rapid cloning of genes which encode interactive proteins to a specific protein (i.e. the yeast two-hybrid system, Chien et al., 1989; "southwest" screening, Matsui et al., 1995), classic biochemical approaches provide several advantages (Table 1). Accordingly, we decided to biochemically purify the COP9 complex, assaying for the complex with our affinity-purified antibodies.

In order to determine the stability of the COP9, we monitored the gel filtration properties of under various conditions. The complex was stable to treatments with 1 M NaCl, 1.6 M NH_4SO_4, 4 M urea, and 0.1% Triton X-100, but dissociated in 1% SDS, 6 M urea, and 100 mM β-mercaptoethanol. These results indicate that the complex is relatively stable and thus should maintain its structure through a biochemical purification

The molecular and initial biochemical studies on COP9 and the COP9 complex were done on Arabidopsis seedlings. As Arabidopsis seedlings are not the prime tissue for protein purification owing to its small size (≈ 2 mg per seedling). Several commercially available plant sources were assayed for COP9 by immuno-blot

Table 1. Comparison of molecular approaches (i.e. the yeast two hybrid system and "southwest" screening) and biochemical approaches to identifying protein-protein interactions

Molecular approaches	Biochemical approaches
rely on simple protein-protein affinity	rely on physiological interactions
identify interactions between two proteins	identify interactions between a number of proteins
identify transient interactions	identify biochemical properties of desired protein
prone to artifacts	identify stable interactions

analysis (Figure 1). While our anti- COP9 antibodies recognized a protein of similar size to Arabidopsis COP9 in all tissues checked, COP9 was most abundant in heads of broccoli and cauliflower. Heads of *Brassica oleracea* L. (cauliflower), could be an excellent source for complex purification as cauliflower heads are composed primarily of floral and inflorescence meristems and do not contain chloroplasts, which are a potential source of major protein contaminants (see below). Most importantly, as cauliflower is a crucifer like Arabidopsis, a high degree of nucleic acid conservation is expected, which will aid in the eventual cloning of the various complex subunits.

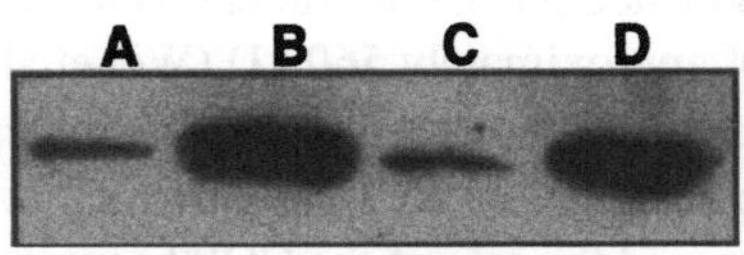

Figure 1. "Garden" immuno-blot detection of COP9 in various plants. A - bean sprouts; B - broccoli heads; C - alfalfa sprouts; D - cauliflower heads. Soluble proteins were extracted from each tissue and analyzed as described (Wei et al., 1994).

To determine if COP9 is present in cauliflower in a complexed form, cauliflower soluble proteins were fractionated on an analytic Superdex 200 gel filtration column. As was found previously for Arabidopsis (Wei et al., 1994), cauliflower COP9 elutes solely at a position corresponding to a molecular mass of approximately 500 kD (Figure 2A). Similar to Arabidopsis, no COP9 monomers were detected in the cauliflower extract, suggesting that COP9 functions in cauliflower only in a complexed form. The COP9 complex form Arabidopsis is very similar in size to the highly abundant ribulose bisphosphate carboxylase (RUBISCO) (Wei et al., 1994). Cauliflower heads essentially lack RUBISCO, as indicated by the silver stained profile of the gel filtration fractions, a property which will aid in the purification of the complex (Figure 2B).

The COP9 complex was purified from cauliflower head extracts through an empirically determined purification scheme (Chamovitz et al, 1996, Table 2, Figure 3). In lieu of any supporting biochemical data which would aid us in planning the

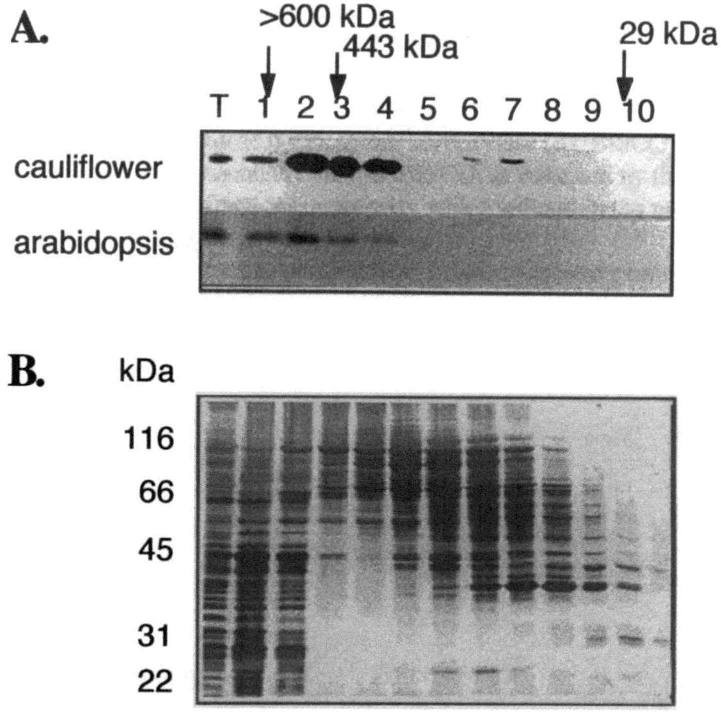

Figure 2. The COP9 complex in cauliflower is similar to that in Arabidopsis. A. Gel filtration profile of COP9 from cauliflower heads and Arabidopsis seedlings. The bands seen in lanes 6 and 7 are background. B. Silver-stained profile of the gel filtration fractions from cauliflower. The elution points of three size standards are noted: >600 kD - thyroglobulin, 443 kD - apoferritin, 29 kD - carbonic anhydrase. Protein extraction, gel filtration and silver staining according to Chamovitz et al., 1996 and Wei et al, 1994.

biochemical purification, a brute force approach was taken and various purification media and techniques were attempted, including, ion (anion and cation) exchange chromatography, hydrophobic interaction chromatography, various affinity chromatography resins, selective precipitation techniques and dialfiltration (for general guidelines to protein purification, see Deutscher, 1990, Janson and Ryden, 1989). Following each trial, the partially purified COP9 was further analyzed for complex integrity by Superdex 200 gel filtration chromatography.

We ascertained that we had succeeded in purifying the COP9 complex to near homogeneity when the fold purification of COP9 monomer in the purified fraction was similar to that anticipated for the purified complex (Chamovitz et al., 1996), and when the major protein bands were determined by amino acid sequence analysis to be COP9 and other components of the complex (see below).

COP9 eluted in the final gel filtration step over a Superose 6 column as a large protein complex of 560 kD, which copurified with approximately 11 other proteins. Amino acid sequence analysis of the 22 kD subunit confirmed its identify as COP9, while amino acid sequence analysis of the 50 kD and 52 kD copurifying proteins

Table 2. Outline of COP9 Complex Purification Protocol. More details are provided in Chamovitz et al., 1996.

Step	Media	Notes
1. total soluble extract		
2. anion exchange	Sepharose FF - pH 7.0	step separation and concentration
3. dilafiltration	Filtron 300k Ultrasette	buffer exchange concentrate large proteins
4. anion exchange	Sepharose HR - pH 7.0	gradient
5. affinity	heparin	step separation
6. anion exchange	Mono Q - pH5.7	gradient
7. concentration	Filtron 100k Microsep	to 100 μl
8. gel filtration	Superose 6	elutes at 560 kD

confirmed their identity as FUS6 (Chamovitz et al., 1996). As mentioned above, plants mutant in *FUS6* or *COP9* are phenotypically identical, and the COP9 complex is absent in *fus6* mutant seedlings. The identification of FUS6 as a component of the COP9 complex was further confirmed by showing that antibodies against either COP9 or FUS6 selectively coimmunoprecipitate both proteins from total Arabidopsis cellular extracts (Staub et al., 1996).

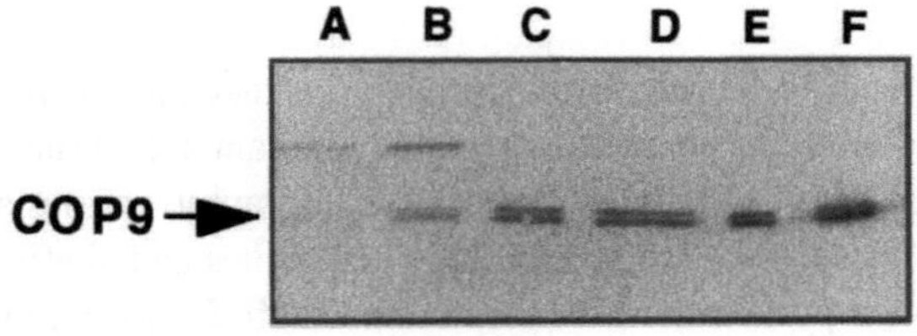

Figure 3. Enrichment of COP9 during purification of the complex. A - 12 μg of total extract (step 1 in Table 2); B - 10 μg following 1st ion exchange (step 2 in Table 2); C - 5 μg following 2nd ion exchange (step 4 in Table 2); D - 2 μg of heparin fraction (step 5 in Table 2); E - 1 μg following 3rd ion exchange (step 6 in Table 2); F - 1 μg following final gel filtration (step 8 in Table 2). According to Chamovitz et al., 1996

The purification of the COP9 complex is an important and essential step in understanding the function of the COP9 complex in regulating plant development. We have succeeded in proceeding from genetics (a mutant phenotype) (Wei and Deng, 1992), to molecular biology (cloning of *COP9*) (Wei et al., 1994), to the biochemical purification of the multi-subunit COP9 protein complex (Chamovitz et al., 1996). This approach should be applicable to studying other proteins which have been identified through genetic and molecular means, but whose activities are not known.

Peptide sequences have been obtained for all the putative complex subunits, and in most cases these peptides have identified cDNA clones in the Arabidopsis EST databases. This purification now provides the tools which will allow us to further dissect the COP9 complex at the biochemical and molecular levels. As the complex is conserved between the Plant and Animal kingdoms (Chamovitz and Deng, 1995; 1997), these studies will have ramifications for understanding developmental regulation also in animal systems.

4. Acknowledgments

Our research on the COP9 complex is supported by a grant from the National Science Foundation, a National Science Foundation Presidential Faculty Fellow award (to X.-W.D.), and a Human Frontiers Research Science Program Long Term Fellowship and an Israel Council for Higher Education Yigal Alon Award (to D.A.C.).

5. References

von Arnim, A. G., and X. W. Deng (1994). Light inactivation of Arabidopsis photomorphogenic repressor COP1 involves a cell-specific regulation of its nucleoplasmic partitioning. Cell 79, 1035-1045.

von Arnim, A.G. and Deng, X.-W. (1996). Light control of seedling development. Annu. Rev. Plant Physiol. Plant Mol. Biol. 47, 215-243.

Castle, L. and Meinke, D. (1994). A *FUSCA* gene of Arabidopsis encodes a novel protein essential for plant development. Plant Cell 6, 25-41.

Chamovitz, D. and Deng, X.-W. (1995). The novel *COP,DET* genes of the Arabidopsis light signaling pathway may define a group of general developmental regulators shared by both animal and plant kingdoms. Cell 82, 353-354.

Chamovitz. D.A., Wei, N., Osterlund, M.T., von Arnim, A.G., Staub, J.M., Matsui, M. And Deng, X.-W. (1996). The COP9 Complex, a novel multisubunit nuclear regulator involved in light control of a plant developmental switch. Cell 86, 115-121.

Chamovitz, D.A. and Deng, X.-W. (1996). Light signaling in plants. Crit. Rev. Plant Sci. 15, 455-478.

Chamovitz, D.A. and Deng, X.-W. (1997). The COP9 Complex: A link between photomorphogenesis and general developmental regulation? Plant, Cell and Environ. 20, 734-739.

Chen, J.-L., Attardi. L.D., Verrizjer, C.P., Yokomori, K., Tjian, R. (1994). Assembly of recombinant TFIID reveals differential coactivator requirements for distinct transcriptional activators. Cell 79, 93-105.

Chien, C.-T., Bartel, P.-L., Sternglanz, R. and Fields, S. (1989). The two-hybrid system: a method to identify and clone genes for proteins that interact with a protein of interest. Proc. Nat. Acad. Sci. USA 88, 9578-9582.

Chory, J. (1992). A genetic model for light-regulated seedling development in Arabidopsis. Development 115, 337-354.

Chory, J. (1993). Out of darkness: mutants reveal pathways controlling light-regulated development in plants. Trends in Genetics 9, 167-172.

Deng, X.-W. and Quail, P.H. (1992). Genetic and phenotypic characterization of *cop1* mutants of *Arabidopsis thaliana*. Plant J. 2, 83-95.

Deng, X.-W. (1994). Fresh view on light signal transduction in plants. Cell, 76, 423-426.

Deng, X.-W., Matsui, M., Wei, N., Wagner, D., Chu, A.M., Feldmann, K.A. and Quail, P.H. (1992). *COP*1, an Arabidopsis regulatory gene, encodes a novel protein with both a Zn-binding motif and a Gβ homologous domain. Cell 71, 791-801.

Deutscher, M.P. (1990). Guide to Protein Purification. Methods in Enzymology 182.

Dikstein, R., Ruppert, S., Tjian, R. (1996). TAFII250 is a bipartite protein kinase that phosphorylates the basal transcription factor RAP74. Cell 84, 781-790.

Dynlacht, B.D., Hoey, T., Tjian, R. (1991). Isolation of coactivators associated with the TATA-binding protein that mediate transcriptional activation. Cell 66, 563-576.

Janson, J.C. and Ryden, L. (1989). Protein Purification. New York: VCH.

Kwok, S.F., Piekos, B., Miséra, S., and Deng, X.-W. (1996). A complement of ten Arabidopsis pleiotropic *COP/DET/FUSCA* loci are essential for plant development and necessary for repression of photomorphogenesis. Plant Physiol. 110, 731-742.

Matsui, M., Stoop, C. D., von Arnim, A. G., Wei, N., and X. W. Deng (1995). Arabidopsis COP1 specifically interacts in vitro with a novel cytoskeleton associated protein. Proc. Natl. Acad. Acd. USA 92, 4239-4243.

McNellis, T.W. and Deng, X.-W. (1995). Light control of seedling morphogenetic pattern. Plant Cell 7, 1749-1761.

Miséra, S., Müller, A.J., Weiland-Heidecker, U., and Jürgens, G. (1994). The *FUSCA* genes of Arabidopsis: negative regulators of light responses. Mol. Gen. Genet. 244, 242-252.

Pepper, A., Delaney, T., Washburn, T., Poole, D., and Chory, J. (1994). *DET1*, a negative regulator of light-mediated development and gene expression in Arabidopsis, encodes a novel nuclear-localized protein. Cell 78, 109-116.

Staub, J.M., Wei, N. and Deng, X.-W. (1996). Evidence for FUS6 as apart of the nuclear localized COP9 complex in Arabidopsis. Plant Cell, in press.

Thut, C.J., Chen, J.-L., Klemm, R., and Tjian, R. (1995). p53 transcriptional activation mediated by coactivators TAFII40 and TAFII60. Science 267, 100-104.

Wampler, S.L., Tyree, C.M., Kadonaga, J.T. (1990). Fractionation of the general RNA polymerase II transcription factors from Drosophila embryos. J. Biol. Chem. 265, 21223-21231.

Wei, N. and Deng, X.-W. (1992). COP9: a new genetic locus involved in light-regulated development and gene expression in Arabidopsis. Plant Cell 4, 1507-1518

Wei, N., Kwok, S.F., von Arnim, A.G., Lee, A., McNellis, T., Piekos, B. and Deng, X.-W. (1994a). Arabidopsis *COP*8, *COP*10 and *COP*11 genes are involved in repression of photomorphogenic developmental pathway in darkness. Plant Cell 6, 629-643.

Wei, N., Chamovitz, D.A., and Deng, X.-W. (1994b). Arabidopsis COP9 is a component of a novel signaling complex mediating light control of plant development. Cell 78, 117-124.

Silencing of Gene Expression in the Anthocyanin Regulatory Gene Families

Angela Ronchi, Roberto Pilu and Chiara Tonelli

Dipartimento di Genetica e di Biologia dei microrganismi
Università degli Studi di Milano
Via Celoria 26, 20133 Milano, Italy

1 Introduction

An inference that can be drawn from Mendel's first law is that contrasting alleles emerge from the heterozygotes without having influenced each other. Thus the pattern of expression of a gene with intact regulatory and coding regions is generally transmitted unchanged to the progeny upon segregation. However, exceptions to this rule have been observed in a number of organisms and include 'epigenetic' modifications of gene expression that are heritable and do not involve changes to the nucleotide sequence.

Genomic imprinting is the best characterized exception to classical Mendelian inheritance. In this widespread process, examples of which can be found in mammals and plants, the expression of a gene depends on the sex of the parent from which it has been inherited and arises from epigenetic mechanisms which modify DNA (i.e. methylation) and/or chromatin structure (Matzke and Matzke 1993; Fundele and Surani 1994).

Epigenetic silencing of gene expression has been widely documented in different organisms. In general, gene inactivation is mediated both by sequences that are foreign to the genome, such as transgenes and mobile genetic elements, and by endogenous DNA sequences bearing varying degrees of similarity to each other. In plants, the inactivation of transposons has been shown to be associated with their methylation (Chandler and Walbot 1986; Chomet et al. 1987; Fedoroff et al. 1996; Martienssen 1996). Whilst it has been suggested that methylation serves to restrict the movement of mobile elements by inactivating them (Chandler and Walbot 1986, Bestor 1990; Brutnell and Dellaporta 1994), the causal relationship between DNA methylation and transposon inactivation is still unclear. Considering gene inactivation to be a consequence of DNA methylation, the inactivation could result from direct interference with protein binding by the methyl adduct, or indirectly through the presence of methyl-DNA binding proteins (Meehan et al. 1989; Boyes and Bird 1991). Conversely, DNA methylation may be a consequence of gene inactivation.

Silencing phenomena have also been encountered in transgene experiments. In *Neurospora crassa*, multiple copies of a gene introduced via DNA-mediated transformation can result in the loss of expression of all copies of that gene during

NATO ASI Series, Vol. H 104
Cellular Integration of Signalling Pathways in Plant Development
Edited by F. Lo Schiavo, R. L. Last, G. Morelli, and N. V. Raikhel

vegetative growth (Pandit and Russo 1992; Romano and Macino 1992). The duplicated sequences are modified by methylation and by C-to-T transitions, a phenomenon called RIP (Repeat-Induced Point mutation) (Selker 1990). In *Ascobolus* and *Coprinus cinereus* however, duplicated sequences only undergo methylation (Rhounim et al. 1992; Freedman and Pukkila 1993). In transgenic plants, mechanisms that silence or alter gene expression in the presence or in the absence of methylation have been documented, e.g. co-suppression and trans-inactivation, (Matzke and Matzke 1993), in which the ectopic transgene can suppress/modify its own expression and that of the endogenous gene (Jorgenssen 1994; Flavell 1994; Meyer and Saedler 1996).

All these phenomena leading to the silencing or modification of gene expression seem to share a common feature, the presence of duplicated sequences in the genome which may be potentially deleterious (e.g. transposons). It would appear that transgenes become silenced because they are present in multiple copies, while endogenous genes may be incorrectly identified as 'foreign' because they become members of a repetitive set or because they contain fragments of transposable elements that are widespread, i.e. duplicated, in the genome.

Eukaryotes have efficient but poorly understood genome-wide searching mechanisms for homologous sequences. Homologues may be located only a few hundred nucleotides apart or on non-homologous chromosomes. Evolutionary considerations suggest that this gene copy-number scanning machinery has evolved as a defense system that disables invasive or foreign DNA, as well as preventing overexpression of endogenous genes. It is evident that proliferating sequences give rise to serious problems of insertional mutagenesis and ectopic recombination that can disrupt chromosome structure (Kricker et al. 1992; Matzke et al. 1996).

2 Paramutation and Related Processes of Plant Epigenetic Regulation

Several unusual epigenetic phenomena involving interactions between alleles or heritable and metastable changes of the genome have been described in plants by Brink (1958; 1973) and McClintock (1967). The advent of molecular work has renewed interest in these phenomena since unravelling the mechanisms underlying these events may be of significant value in the general context of understanding the control of gene expression and cell differentiation.

Paramutation, as defined by Brink (1958), is "an interaction between alleles that leads to directed, heritable change at the locus with high frequency and sometimes invariably within the time span of a generation". The term implies a phenomenon distinct from mutation where the latter occurs sporadically and undirected. Paramutation regularly occurs under specific conditions and the resulting change in phenotype is specific and unidirectional. In other words, the expression of one allele (the recipient, termed paramutable) has been heritably changed through the interaction, during the heterozygous state, with another allele (the inducer, termed paramutagenic). This interaction is dependent on homology but is not generally dependent on meiotic pairing.

The known examples of paramutation come mainly from higher plants and, with the exception of the *sulf* locus in tomato (Hagemann 1969), the phenomenon has been characterized in genes that regulates the anthocyanin pathway in maize. This pathway is an excellent model system with which to monitor epigenetic changes since their effects on color intensity, distribution and tissue specificity can be easily visualized. The enzymes involved in pigment biosynthesis are expressed in a coordinated manner (reviewed by Dooner et al. 1991). Genetic and molecular analyses indicate that the structural pigmentation genes are temporally and spatially regulated by two families of transcription factors : the *R/B* family, encoding related proteins with sequence homologies to the basic helix-loop-helix (bHLH) DNA binding/dimerization domain found in the myc oncoproteins (Chandler et al. 1989; Ludwig et al. 1989; Tonelli et al. 1991; Consonni et al. 1993) and the *C1/Pl* family, encoding related proteins with sequence homologies to the DNA binding domains of the myb oncoproteins (Paz-Ares et al. 1987; Cone et al. 1986, 1993). One member from each family must be expressed for transcriptional activation of the pigment biosynthetic genes to occur. Numerous studies indicate that each family shows extensive allelic variability, comprising multiple homologous genes that control pigmentation in specific tissues. All paramutation phenomena observed in maize involve the regulatory genes of the anthocyanin biosynthetic pathway (Table 1 and 2). The occurrence of epigenetic mechanisms that down-regulate the anthocyanin genes may be due to the fact that the regulatory genes involved share sequence homology.

2.1 Paramutation at the *R* Locus

The most extensive investigation of paramutation is that of the *R* locus made by Brink (1973). Paramutation can be described as a quantitative change in the phenotypic expression of an *R* allele. The change, which leads to a lower level of pigmentation, is genetically transmissible. The standard *R-r* allele, a complex locus that contains multiple homologous *r* genes, determines the pigmentation of the aleurone, the outer layer of the endosperm, and its level of expression can be monitored by assaying the amounts of anthocyanins accumulated in this tissue. Paramutation was first observed in the progeny of *R-r/R-st* heterozygotes crossed to a null allele *r-g*. While the *R-st* allele is not altered with respect to its original phenotype, the *R-r* allele is heritably changed to yield a weakly pigmented phenotype. The *R-r* allele is said to be paramutable and is designated *R-r'*, while the *R-st* allele that suppresses the level of pigmentation of *R-r*, is termed paramutagenic. Paramutation is progressive through generations, the level of repression depending on the number of generations in which the *R-r* allele has been heterozygous with *R-st*. Upon separation of the alleles by segregation, *R-r'* tends to recover its pigmenting potential after a few generations but stabilizes at a level below that of the original *R-r* (Brink 1973; Kermicle 1973). Another intriguing observation is that all *R-r'* derivatives tend to be paramutagenic themselves when put in trans with another *R-r*, though this secondary paramutation has a lesser suppressive effect than *R-st*. Changes in the structure of the chromosome where *R-r* is located, such as insertion of *R-r* into a reciprocally translocated chromosome or into a chromosome which carries a large heterochromatic segment at the end, reduces the sensitivity of *R-r* to paramutation in *R-st* heterozygotes. Recently it has been reported that changes in *R-st* structure alter the level of its paramutagenicity (Kermicle et al. 1995). The *R-st* locus is comprised of four *r* genes arranged in direct orientation. Deletions of *R-st* in which between one to three *r* genes are retained have been obtained through unequal crossing over and it has been shown that paramutagenic strength decreases in parallel with copy number, the single *R-st* gene no longer being paramutagenic. Subsequent re-addition of *r* genes by intragenic recombination restores paramutagenicity in relation to the total *r* gene number.

Table 1: Paramutation of the anthocyanin regulatory genes

Paramutable allele	Paramutagenic allele	Percentage of paramutation[a]	Paramutation observed in F1	Reversion to w.t.	Secondary paramutation	Reference
R-r	*R-st*	100	+/-	+[b]	+	Brink 1973
B-I	*B'*	100	+	-	++	Coe 1966
Pl-Rh	*Pl'-mah*	100	+	+	++	Hollick et al. 1995
pl-W22	*pl-bol3*	100	n.d.[c]	+	+/-	Tonelli and Pilu (unpublished)

(a) Frequencies calculated in the progeny of test-crosses with null or contrasting alleles.
(b) *R-r''*, which derives from two generations of exposure to a paramutagenic allele, does not spontaneously revert.
(c) n.d., not determined

2.2 Paramutation at the *B* Locus

In the case of *Booster* (*B*), studied by Coe (1966), alleles of *B* and *b* determine the presence and absence respectively, of anthocyanin pigmentation in the vegetative parts of the plant, where *B* is dominant to *b*. Homozygous plants carrying the *B-I* allele are intensely colored but occasionally *B-I* mutates to *B'* which results in reduced pigment accumulation in the same plant tissues. *B'* is extremely paramutagenic. Heterozygous *B-I/B'* plants are weakly pigmented, *B'* is almost completely dominant over *B-I*, and all the progeny arising either from self pollination or from test-crosses with the non-functional *b* allele show the *B'* phenotype. Paramutation is, in this case, 100% efficient. In fact *B-I* is never transmitted from a *B-I/B'* heterozygote, it is always paramutated to *B'*. This lack of segregation is not due to segregation distortion since markers linked to *B-I* segregate normally. The new *B'* allele becomes paramutagenic and is able to paramutate new *B-I* alleles that were never previously exposed to *B'*.

Molecular data (Patterson et al. 1993) shows that *B'* acts in trans to suppress the transcription of *B-I* which consequently remains low also in subsequent generations, even when the paramutable *B'* allele is segregated away. It was also reported that despite the dramatic differences in phenotype and transcription of *B'* and *B-I*, no changes in DNA sequence or methylation pattern where found between the two alleles. On the basis of these observation the authors suggested a model in which an allelic interaction promotes heritable changes in chromatin structure.

Table 2 : Characteristics of the anthocyanin regulatory genes involved in Paramutation

Allele	Derivation	Molecular features	Change into paramutagenic derivative	Phenotype[(a)]	References
R-r	Natural, paramutable	complex		Dark red aleurone	Dooner and Kermicle 1971
R-st	Natural, paramutagenic	complex	1	Stippled aleurone	
B-I	Natural	simple	10^{-1}-10^{-2}	Dark purple plant	Coe 1966, Patterson et al. 1995
B'	Paramutagenic derivative of B-I	simple		Light purple plant	
Pl-Rh	Natural	simple	10^{-1}	Dark red plant	Hollick et al. 1995
Pl'-mah	Paramutagenic derivative of Pl-Rh	simple		Mahogany plant	
pl-W22	Natural, paramutable	simple		Light red mesocotyl	Tonelli and Pilu, (unpublished)
pl-bol3	Natural, paramutagenic	complex	1	Strong red mesocotyl	

(a) Phenotype of organs scored for paramutation

2.3 Paramutation at the *pl* Locus

Paramutation also occurs in *pl*, a member of the myb family of anthocyanin regulatory genes (Cone et al. 1993; Hollick et al. 1995). The *Pl-Rh* allele results in dark purple anthocyanin pigmentation in several tissues. Spontaneous derivatives of *Pl-Rh*, termed *Pl'-mah*, have been isolated on the basis of an altered phenotype termed 'mahogany' that confers reduced and sunlight dependent pigmentation. These derivatives influence *Pl-Rh* such that only the *Pl'-mah* phenotype is transmitted from a *Pl-Rh/Pl'-mah* heterozygote. It has been demonstrated that the phenotypic expression of *Pl'-mah* is inversely correlated to paramutagenic strength and that the paramutagenic actvity is genetically linked to the *pl* locus. The behavior of the *pl* system is unique, yet it shares characteristics with paramutation at both *b* and *r* loci (Table 1).

Another particular case of epigenetic regulation is represented by the *Pl-Bh* gene. This allele of *pl* differs from the wild type in both its pattern of expression and its organ specificity. In contrast to the uniformed pigmentation observed in *Pl* plants, the *Pl-Bh* gene determines a variegated pigmentation in all tissues of the plants, including the aleurone, a tissue not pigmented by other *Pl* alleles. However, only a single base pair has been detected to differ between this allele and the wild type gene, while the lower level of mRNA expression found in the variegated *Pl-Bh* tissues was correlated with hypermethylation of *Pl-Bh* DNA (Cocciolone and Cone, 1993).

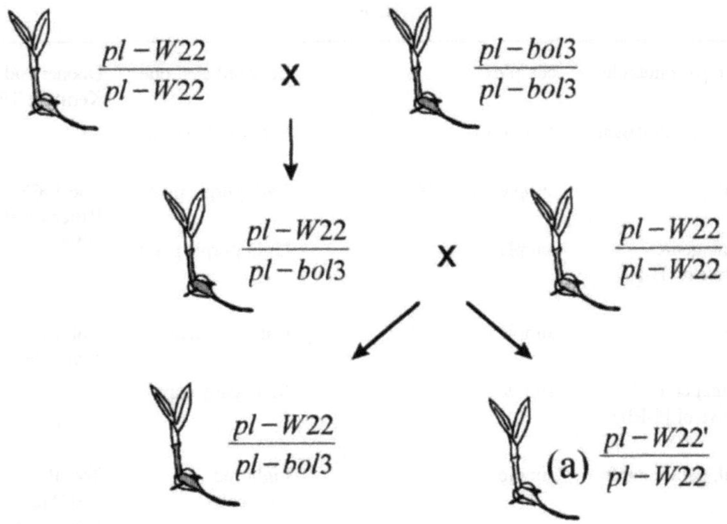

Figure 1. *pl* gene paramutation. The *pl-bol3* allele produces strong pigmentation. The *pl-W22* produces weak pigmentation and is affected by *pl-bol3*. (a) The *pl-W22'* is the dominant paramutated allele that produces four time less pigment than the parental *pl-W22*.

Recently we have discovered a new case of allelic interaction between two alleles of the *pl* family (Tonelli and Pilu, manuscript in preparation). We have isolated a *pl* allele, named *pl-bol3*, that is able to negatively influence the phenotypic expression of another *pl* allele, resident in the W-22 line termed *pl-W22*. Both these alleles are recessive to *Pl-Rh* but they exhibit tissue specific pigmentation (e.g. mesocotyl) which differs from that of *Pl-Rh*. In addition the light requirement of the *pl-bol3* and *pl-W22* alleles is different from that of *Pl-Rh*, being their expression light dependent. Preliminary genetic data indicate that this paramutation phenomenon, observed in all the juvenile tissues of the maize seedling, occurs with a frequency of 100%. *pl-W22* emerging from an heterozygous state with *pl-bol3* shows a four fold decrease in pigment content and this reduction is transmitted to subsequent progeny (Figure 1).
Other characteristics of this phenomenon that are hallmarks of paramutation, are the progressive reversion to the wild type phenotype following three/four generations of test-crosses and a weak ability of the *pl-W22'* to exert secondary paramutation.

Furthermore, unlike the *B'/B-I* and *Pl-Rh/Pl'-mah* systems, *pl-bol3* is not a spontaneous derivative of the *pl-W22* allele, but is instead a distinctive allele, similar to *R-r* and *R-st*. At the molecular level, *pl-bol3* is a complex of four genes, while *pl-W22* is a simple gene. This is particularly interesting in view of recent molecular and genetic studies indicating paramutation as a silencing process whose strength decreases in parallel with gene copy number and sequence similarity (Eggleston et al. 1995; Kermicle et al. 1995). It is worth noting that *Pl-Rh* and *Pl-mah*, although simple loci consisting of only one gene each, both contain a long duplication at the 3'end (Cone et al. 1993). The *B* gene is also a single gene but it probably represents an ancestral duplication of the *r* gene (Chandler et al. 1989; Consonni et al. 1993). Similarly, *pl* shows a high degree of sequence homology with *C1* (Cone et al. 1993) as well as with numerous other myb genes widespread in the genome (Chiara Tonelli, unpublished results).

Analogies with paramutation can be found in a case of Arabidopsis plants transformed with multiple copy constructs where unequal crossing-over has been used to analyze transgene silencing (Assaad et al. 1993). The changes in copy number were correlated with changes in transgene expression and promoter region methylation and, similarly to *r* paramutation, the level of gene silencing was progressive over generations and the effects potentially reversible.

To date, the specific behavior of the *pl* system involving *Pl-Rh/Pl'-mah*, disclosed similarities to those seen at either the *b* and *r* loci, but not both. The finding of a new paramutation case involving other *pl* alleles that shows similar characteristics to the *r* system indicates that paramutation at the *pl* locus may not necessarily present unique features. This observation supports the hypothesis that the basic process of paramutation may be mechanistically similar in all three loci of maize accessible to analysis.

2.4 REED : Reduced Expression of Endogenous Duplications

The mechanisms underlying gene inactivation are still unclear, though ectopic pairing (Jorgenssen, 1994) and interactions between sense and antisense RNA (Grierson et al. 1991) have been suggested. Nevertheless, the several cases of repression of gene expression which correlate with elevated sequence homology indicate that a cellular system exists that is able to count the copy number of sequences in the genome.

The term *REED* (Reduced Expression of Endogenous Duplications) has been coined for the interaction observed between the duplicated endogenous genes *R* and *Sn*, regulatory genes of the anthocyanin pathway (Ronchi et al. 1995). As a result of their coupling in the genome, *Sn* is partially silenced giving rise to weakly pigmented derivatives. Seedlings bearing a normal level of red pigmentation have been termed Sn-I, while the weak derivatives are termed Sn-Tr. No gross DNA rearrangement has been found to be the basis of the interaction, while differences in pigmentation were inversely correlated with differences in the degree of methylation of the *Sn* promoter. These results have been supported by data from treatment of plants with 5-azacytidine (AZA), a demethylating agent, that restored a strong level of pigmentation that was transmitted to the progeny and correlated with an increased expression of the *Sn* and structural gene transcripts. Genomic sequencing showed a negative correlation between the level of *Sn* expression in Sn-I, Sn-tr and Sn-AZA plants and the presence of methyl-C-residues within the promoter. The degree of cytosine methylation has been shown to be very low in the AZA-treated *Sn* promoter compared with Sn-I and Sn-tr promoters, while the latter showed the highest level of methyl-C residues. Interestingly, the presence of methyl groups for some C residues is not restricted to the canonical CpG or CpNpG context. In addition some putative binding sites for transcription factors are present in regions of the *Sn* promoter in which methylation has been observed. However, despite the clear correlation between methylation and the *Sn REED* phenomenon, the question of whether methylation is a consequence or a cause of the altered gene expression has not yet been established. Other factors could exist which coordinate both the expression level and the methylation state of a gene promoter. What seems to be important is that duplicated sequences, particularly those exerting a critical role in regulation, must be maintained in the genome under a strict dosage control. It has been found recently that for some of these regulatory genes (e.g. *pl*) extremely low levels of transcription are sufficient for the activation of structural gene expression (Procissi et al. 1997). In contrast, their overexpression in transgenic systems often leads to pleiotropic effects on development (Lloyd et al. 1992). Thus silencing systems may be a necessary level of control for those genes whose expression needs to be very tightly modulated and the understanding of their mechanism of action may well have broad implications for gene regulation within eukaryotic genomes.

Acknowledgements

We are grateful to Andrew MacCabe for critical reading of the manuscript. This work was supported by EC-BIOTECH Grant No. BIO2 CT93 0101 and by MIRAAF P.F.N. 'Biotecnologie vegetali' (Area 1 Progetto No.2) to C.T.

References

Assaad F.F., Tucker K.L., Signer E.R. (1993) Epigenetic repeat-induced gene silencing (RIGS) in *Arabidopsis*. Plant Mol. Biol. 22 :1067-1085.

Bestor, T. (1990) DNA methylation: evolution of a bacterial immune function into a regulator of gene expression and genome structure in higher eukariotes. Philos. Trans. R. Soc. London Ser. B 326, 179-187.

Boyes, J., Bird, A. (1991) DNA methylation inhibits transcription indirectly via a methyl-CpG binding protein. Cell, 64: 1123-1134.

Brink, R.A. (1958) Paramutation at the *R* locus in maize. Cold Spring Harbour Symp. Quant. Biol. 23: 379-391.

Brink, R.A. (1973) Paramutation. Annu. Rev. Genet. 7: 129-152.

Brutnell T.P., Dellaporta S.L. (1994) Somatic inactivation and reactivation of *Ac* associated with changes in cytosine methylation and transposase expression. Genetics 138:213-225

Chandler, V.L., Radicella, J.P., Robbins, T.P., Chen, J., Turks, D. (1989) Two regulatory genes of the maize anthocyanin pathway are homologous: isolation of *b* using *r* genomic sequences. Plant Cell 1: 1175-1183.

Chandler, V.L., Walbot, V. (1986) DNA modification of a maize transposable element correlates with loss of activity. Proc. Natl. Acad. Sci. USA, 83: 1767-71

Chomet, P.S., Wessler, S., Dellaporta, S. (1987) Inactivation of the maize transposable element activator (*Ac*) is associated with DNA modification. EMBO J. 6: 295-302.

Cocciolone, M.S., Cone, K.C. (1993) *Pl-Bh*, an anthocyanin regulatory gene of maize that leads to variegated pigmentation. Genetics 153: 575-588.

Coe, E.H. (1966) The properties, origins and mechanism of conversion-type inheritance at the *B* locus in maize. Genetics 53: 1035-1063.

Cone, K.C., Burr F.A., Burr B. (1986) Molecular analysis of the maize anthocyanin regulatory locus *C1*. Proc. Natl. Acad. Sci. USA, 83: 9631-9635.

Cone, K. C., Cocciolone, S.M., Burr A.F., Burr B. (1993) Maize anthocyanin regulatory gene pl is a duplicate of *c1* that function in the plant. Plant Cell, 5: 1795-1805.

Consonni, G., Geuna, F., Gavazzi, G., Tonelli, C. (1993) Molecular homology among members of the *R* gene family in maize. Plant Journal 3: 335-346.

Dooner H.K., Robbins T.P., Jorgensen R.A. (1991) Genetic and developmental control of anthocyanins biosynthesis. Ann. Rev. Genet. 25: 173-199.

Dooner, H.K., Kermicle, J.L. (1971) Structure of the R^r tandem duplication in maize. Genetics 67: 427-436.

Eggleston, W.B., Alleman, M., Kermicle, J.L. (1995) Molecular organization and germinal instability of *R-stippled* in maize. Genetics, 141: 347-360.

Fedoroff, N., Schlappi, M., Raina, R. (1996) Epigenetic regulation of the maize *Spm* transposon. Bioessays 17: 291-294.

Flavell, R.B. (1994) Inactivation of gene expression in plants as a consequence of novel sequence duplication. Proc. Natl. Acad. Sci. USA, 91: 3490-3496.

Freedman, T., Pukkila, P.J. (1993) *De novo* methylation of repeated sequences in *Coprinus cinereus*. Genetics, 135: 357-366.

Fundele, I.H.R., Surani, A. M. (1994) Experimental embryological analysis of genetic imprinting in mouse development. Develop. Genet. 15: 515-522.

Grierson, D., Fray, R.G., Hamilton, A.J., Smith, C.J.S., Watson, C.F. (1991) Does co-suppression of sense genes in transgenic plants involve antisense RNA? Trends Biotech. 9: 122-123.

Hagemann, R.(1969) Somatic reversion (paramutation) at the *Sulfurea* locus of *Lycopersicon esculentum*. III Studies with trisomics. Can.J.Genet.Cytol. 11: 346-358.

Hollick, J.B., Patterson, G:I:, Coe, E.H.Jr., Cone, C.K., Chandler, V.L. (1995) Allelic interactions heritably alter the activity of a metastable maize *pl* allele. Genetics, 141: 709-719.

Jorgenssen, R. (1994) Developmental significance of epigenetic impositions on the plant genome: a paragenetic function for chromosomes. Develop. Genet., 15: 523-532.

Kermicle, J.L. (1973) Organization of paramutational components of the *R* locus in maize. Basic mechanisms in plant morphogenesis. Brookhaven Symposia in Biology, 25: 262-278.

Kermicle, J.L., Eggleston, W.B., Alleman, M. (1995) Organization of paramutagenicity in *R-stippled* maize. Genetics, 141: 361-372.

Kricker, M.C., Drake,J.W., Radman, M.(1992) Duplication-targeted DNA methylation and mutagenesis in the evolution of eukaryotic chromosomes. Proc. Natl. Acad. Sci. USA 89, 1075-79.

Lloyd, A.M., Walbot, V., Davis, R.W. (1992) Arabidopsis and Nicotiana anthocyanin production activated by maize regulators *R* and *C1*, Science 258: 1773-1775.

Ludwig. S.R., Habera, L.F., Dellaporta, S.L., Wessler, S.R. (1989) *Lc*, a member of the maize *R* gene family responsible for tissue-specific anthocyanin production, encodes a protein similar to transcriptional activators and contains the *myc*-homology region. Proc. Natl. Acad. Sci. USA, 86: 7092-7096.

Martienssen, R. (1996) Paramutation and gene silencing in plants. Curr.Biol. 6: 810-813.

Matzke, M.A., Matzke, A.J. M. (1993) Genomic imprinting in plants : parental effects and trans-inactivation phenomena. Annu. Rev. Plant Physiol. Plant. Mol. Biol. 44: 53-76.

Matzke M.A., Matzke A.J.M., Eggleston, B. (1996) Paramutation and transgene silencing : a common response to invasive DNA ? Trends Plant Sci. 1: 382-388.

McClintock, B. (1967) Regulation of pattern of gene expression by controlling elements in maize. Carnegie Inst. Yearb. 65: 568-78.

Meehan R.R., Lewis J.D., McKay S., Kleiner E.L., Bird A.P. (1989) Identification of a mammalian protein that binds specifically to DNA containing methylated CpGs. Cell 58:499-507.

Meyer, P., Saedler, H. (1996) Homology-dependent gene silencing in plants. Annu. Rev. Plant Physiol. Plant Mol. Biol. 47: 23-48.

Pandit, N.N., Russo, V.E.A. (1992) Reversible inactivation of a foreign gene, ***hph***, during the sexual cycle of *Neurospora crassa* transformants. Mol. Gen. Genet. 234: 412-422.

Patterson, I. G., Thorpe, J.C., Chandler, L. V. (1993) Paramutation, an allelic interaction, is associated with a stable and heritable reduction of transcription of the maize *b* regulatory gene. Genetics 135:881-894

Paz-Ares, J., Ghosal D., Wienand, U., Peterson, P.A., Saedler, H. (1987) The regulatory *C1* locus of *Zea mays* encodes a protein with homology to myb proto-oncogene products and with structural similarities to transcriptional activators. EMBO J. 6: 3553-3558.

Procissi, A. ,S. Dolfini, A. Ronchi, C.Tonelli. (1997) Light-dependent spatial regulation and temporal expression of pigment regulatory genes in developing maize seeds The Plant Cell 9 : (in press)

Rhounim, L., Rossignol, J.L., Faugeron, G. (1992) Epimutation of repeated genes in *Ascobolus immersus*. EMBO J. 11: 4451-4457.

Romano, N., Macino, G. (1992) Quelling: transient inactivation of gene expression in *Neurospora crassa* by trasformation with homologous sequences . Mol. Mic. 6: 3343-53.

Ronchi, A., Petroni, K., Tonelli, C. (1995) The reduced expression of endogenous duplications (REED) in the maize R gene family is mediated by DNA methylation. EMBO J., 14: 5318-5328.

Selker, E.U. (1990) Premeiotic instability of repeated sequences in *Neurospora crassa*. Annu. Rev. Genet. 24: 579-613.

Tonelli, C. Consonni, G., Faccio Dolfini, S. Dellaporta, S.L., Viotti, A., Gavazzi, G. (1991) Genetic and molecular analysis of *Sn*, a light-inducible, tissue-specific regulatory gene in maize. Mol. Gen. Genet. 225: 401-410.

Quelling: transgene-induced gene silencing in *Neurospora crassa*

Carlo Cogoni and Giuseppe Macino

Dipartimento di Biotecnologie Cellulari ed Ematologia, Sezione di Genetica Molecolare, Policlinico Umberto I, Università di Roma "La Sapienza", Viale Regina Elena, 324, OO161 Roma, Italy.

Introduction

During recent years evidence of the capability of some fungi and plants to recognise and modify duplicated sequences of integrated transforming DNA have been provided. In *Neurospora crassa* it has been shown that duplicated genes are irreversibly inactivated by a mechanism called RIP (Repeat-Induced Point mutation). Duplicated sequences are extensively modified by point mutation, namely cytosine-timine transitions (Selker *et al.*, 1987). *N. crassa* is also able to inactivate duplicated sequences introduced by transformation through a phenomenon termed quelling (Romano and Macino, 1992). Unlike RIP, quelling occurs in the vegetative phase of the *Neurospora* life cycle and consists of a transient inactivation of the gene expression as consequence of transformation with homologous sequences. A substantial difference between RIP and quelling is that the former takes place in dikaryotic cells resulting from fertilisation, and the latter in coeneocytic cells where the exogenous DNA is introduced by transformation. Gene inactivation by quelling is not the result of mutagenesis as in RIP, instead it is the result of the suppression of gene expression, through a reduction of the steady-state mRNA level of the duplicated gene. Quelling is reversible and the function of the inactivated gene may be restored after prolonged culturing time.

Similar phenomena of reversible gene inactivation have been reported in plants and have been termed RIGS (Repeat Induced Gene Silencing) (Assaad *et al.*, 1993) or "homology-dependent gene silencing" (Matzke *et al.*, 1994) to emphasise the apparent requirement for multiple copies of the silenced genes. In plants gene silencing has been found to occur at two levels, transcriptional and post-transcriptional (reviewed in Flavell, 1994) Generally, transcriptional inactivation or trans-inactivation requires homology between promoters, while, homology in transcribed regions may induce post- transcriptional gene silencing or co-suppression (reviewed in Stam *et al.*, 1997). The concurrence of several common features shared between quelling in *Neurospora* and co-suppression in plants suggests that the mechanisms for co-suppression in plants and fungi may have evolved from a common ancestral mechanism. Since *Neurospora* is a relatively simple and genetically well characterised organism, it provides a important tool to dissect the molecular mechanisms of gene silencing which may have universal features that are evolutionarely conserved between organisms.

NATO ASI Series, Vol. H 104
Cellular Integration of Signalling Pathways in Plant Development
Edited by F. Lo Schiavo, R. L. Last, G. Morelli, and N. V. Raikhel

The albino genes as reporter system to study the gene silencing in *Neurospora crassa*

In *Neurospora crassa* the first reported observations of gene silencing induced by sequence duplication referred to as "quelling" are the silencing of the *al-1* and *al- 3* genes (Romano and Macino, 1992) and the *hph* transgene (Pandit and Russo, 1992). In both cases the silencing was found to be reversible during the vegetative growth phase. The *hph* bacterial transgene can undergo through inactivation-activation cycles, while in the case of silencing of the *al-1* and *al-3 Neurospora* genes, the reactivation of gene expression is unidirectional i.e. inactive gene becomes activated. After these two observations a number of cases of quelling for different gene have been reported and quelling appears to be a general phenomenon occurring in *Neurospora*. Quelling was observed in transgenic *Neurospora* strains for a number of genes including: *wc-1* (Ballario *et al.*, 1996), *wc-2* (Linden and Macino, 1997), *qa-2* (Cogoni, unpublished) *ad-9* (Schmidhauser T.J., personal communication). These genes codify products involved in completely different functions in the cell: *wc-1* and *wc-2* are transcription factors involved in the light transduction pathway, *qa-2* is a quinic acid dehydrogenase and *ad-9* is a gene involved in adenine biosynthesis. Thus, the occurrence of silencing should be explained on the basis of sequence specificity rather than as perturbation of specifically regulated metabolic functions, as a consequence of the introduction of transgenes.

Quelling in *Neurospora* has been studied by using as a model system the silencing of the albino genes. Wild type *Neurospora crassa* is bright orange in colour due to the presence of carotenoids. These pigments are synthesised by different enzymes encoded by three structural genes: albino-1 (*al-1*) encoding phytoene dehydrogenase (Schmidhauser *et al.*, 1990), albino-2 (*al-2*) encoding phytoene synthetase (Schmidhauser *et al.*, 1994) and albino-3 (*al-3*) which codes for geranylgeranyl pyrophosphate synthase (Carattoli *et al.*, 1991). The advantage of using carotenoid biosynthetic genes to study gene silencing is that inactivation of these genes produce an easily visible phenotype. Moreover, by simple visual inspection it is also possible to evaluate the level of the albino gene silencing by quantification of carotenoid content.

DNA sequence requirements for silencing

Studies using several deletion constructs of the *al-1* gene in transformation experiments, indicate that a duplication of 170 bp of the coding region is sufficient to induce silencing, whereas the duplication of the promoter only do not cause quelling (Cogoni *et al.*, 1996). The observation that transgenic fragments containing any region of *al-1* coding sequences of sufficient size (170 bp) are capable of inducing quelling, combined with the observation that inclusion of at least some coding sequences in the transgene is required for quelling, suggest a requirement for a minimal degree of homology at the RNA level. Specific sequences are probably not required for quelling, since different gene regions all produced quelling with comparable efficiencies (Fig. 1). As in quelling in *Neurospora*, co-suppression in plants has been shown to require duplication of the sequences belonging to the transcribed region. However, in co-suppression in plants the requirements of homology of specific sequences has not been well defined. Recently, in *chs* silencing in *Petunia* it has been proposed that the transgenic RNA must be contain a repetitive portion of the 3' end of the *chs* gene in order to trigger silencing by interference in mRNA metabolism (Metzlaff *et al.*, 1997). In another case, transgenic plants carrying silenced *uidA* transgene did not accumulate the chimeric virus containing a sequence of the *uidA* gene. Viral RNAs are degraded only if they carrying the 3' region of the *uidA* gene (English et al., 1996). However, no specific sequence or RNA structure have been identified as yet no general picture is emerging regarding sequence homology required in co-suppression phenomena.

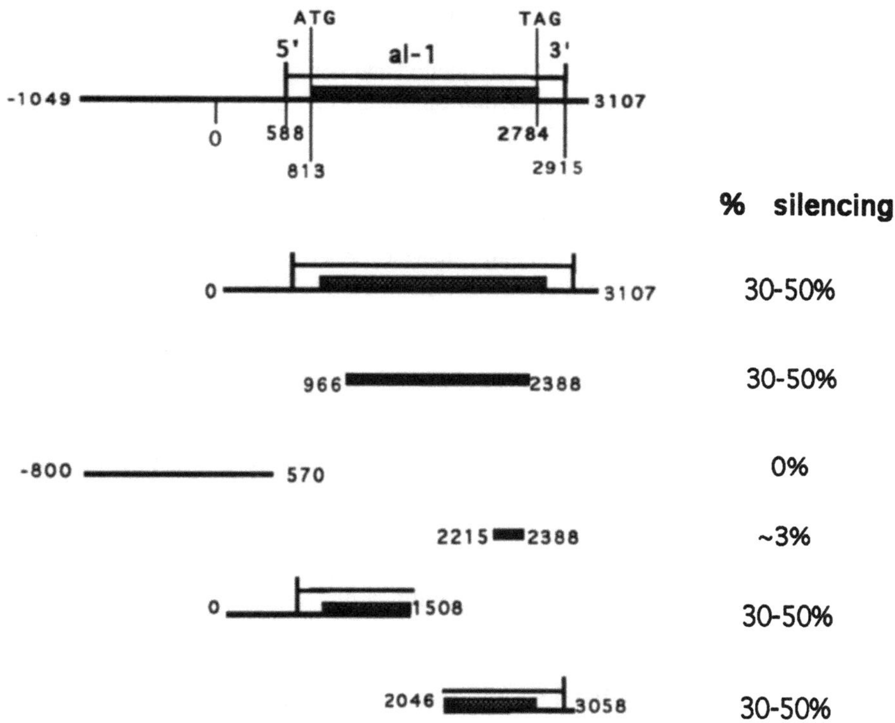

Fig 1. Above: physical map of the *N. crassa al-1* gene region. The *al-1* transcription start site is at position 588 and the stop codon is located at position 2784. The shaded box indicates the coding region of the *al-1* mRNA. Below: *al-1* 5' and 3' deletion derivatives used in transformation experiments. The silencing efficiency for each construct was measured as the percentage of transformants that exhibited an albino phenotype

Quelling results in specific inhibition at a post-transcriptional level of the accumulation of mRNA of the duplicated gene

RNA analysis of silenced albino transformants revealed that quelling acts through a heavy reduction of the steady-state level of mRNA of the genes duplicated by transformation. The reduction of the mRNA level is absolutely sequence specific as only the expression of the duplicated gene is affected (Fig. 2). This reduction could result from effects at either a transcriptional or post-transcriptional level. By using an RNase Protection Assay the relative amounts of the endogenous precursor *al-1* mRNA and mature *al-1* mRNA in both quelled and non quelled transformant strains have been determined. It has been shown that quelled strains produce the same amount of *al-1* primary transcript as the wild-type unquelled strain. By contrast, the level of mature spliced *al-1* mRNA is greatly reduced in quelled strains (Fig. 3). Such a reduction may be due to several causes, such as i) defective splicing, resulting in degradation of partially spliced intermediates and in the production of only small amounts of correctly matured mRNA; ii) reduced cytoplasmic transport and/or accelerated nuclear degradation of mature mRNA; iii) cytoplasmic instability and accelerated degradation of mature mRNA. In view of the apparently unchanged steady-state levels of precursor mRNA in both quelled and non quelled strains, this would be the simplest explanation, indicating that quelling in *Neurospora* takes place at the post-transcriptional level in the cytoplasm.

A specific decrease of steady-state mRNA levels of silenced genes has also been observed in all cases of gene suppression in plants. Based on transcription and RNA turnover analysis the gene silencing phenomena in plants have been divided into post-transcriptional and transcriptional categories. Post-transcriptional gene silencing has been also defined as co-suppression to point out the fact that silencing acts at the same time on both the transgenes and endogenous genes. Also in quelling the introduced transgenes are silenced contemporaneously with the resident gene. Co-suppression was first observed with gene involved flower pigmentation in petunia (van der Krol *et al.*, 1990; Napoli *et al.*, 1990) and fruit ripening in tomato (Smith *et al.*, 1990). Subsequently cases of co-suppression have been shown to occurs in a variety of plant species and for a number of genes including: ß-1,3 glucanase (de Carvalho *et al.*, 1992), ß glucuronidase (Elmayan and Vaucheret, 1996), Neomycin phospotransferase (Ingelbrecht, *et al.* 1994), Chitinase (Hart, *et al.*, 1992) , *rolB* (Dehio and Schell, 1994) and Nitrite reductatase (Vaucheret *et al.*, 1995), as well as for viral transgenes (Linbdo *et al.*, 1993; Smith *et al.*, 1994).

Gene silencing stability

The albino transformants produced by quelling have an unstable phenotype and were observed to revert progressively to wild type or an intermediate phenotype over prolonged culturing time. The reversion of quelling of the *al-1* gene appears to be unidirectional since once it has occurred, gene silencing cannot take place again. The release of silencing took place in a fraction of growing cells and was found to be clonal as expected for cell-heritable variation and it was observed that revertant strains showed a strong reduction of copy number of exogenous sequences (Romano and Macino, 1992). The mitotic instability of ectopic sequences seems to be a naturally occurring phenomenon in *Neurospora crassa* independent from quelling. A case of instability of exogenous sequences has been observed in a *fluffy* strain transformed with a plasmid containing the benomyl resistance gene (Rossier *et al.*, 1992). Loss of tandem arranged transforming sequences has been observed (Selker *et al.*, 1987) during vegetative growth. Loss involving deletion of plasmid sequences, is probably the result of homologous recombination occurring in the vegetative phase. The observation that reversion is associated to the loss of exogenous copies suggests that the instability of gene silencing may simply reflect the instability of the exogenous (ectopic) DNA in *Neurospora* rather than an intrinsic characteristic of quelling. This conclusion is also supported by the observation that quelling correlates with the arrangement in tandem of the exogenous DNA, which may be more prone to recombination leading to excision of the DNA. A similar case of reactivation of a gene as a consequence of intrachromosomal recombination during meiosis leading to a reduction of transgenic copy number has been described in *Arabidopsis*. In a plant deriving from a cross of lines carrying tandem repeats of the inactive *hpt* gene a correlation between reduction of transgenic copy number and reactivation of the *hpt* gene was observed (Mittelstein *et al.*, 1994). Also, in *Petunia* a case somatic instability as consequence of mitotic recombination has been described. Thus, in quelling and in some cases of co-suppression in plants the persistence of transgene copies seems to be required for maintenance of the silenced state. Nevertheless, the maintenance and inheritance of silencing results more complex in most cases described in plants and cannot always be predicted on the basis of the transgene presence.

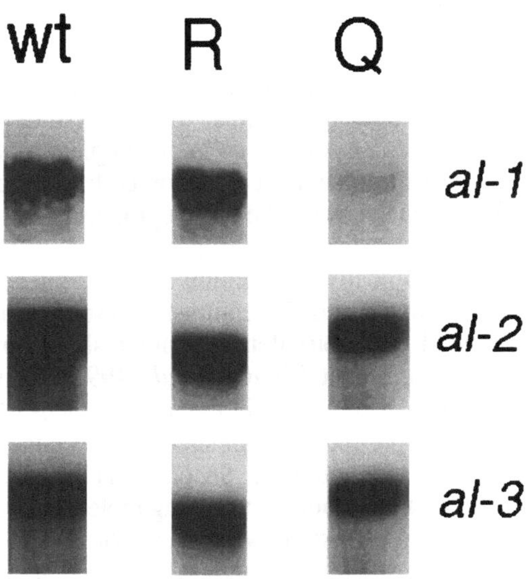

Fig 2. Northern analysis was performed on total RNA extracted from an *al-1* quelled strain (Q), from its revertant (R) and from a wild-type strain (wt).

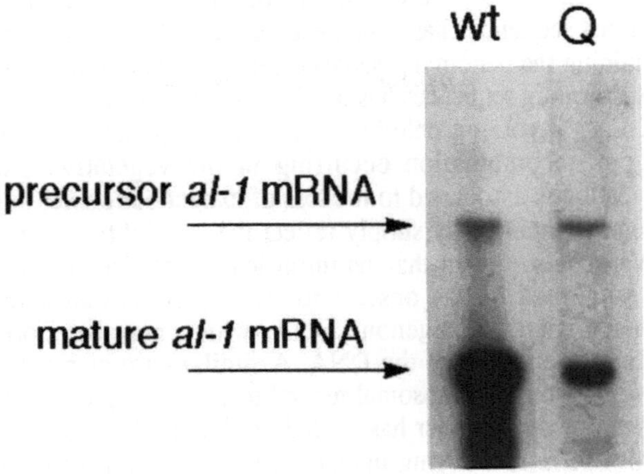

Fig 3. RNase Protection Assay (RPA) performed on total RNA extracted from an *al-1* quelled transformant (Q) and from a wild-type untransformed strain (wt). The level of precursor al-1 RNA is similar in both quelled and non quelled strains while the amount of mature mRNA appears to be greatly reduced in the quelled strain with respect to the non-quelled strain

Quelling is not nucleus limited.

The vegetative growth phase of the filamentous fungus *Neurospora crassa* is characterised by a system of branched hyphae which constitute the mycelium. The hyphae, the basal cellular type, are segmented by incomplete septa. Using opportune selection markers two *Neurospora* strains can be forced to grow as a heterokaryon in which genetically different nuclei share the same cytoplasm. Since *Neurospora* is a haploid organism, heterokaryosis is commonly used to test the dominance or recessivity of a mutation. Typically, a mutation in the carotenogenic genes shows recessive behaviour in heterokaryosis. By contrast, heterokaryotic strains containing a mixture of *al-1* silenced and non-silenced nuclei show an albino phenotype indicating that quelling is dominant (Cogoni *et al.*, 1996). This evidence has two important implications. First, the presence in the same nucleus of both the transgene and endogenous gene is not a pre-requisite for silencing since the resident gene of the non-transformed nucleus is also silenced. Second, quelling is not limited to the nucleus. The involvement of a diffusable trans-acting molecule could be hypothesised. This molecule must be able to diffuse between nuclei or to act in mediating mRNA degradation in the cytoplasm which is common to the different nuclei.

In plants there are indications that co-suppression at least in some cases is not confined to the nucleus, but acts in the cytoplasm. A strong indication that gene silencing can act in the plant cytoplasm derives from the cases in which co-suppressed plants are resistant to viral infection. For example, transgenic tobacco plants containing either a full length or truncated TEV (tobacco etch virus) coat protein (CP) showed a resistance to infection with TEV. Virus resistant plants showed low levels of transgenic CP mRNA and the viral RNA. TEV replicates exclusively in the cytoplasm indicating that gene silencing is a cytoplasmic event in which specific RNA sequences (TEV, CP mRNA) are degraded. Consistent with a cytoplasmic site of action is the observation that in tobacco co-suppression of the ß-1,3 glucanase genes does not affect accumulation of transgene nuclear mRNA (de Carvalho *et al.*, 1995).

An aberrant sense RNA: the dominant signal for quelling?

The observation that quelling is dominant in heterokaryons indicates that quelling results from the production of a trans-acting substance. Since quelling displays gene specificity (Romano and Macino, 1992) and requires portions of the transcribed region of *al-1* in the transgene, it seemed possible that quelling may be mediated by an RNA molecule. The frequency of quelling does not appear to be reduced when a promoter-less transgene, or even fragments thereof, are used for transformation, however. These cases may result from insertion of the transgene in the vicinity of an endogenous promoter, which may trigger the synthesis of a transgene-specific RNA in either the sense or the antisense direction. While the presence of antisense transgenic RNA has not been detected in quelled strains, RNase Protection experiments have established a correlation between quelling and production of a unexpected sense transcript deriving from the transgenic promoter-less construct. Even if the expression of the transgenes was found to be necessary for gene silencing the absolute level of normal transgenic RNA is not important, as indicated by the fact that a promoter-less transgene was found to induce gene silencing as well as a transgene under the control of strong promoter (Cogoni *et al.*, 1996). Thus, the unexpected sense RNA may be qualitatively distinct from the regular RNA and therefore be able to trigger gene silencing.

The existence of "qualitatively different" RNA also referred to as aberrant RNA (aRNA), has been proposed in co-suppression in plants in order to explain some observations in which the highly expressed transgene is not required for co-suppression. For example, the *chs* gene silencing in *Petunia* is induced by a promoter-less construct (van Blokland *et al.*, 1994) and cases in which the degree of viral resistance is not always related to a high level of transgene expression (English *et al.*, 1996). The hypothesis predicts that an aRNA produced directly or induced by the transgene can trigger a degradation mechanism in the cell involving other RNAs (i.e. endogenous RNA or viral RNA) on the basis of sequence identities.

An aRNA may be produced due to the location of the transgenes in the nucleus or some feature of the chromatin of exogenous loci. It has been proposed, based on correlation between the occurrence of silencing and the presence of transgenic DNA repeats, that the DNA repeats may lead to aRNA transcription (Stam *et al.*, 1997b). This observation is also consistent with the correlation between tandem repeats and quelling in *Neurospora*. Repeats and especially inverted repeats could, by DNA-DNA interactions leading to the formation of cruciform structures, be a substrate for aRNA transcription. Lindbo *et al.* (1993) proposed that the aRNA could be recognised by a RNA-dependent RNA polymerase (RdRP), leading the production of cRNA, formation of double stranded RNA, and RNA degradation. The proposed mechanism can explain co-suppression events in which endogenous and transgenic RNA are simultaneously degraded. This mechanism is also consistent with the presence of RdRP activity in plants. The fact, that both in plants and in *Neurospora* there is no clear evidence of cRNAs, could be explained by their instability or by the hypothesis that these cRNAs could be very small and difficult to detect. The aRNAs could be able to induce degradation by other processes. For example the structure of the aRNA could be recognised by a protein complex determining a sort of footprint and RNAs containing an identical structure could be targeted and degraded. Otherwise, the aRNA could interfere directly with the metabolism of the target RNA as suggested by Metzlaf *et al.* (1997). In this case the aRNA produced by the *chs* transgene is not exported efficiently form the nucleus and it is proposed to interact with the endogenous RNA determining its degradation.

Conclusions

The silencing of duplicated gene sequences seems to be a widely diffused phenomenon in plants and in fungi. The evolution of this complex mechanism suggests that it may play an important role in genome organisation and in the control of gene expression. A possible role for co-suppression may be the protection of plant genomes from transposable elements, viruses and other repeated elements that can accumulate in the genome. Similarly, the *Neurospora crassa* genome could be protected from the expansion of transposons and viruses by the evolution of two mechanisms of inactivation of duplicated sequences. Quelling could act in the vegetative phase through the silencing of transposons and viruses genes, inhibiting their expansion and replication, while RIP acts in the sexual phase through irreversible mutagenesis of duplicated sequences preventing their transmission to progeny.

The mechanisms of action of both cosuppression in plants and quelling in *Neurospora* are still unknown. Their understanding can increase our comprehension of the regulation of gene expression and genome organisation. Moreover, understanding the mechanisms involved in this process may be useful in biotechnological applications to enhance or avoid gene silencing in transgenic plants and fungi.

Acknowledgements

We thank Annette Pickford for her assistance with the manuscript. This work was supported by grants from the Ministero Agricoltura e Foreste, Piano Nazionale Tecnologie Applicate alle Piante, the Progetti Finalizzati, Ingegneria Genetica del Consiglio nazionale delle Ricerche and the Istituto Pasteur Fondazione Cenci Bolognetti.

References

Assaad, F.F. *et al.* (1993) Epigenetic repeat-induced gene silencing (RIGS) in Arabidopsis. *Plant Mol. Biol.*, 22: 1067-1085.

Ballario, P. *et al.* (1996) White collar-1, a central regulator of the blue-light responses in Neurospora crassa, is a Zinc-finger protein. *EMBO J.* 15: 1650-1657

Carattoli, A. *et al.* (1991) The *Neurospora crassa* carotenoid biosynthetic gene (albino-3) reveals highly conserved regions among prenyltransferases. *J. Biol. Chem.* 266: 5854-5859

Cogoni, C. *et al.* (1996) Transgene silencing of the al-1 gene in vegetative cells of Neurospora is mediated by a cytoplasmic effector and does not depent on DNA-DNA interactions or DNA methylation, *EMBO J.* 15: 3153-3163

de Carvalho, F. *et al.* (1992) Suppression of b-1,3-glucanase transgene expression in homozygous plants. *EMBO J.* 11: 2595-2602

Dehio, C. and Schell, J. (1994) Identification of plant genetic loci involved in a post-transcriptional mechanism for meiotically reversible transgene silencing. *Proc. Natl. Acad. Sci. USA* 91: 5538-5542

Elmayan, T. and Vaucheret, H. (1996) Single copies of a 35S-driven transgene can undergo post-transcriptional gene silencing at each generation or can be transcriptionally inactivated in trans by a 35S silencer. *Plant J.* 9: 787-797

English, J.J. *et al.* (1996) Suppression of virus accumulation in transgenic plants exhibiting silencing of nuclear genes. *Plant Cell* 8: 179-188

Flavell,R.B. (1994) Inactivation of gene expression in plants as a consequence of specific sequence duplication. *Proc. Nat. Acad. Sci. USA* 91: 3490-3496.

Hart, C.M. *et al.* (1992) Regulated inactivation of homologous gene expression in transgenic *Nicotiana sylvestris* plants containing a defence-related tobacco chitinase gene. *Mol. Gen. Genet.* 235: 179-188

Ingelbrecht, I. *et al.* (1994) Post-transcriptional silencing of reporter transgenes in tobacco correlates with DNA methylation. *Proc. Natl. Acad. Sci. USA* 91: 10502-10506

Linbdo, J.A. *et al.* (1993) Induction of a highly specific antiviral state in transgenic plants: implications for regulation of gene expression and virus resistance. *Plant Cell* 5: 1749-1459

Linden, H. and Macino, G. (1997) White collar 2, a partner in blue-light signal trasduction, controlling expression of light-regulated genes in Neurospora crassa. *EMBO J.* 16: 98-109

Matzke, A.J.M. *et al.* (1994) Homology-dependent gene silencing in transgenic plants: epistatic silencing loci contain multiple copies of methylated transgenes. *Mol. Gen. Genet.* 244: 219-229.

Metzlaff, M. *et al.* (1997) RNA-mediated RNA degradation and chalcone synthase A silencing in Petunia. *Cell* 88: 845-854

Mittelstein Scheid, O. *et al.* (1994) Gene inactivation in *Arabidopsis thaliana* is not accompanied by an accumulation of repeat-induced point mutations, *Mol. Gen. Genet.* 244: 325-330

Napoli, C. *et al.* (1990) Introduction of a chimeric chalcone synthase gene into Petunia results in reversible co-suppression of homologous genes in trans. *Plant Cell* 2: 279-289.

Romano, N. and Macino, G. (1992) Quelling: transient inactivation of gene expression in *Neurospora crassa* by transformation with homologous sequences. Molec. Microbiol. 6: 3343-3353

Rossier, C. *et al.* (1992) Mitotic instability in benomyl-resistant transformants of a fluffy strain of *Neurospora crassa. Fungal Gen.Newslett.* 39: 73-75

Pandit, N.N. and Russo, V.E.A. (1992) Reversible inactivation of a foreign gene, hph, during the asexual cycle in Neurospora crassa transformants. *Mol.Gen.Genet.* 234: 412-422

Schmidhauser, T.J. *et al.* (1990) Cloning, sequence and photoregulation of al-1, a carotenoid biosynthetic gene of *Neurospora crassa. Mol. Cell. Biol.* 10: 5064-5070

Schmidhauser, T. J., *et al.* (1994) Characterization of al-2, the phytoene synthase gene of Neurospora crassa. Cloning, sequence analysis, and photoregulation. *J. Biol. Chem.* 269: 12060-12066.

Smith, C.J.S. *et al.* (1990) Expression of a truncated tomato polygalacturonase gene inhibits expression of the endogenous gene in transgenic plants. *Mol. Gen. Genet.* 224: 477-481

Smith, H.A. *et al.* (1994) Transgenic plant virus resistance mediated by untranslatable sense RNAs: expression, regulation, and fate of nonessential RNAs. *Plant Cell* 6: 1441-1453

Selker, E.U. *et al.* (1987) Rearrangement of duplicated DNA in specialized cells of Neurospora. *Cell* 51: 741-752

Stam, M.*et al.* (1997a) The silence of gene in transgenic plants. *Annals of botany* 79: 3-12

Stam, M. *et al.* (1997b) Post-transcriptional silencing of chalcone synthase in Petunia by inverted transgene repeats. *Plant J. in press*

Vaucheret, H. *et al* (1995) Molecular and genetic analysis of nitrite reductase co-suppression in transgenic tobacco plants. *Mol. Gen. Genet.* 248: 311-317

van Blokland, R. *et al.* (1994) Transgene-mediated suppression of chalcone synthase expression in *Petunia hybrida* results from an increase in RNA turnover. *Plant J.* 6: 861-877

van der Krol, A.R. *et al.* (1990) Flavonoid genes in petunia: Addition of a limited number of gene copies may lead to a suppression of gene expression. *Plant Cell* 2: 291-299

Import and editing of plant mitochondrial transfer RNAs

Laurence Maréchal-Drouard, Ian Small, Anne Cosset, Anne-Marie Duchêne, Raman Kumar, Hakim Mireau, Nemo Peeters, Vera Carneiro, Daniel Ramamonjisoa, Claire Remacle, Ginette Souciet, Henri Wintz, Jacques-Henry Weil and André Dietrich

Institut de Biologie Moléculaire des Plantes, UPR A0406 du CNRS, Université Louis Pasteur, 12, rue du Général Zimmer, 67084 Strasbourg cedex, France

Station de Génétique et d'Amélioration des Plantes, INRA, Route de St-Cyr, 78026 Versailles cedex, France

The recombinations and mutations that higher plant mitochondrial (mt) DNA has undergone during evolution have led to the inactivation or complete loss of a number of the native transfer RNA (tRNA) genes deriving from the genome of the ancestral endosymbiont. Whereas the mt genome of the bryophyte *Marchantia polymorpha* still contains almost all these native tRNA genes (Oda *et al.* 1992a and b), one third to one half of them have disappeared from higher plant mitochondria and the remaining genes cannot provide for the needs of the mt translation system (Ceci *et al.* 1996, Sangaré *et al.* 1990, Unseld *et al.* 1997, Weber-Lotfi *et al.* 1993).

Due to the insertion of chloroplast (cp) DNA sequences into plant mt genomes during evolution, the loss of native tRNA genes has been accompanied in a number of cases by their replacement with functional and expressed "chloroplast-like" (cp-like) tRNA genes. In contrast to the mt genomes of lower plants where no sequences homologous to cp DNA have been detected (Michaelis *et al.* 1990, Oda *et al.* 1992b, Wolff *et al.* 1994), higher plant mt genomes contain several cp-derived DNA fragments varying in size and/or in identity from species to species. In most cases, frameshift mutations or truncations have occurred in the transferred cp genes leading to non-functional genes and, so far, the cp-like tRNA genes constitute the only examples of transferred cp sequences expressed in higher plant mitochondria (for a review see (Maréchal-Drouard *et al.* 1993b). The existence of both native and cp-like tRNA genes in plant mitochondria raises the problem of their transcriptional regulation.

Taken together, the native tRNAs and the cp-like tRNAs still do not account for the 20 amino acids found in proteins and it is now clear that a number of nucleus-encoded tRNAs are imported into plant mitochondria, in order to complete the set of tRNAs needed for protein synthesis in these organelles. Evidence for the presence of nucleus-encoded tRNAs

NATO ASI Series, Vol. H 104
Cellular Integration of Signalling Pathways in Plant Development
Edited by F. Lo Schiavo, R. L. Last, G. Morelli, and N. V. Raikhel

in plant mitochondria was first indirect (incomplete set of tRNA genes in the mt genome, discovery of mt tRNAs with primary sequences identical to those of their cytosolic counterparts and hybridizing only to nuclear DNA) and a direct proof was finally obtained from experiments using transgenic plants (for a review see Dietrich *et al.* 1992). Mitochondria of other organisms also import nucleus-encoded tRNAs, although the processes probably have independent origins and different tRNAs are imported in plants, yeast and protozoa (for a review, see Schneider 1994). In the present article we will review the progress made in our groups during the last few years in order to better understand the evolutionary and mechanistic aspects of plant mt tRNA import.

A further peculiar feature of plant mt tRNAs has recently been characterized: following limited sequence divergence in their genes, some of the higher plant native mt tRNAs are rescued by editing. In higher plant mitochondria, RNA editing is a post-transcriptional process which generally results in the conversion of cytidines to uridines and affects almost exclusively messenger RNAs (for a review, see Maier *et al.* 1996). For proteins, it contributes to the conservation of functional sequences, whereas for tRNAs editing is a prerequisite for efficient processing. Transfer RNA editing has now been observed in various organisms and the recent data obtained in higher plant mitochondria will be discussed below.

1. tRNA import into higher plant mitochondria

1.1 Evolutionary aspects

Since the first report on the presence of a cytosolic-like $tRNA^{Leu}$(C*AA) in bean (*Phaseolus vulgaris*) mitochondria (Green *et al.* 1987), an increasing number of other tRNAs (e. g. $tRNA^{Ala}$, $tRNAs^{Arg}$, $tRNA^{Gly}$, $tRNAs^{Leu}$, $tRNAs^{Thr}$ and $tRNAs^{Val}$) have also been shown to be encoded in the nucleus and imported into plant mitochondria (Dietrich *et al.* 1992, Kumar *et al.* 1996). Although this process appears to be a general phenomenon in higher plant mitochondria, a systematic comparison of the tRNAs imported into the mitochondria of larch, maize and potato revealed striking differences in the number and/or in the identity of the imported tRNAs between these three species (Figure 1).

It is worth noting, that despite having a genome at least twice as large as that of most angiosperms, mitochondria of larch (a gymnosperm) import about twice as many tRNAs as angiosperms. There are also slight differences in import between closely related plants; wheat mitochondria, unlike maize mitochondria, import $tRNA^{His}$, sunflower and *Arabidopsis thaliana* mitochondria, unlike mitochondria from other angiosperms tested, import $tRNAs^{Ser}$ and $tRNA^{Phe}$, respectively. The patterns of tRNA import in different plant species gives some clues as to the evolution of the phenomenon and to the mechanisms involved. The ability to import each tRNA (with the possible exception of isoacceptors for the same amino acid) has very unlikely been acquired independentely at different times during evolution of higher plants.

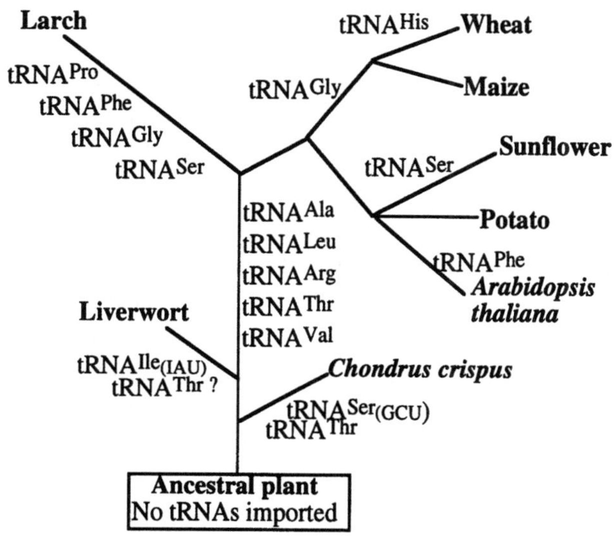

Figure 1. Evolution of tRNA import into plant mitochondria (references in Kumar *et al.* 1996). The figure shows a phylogenetic tree of the plant species in which mt tRNA import has been investigated. In the case of the liverwort *Marchantia polymorpha*, the alga *Chondrus crispus,* and the angiosperm *Arabidopsis thaliana*, the tRNAs assumed to be imported are those needed to have a complete set for the translation of the genetic information, but which are missing from the corresponding mt genomes (Leblanc *et al.* 1995, Oda *et al.* 1992a and b, Unseld *et al.* 1997). On each branch of the tree are indicated the tRNAs which were first imported during this period. Branch lengths are arbitrary.

1.2 tRNA import mechanism

1.2.1 General considerations

Alignment of a number of cytosolic tRNAs and nuclear tRNA genes, corresponding to imported or non-imported plant tRNAs, has not revealed any conserved structural feature specific to the tRNAs which are targeted to mitochondria, either in the tRNAs themselves or in the flanking regions of their genes (Ramamonjisoa, D., Kauffmann S., Choisne, N.,

Maréchal-Drouard, L, Green G., Wintz, H., Small, I.., Dietrich A., submitted). The methylation of the G at position 18, which constitutes the only difference between the cytosolic and mt forms of sequenced $tRNAs^{Leu}$ is not specific to all imported tRNAs and is therefore unlikely to be a signal for mt import (Maréchal-Drouard *et al.* 1990, Maréchal-Drouard, L, unpublished results).

The apparent absence of a conserved import signal, together with the differences in the number and identity of RNAs imported in various plant species, suggest that each imported tRNA is recognized specifically by one or several factors. There are few apparent restrictions on whether a tRNA can or cannot be imported into higher plant mitochondria and it is very unlikely that there is a specific import channel for each tRNA. A more likely hypothesis is the involvement of specific carrier proteins capable of directing tRNAs into a non-specific import system. Such a non-specific import system could be the protein import channel. The search for tRNA carriers has centered on proteins capable of binding specifically to tRNAs. Mt aminoacyl-tRNA synthetases, which are all encoded in the nuclear genome and imported into mitochondria, can be considered as ideal candidates for such a role. For example, just as the $tRNAs^{Leu}$ are distributed between the cytosol and the mitochondria in plants, it seems (on the basis of biochemical and immunological data) that the same leucyl-tRNA synthetase is present in the cytosol and in the mitochondria, at least in bean (*Phaseolus vulgaris*), (Maréchal-Drouard *et al.* 1988, Guillemaut, P, Maréchal-Drouard, L, Dietrich, A, unpublished results). A more detailed study carried out recently with the alanyl-tRNA synthetase (AlaRS) clearly demonstrated that the *Arabidopsis thaliana* AlaRS gene is bifunctional, coding for both the cytosolic and mt forms of the enzyme, depending on which of the two translation initiation codons is used (Mireau *et al.* 1996).

1.2.2 Recognition by the cognate aminoacyl-tRNA synthetase seems necessary but not sufficient for tRNA import into higher plant mitochondria

In the absence of an *in vitro* system for studying tRNA import into isolated mitochondria, the work aimed at testing the implication of the aminoacyl-tRNA synthetases in the process has essentially relied on *in vivo* experiments by investigating the import of modified tRNAs in transgenic plants.

In bacteria, animals and plants, the major identity element in $tRNA^{Ala}$ recognized by AlaRS is the G_3:U_{70} wobble pair in the acceptor stem. The alteration of U_{70} to C_{70}, which restores normal Watson-Crick base-pairing, abolishes binding of AlaRS to $tRNA^{Ala}$ and thus aminoacylation (Carneiro *et al.* 1994, Hou and Schimmel 1988, Hou and Schimmel 1989). Mt $tRNA^{Ala}$ analyzed so far has been shown to be a nucleus-encoded species in all higher plants. In order to test the implication of AlaRS in addressing the nucleus-encoded $tRNA^{Ala}$ into plant mitochondria, constructs carrying wild-type (U_{70}) or mutant (C_{70}) forms of the *Arabidopsis thaliana* $tRNA^{Ala}$ gene were used to generate transgenic tobacco plants (Dietrich *et al.* 1996a). Whereas both forms of $tRNA^{Ala}$ were present in cytosolic fractions from the transgenic plants, only the wild-type $tRNA^{Ala}$ was found in mitochondria. Thus a single nucleotide change is sufficient to abolish both aminoacylation and import of $tRNA^{Ala}$, strongly supporting the hypothesis that aminoacyl-tRNA synthetases are involved in the tRNA import process in plants. Furthermore, insertion of four bases into the anticodon loop of $tRNA^{Ala}$ did not prevent mitochondrial import, implying that the tRNA might not need to participate in translation to be imported. These experiments suggest that aminoacyl-tRNA synthetases are involved in tRNA import, but their precise role remain to be defined.

To determine whether an aminoacyl-tRNA synthetase alone can drive a tRNA into mitochondria, transgenic potato plants were transformed with a yeast aminoacyl-tRNA synthetase able both to be imported into mitochondria and to recognize a non-imported cytosolic tRNA species. The cytosolic aspartyl-tRNA synthetase (AspRS) from yeast (*Saccharomyces cerevisiae*) was chosen for the experiment, as the $tRNA^{Asp}$ present in potato mitochondria is mitochondrially encoded and potato cytosolic $tRNA^{Asp}$ is a good substrate for the yeast enzyme *in vitro* (Dietrich *et al.* 1996b). A translational fusion consisting of the first 90 amino acids (including the mt targeting transit peptide) of the β subunit of the *Nicotiana plumbaginifolia* mt ATP synthase followed by the yeast cytosolic AspRS was constructed. Expression and mt import of the yeast AspRS fusion protein in the transgenic plants was demonstrated by probing western blots with antibodies against the yeast enzyme. However, the endogenous plant cytosolic $tRNA^{Asp}$ could not be detected in the mt fractions of the same plants, demonstrating that import of the AspRS did not lead to import of its tRNA substrate. Although it cannot be excluded that the expression level of the yeast AspRS construct in the potato transgenic plants was too low to allow a detectable $tRNA^{Asp}$ import or that the chimaeric precursor form of this enzyme does not efficiently recognize the plant cytosolic $tRNA^{Asp}$, these experiments support the idea that recognition by a mitochondrially imported aminoacyl-tRNA synthetase is not sufficient to promote efficient import of a tRNA.

1.2.3 Mt import of tRNAs in other organisms and possible models for the tRNA import mechanism

As already mentioned, the import of tRNAs into mitochondria is not limited to the plant kingdom. In particular, similar observations were reported for yeast, *Trypanosoma*, *Leishmania*, *Tetrahymena*, *Paramecium* and a few other organisms (Akashi *et al.* 1996, Feagin *et al.* 1992, Hancock and Hajduk 1990, Kumar et al. 1996, Laforest *et al.* 1997,

Leblanc et al. 1995, Lye *et al.* 1993, Martin *et al.* 1979, Michaelis et al. 1990, Oda et al. 1992b, Pritchard *et al.* 1990, Simpson *et al.* 1989, Suyama 1986). The experimental approaches providing direct proof of tRNA import fall into two types: for plants, *Trypanosoma* and *Tetrahymena*, mt import was demonstrated *in vivo* using transgenic material (Chen *et al.* 1994, Dietrich et al. 1996a, Hauser and Schneider 1995, Lima and Simpson 1996, Rusconi and Cech 1996a, Schneider *et al.* 1994, Small *et al.* 1992), while for yeast and *Leishmania*, *in vitro* import of RNA into isolated mitochondria has been obtained (Mahapatra and Adhya 1996, Mahapatra *et al.* 1994, Tarassov *et al.* 1995a and b, Tarassov and Entelis 1992). Important differences have been discovered between these organisms concerning the number of tRNAs imported and there are reasons to believe that the mechanisms of import might differ too. For example, yeast mitochondria import a single tRNA, $tRNA^{Lys}$(CUU), whereas mitochondria from *Typanosoma* and *Leishmania* import all their tRNAs; other organisms fall between these two extremes.

Supported by a number of general considerations and experimental results, two different mechanisms for tRNA import have been proposed: a co-import of the tRNA with specific protein factors and a direct import of the tRNA after recognition by membrane-bound receptor(s). Concerning the first model, there is strong evidence for the involvement of soluble protein factors, including both the cytosolic lysyl-tRNA synthetase (LysRS) and the precursor to the mt LysRS, in the import of $tRNA^{Lys}$(CUU) into yeast mitochondria (Tarassov *et al.* 1995). In plants, the *in vivo* experiments described above also suggest the requirement of some kind of interaction between the tRNA to be imported and the corresponding aminoacyl-tRNA synthetase. However, both in yeast and in plants, additional factors appear to be involved. In protozoa, proteins other than aminoacyl-tRNA synthetases may play the role of specific carriers. For instance in *Tetrahymena*, mitochondria specifically import only one out of the three cytosolic $tRNAs^{Gln}$. This specificity relies on the recognition of the distinctive anticodon (UUG) of the imported tRNA (Rusconi and Cech 1996b). As in *Tetrahymena* neither the mt nor the cytosolic glutaminyl-tRNA synthetases can distinguish between these $tRNAs^{Gln}$, it seems that some other factor must be responsible for the observed import specificity. The second proposed model is proposes direct import of tRNA and could be applicable to *Trypanosoma* and to *Leishmania*. This model is essentially based on the data obtained by Mahapatra *et al.* (Mahapatra and Adhya 1996, Mahapatra *et al.* 1994) who were able to import a short RNA into isolated *Leishmania* mitochondria in the absence of additional soluble protein factors.

It is clear that each of these models requires further confirmation and the factors involved at the different stages of the transport still need to be defined.

2. tRNA editing in higher plant mitochondria

RNA editing is a widespread phenomenon, which is defined as the modification of a transcript such that its sequence differs from that of the gene from which it was transcribed. RNA editing has been observed mainly in mitochondria, to a lesser extent in chloroplasts and in a few cases in transcripts of nuclear genes (for reviews see e. g. Benne 1996, Covello and Gray 1996, Gray 1996, Maier *et al.* 1996, Smith and Sowden 1996). In higher plant mitochondria and in chloroplasts, RNA editing almost always results in the conversion of cytidines (C) into uridines (U). In other organisms, other types of editing exist, such as insertion/deletion of Us in mitochondria of *Trypanosoma brucei,* and the mechanisms almost certainly differ from that operating in plants.

Editing was first found exclusively in mRNAs and improves amino acid sequence similarity with the corresponding mt proteins of other organisms. However, a few examples of editing events affecting ribosomal RNA or tRNA have been described (references in Laforest *et al.* 1997, Maréchal-Drouard *et al.* 1996a). Editing of structural RNAs is difficult to define, in particular for tRNAs because they undergo many post-transcriptional modifications which are important for correct structure and function. When applied to tRNA, the term "RNA editing" is restricted to base changes which greatly resemble those seen in mRNAs of the same organism and which are presumably caused by the same mechanism. By these criteria, three cases of tRNA editing have been found so far in plant mitochondria: in bean, potato and *Oenothera* $tRNA^{Phe}$ (Binder *et al.* 1994, Maréchal-Drouard *et al.* 1993), in *Oenothera* and potato $tRNA^{Cys}$ (Binder *et al.* 1994; Fey, J, Maréchal-Drouard, L unpublished results) and in larch $tRNA^{His}$ (Maréchal-Drouard *et al.* 1996b). In $tRNA^{Phe}$, the C encoded at position 4 in the gene is changed into a U in the mature tRNA correcting the C_4:A_{69} mismatched base-pair which appears when folding the gene sequence into the cloverleaf structure (Figure 2). Similarly, folding the sequence of the native mt $tRNA^{His}$ gene expressed in larch into the cloverleaf structure yields three C:A mismatched base-pairs (in the acceptor stem, D stem and anticodon stem, respectively) which are edited into classical U:A base-pairs. Conversely, in the third example, $tRNA^{Cys}$, the C to U conversion occurring at position 28 in the anticodon stem does not correct a mismatched C:A base-pair but instead converts a C:U mismatch into another mismatch (U:U).

Whereas in the case of mRNAs editing contributes to the conservation of functional proteins, an interesting question concerns the function of the editing observed in these tRNAs. *In vitro* processing assays of *in vitro*-synthesized edited or unedited forms of tRNA precursors clearly demonstrated that editing is a prerequisite for efficient processing of potato and *Oenothera* mt $tRNA^{Phe}$ and larch mt $tRNA^{His}$ (Marchfelder *et al.* 1996, Maréchal-Drouard et al. 1996a and b) and thus for their accumulation as functional tRNAs in the organelles. It can be speculated that, at least for these two tRNAs, editing is essential for maintaining the cloverleaf folding of the tRNA part in the precursor RNA, thereby allowing efficient cleavage by processing enzymes such as RNase P.

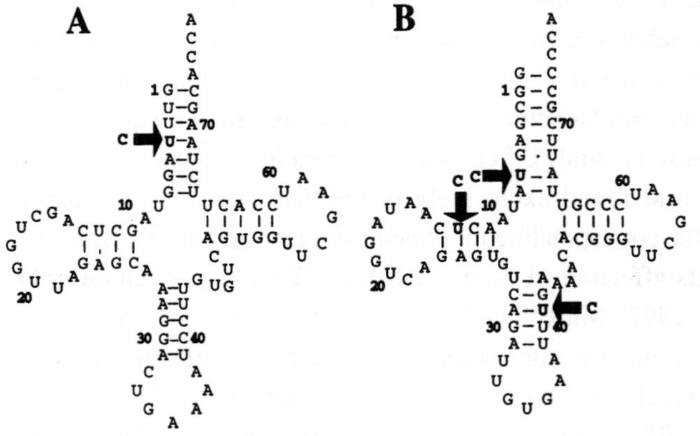

Figure 2. Secondary structure of potato mt tRNA^{Phe} (A) and larch mt tRNA^{His} (B). Numbering is according to Sprinzl *et al.* (1996). The nucleotide changes introduced by editing are written in bold and indicated by large arrows. The sequences are represented as deduced from the genes, no other post-transcriptional modifications than editing are mentioned.

Native mt tRNA^{Phe}, tRNA^{His} and tRNA^{Cys} are found in the bryophyte *Marchantia polymorpha* (Oda *et al.* 1992b). In this non-vascular primitive plant, no editing has ever been detected and a T, instead of a C, is found at the gene level in all the positions mentioned above as editing sites. Editing appears therefore to correct point mutations that have occurred during evolution in some native plant mt tRNAs and to "rescue" them (at least for tRNA^{Phe} and tRNA^{His} which would otherwise be non-functional).

References

Akashi K, Sakurai K, Hirayama J, Fukuzawa H, Ohyama K (1996) Occurrence of nuclear encoded tRNA^{Ile} in mitochondria of the liverwort *Marchantia polymorpha*. Curr. Genet. 30:181-185

Benne R (1996) RNA editing : how a message is changed. Curr. Opin. Genet. Dev. 6:221-231

Binder S, Marchfelder A, Brennicke A (1994) RNA editing of $tRNA^{Phe}$ and $tRNA^{Cys}$ in mitochondria of *Oenothera berteriana* is initiated in presursor molecules. Mol. Gen. Genet. 244:67-74

Carneiro VTC, Dietrich A, Maréchal-Drouard L, Cosset A, Pelletier G, Small I (1994) Characterization of some major identity elements in plant alanine and phenylalanine transfer RNAs. Plant Mol. Biol. 26:1843-1853

Ceci L, Veronico P, Gallerani R (1996) Identification and mapping of tRNA genes on the *Helianthus annuus* mitochondrial genome. DNA sequence 6:159-166

Chen DT, Shi X, Suyama Y (1994) *In vivo* expression and mitochondrial import of normal and mutated $tRNA^{Thr}$ in *Leishmania.* Mol. Biochem. Parasitol. 64:121-133

Covello PS, Gray MW (1996) On the evolution of RNA editing. Trends Genet. 9:265-268

Dietrich A, Maréchal-Drouard L, Carneiro V, Cosset A, Small I (1996a) A single base change prevents import of cytosolic $tRNA^{Ala}$ into mitochondria in transgenic plants. Plant J. 10:101-106

Dietrich A, Small I, Cosset A, Weil JH, Maréchal-Drouard L (1996b) Editing and import: Strategies for providing plant mitochondria with a complete set of functional transfer RNAs. Biochimie 78:518-529

Dietrich A, Weil JH, Maréchal-Drouard L (1992) Nuclear-encoded transfer RNAs in plant mitochondria. Annu. Rev. Cell Biol. 8:115-131

Feagin JE, Werner E, Gardner MJ, Williamson DH, Wilson RJM (1992) Homologies between the contiguous and fragmented rRNAs of the two *Plasmodium falciparum* extrachromosomal DNAs are limited to core sequences. Nucleic Acids Res. 20:879-887

Gray MW (1996) RNA editing in plant organelles: a fertile field. Proc. Natl. Acad. Sci USA. 93:8157-8159

Green AG, Maréchal L, Weil JH, Guillemaut P (1987) A *Phaseolus vulgaris* mitochondrial $tRNA^{Leu}$ is identical to its cytoplasmic counterpart : sequencing and *in vivo* transcription of the gene corresponding to the cytoplasmic $tRNA^{Leu}$. Plant. Mol. Biol. 10:13-19

Hancock K, Hajduk SL (1990) The mitochondrial tRNAs of *Trypanosoma brucei* are nuclear-encoded. J. Biol. Chem. 265:19208-19215

Hauser R, Schneider A (1995) tRNAs are imported into mitochondria of *Trypanosoma brucei* independently of their genomic context and genetic origin. EMBO J. 14:4212-4220

Hou YM, Schimmel P (1988) A simple structural feature is a major determinant of the identity of a transfer RNA. Nature 333:140-145

Hou YM, Schimmel P (1989) Evidence that a major determinant for the identity of a transfer RNA is conserved in evolution. Biochemistry 28:6800-6804

Kumar R, Maréchal-Drouard L, Akama K, Small I (1996) Striking differences in mitochondrial tRNA import among different plant species. Mol. Gen. Genet. 252:

Laforest MJ, Roewer I, Lang BF (1997) Mitochondrial tRNAs in the lower fungus *Spizellomyces punctatus*: tRNA editing and UAG 'stop' codons recognized as leucine. Nucleic Acids Res. 25:626-632

Leblanc C, Boyen C, Richard O, Bonnard G, Grienenberger JM, Kloareg B (1995) Complete sequence of the mitochondrial DNA of the Rhodophyte *Chondrus crispus* (Gigartinales). Gene content and genome organization. J. Mol. Biol. 250:484-495

Lima BD, Simpson L (1996) Sequence-dependent *in vivo* importation of tRNAs into the mitochondrion of *Leishmania tarentolae*. RNA 2:429-440

Lye DF, Chen DT, Suyama Y (1993) Selective import of nuclear-encoded tRNAs into mitochondria of the protozoan *Leishmania tarentolae*. Mol. Biochem. Parasitol. 58:233-246

Mahapatra S, Adhya S (1996) Import of RNA into *Leishmania* mitochondria occurs through direct interaction with membrane-bound receptors. J. Biol. Chem. 271:20432-20437

Mahapatra S, Ghosh T, Adhya S (1994) Import of small RNAs into *Leishmania* mitochondria *in vitro*. Nucleic Acids Res. 22:3381-3386

Maier RM, Zeltz P, Kössel H, Bonnard G, Gualberto JM, Grienenberger JM (1996) RNA editing in plant mitochondria and chloroplasts. Plant Mol. Biol. 32:343-365

Marchfelder A, Brennicke A, Binder S (1996) RNA editing is required for efficient excision of tRNA(Phe) from precursors in plant mitochondria. J. Biol. Chem. 271:1898-1903

Maréchal-Drouard L, Cosset A, Remacle C, Ramamonjisoa D, Dietrich A (1996a) A single editing event is a prerequisite for efficient processing of potato mitochondrial phenylalanine tRNA. Mol. Cell. Biol. 16:3504-3510

Maréchal-Drouard L, Kumar R, Remacle C, Small I (1996b) RNA editing of larch mitochondrial $tRNA^{His}$ precursors is a prerequisite for processing. Nucleic Acids Res. 24:3229-3234

Maréchal-Drouard L, Neuburger M, Guillemaut P, Douce R, Weil JH, Dietrich A (1990) A nuclear encoded potato (*Solanum tuberosum*) mitochondrial $tRNA^{Leu}$ and its cytosolic counterpart have identical nucleotide sequences. FEBS Lett. 262:170-172

Maréchal-Drouard L, Ramamonjisoa D, Cosset A, Weil JH, Dietrich A (1993a) Editing corrects mispairing in the acceptor stem of bean and potato mitochondrial phenylalanine transfer RNAs. Nucleic Acids Res. 21:4909-4914

Maréchal-Drouard L, Weil JH, Dietrich A (1993b) Transfer RNAs and transfer RNA genes in plants. Annu. Rev. Plant Physiol. Plant Mol. Biol.44:13-32

Maréchal-Drouard L, Weil JH, Guillemaut P (1988) Import of several tRNAs from the cytoplasm into the mitochondria in bean *Phaseolus vulgaris*. Nucleic Acids Res. 16:4777-4788

Martin RP, Schneller JM, Stahl A, Dirheimer G (1979) Import of nuclear deoxyribonucleic acid coded lysine accepting transfer ribonucleic acid (anticodon CUU) into yeast mitochondria. Biochemistry 18:4600-4605

Michaelis G, Vahrenholtz C, Pratje E (1990) Mitochondrial DNA of *Chlamydomonas reinhardtii*: the gene for apocytochrome b and the complete functional map of the 15.8 kb DNA. Mol. Gen. Genet. 223:211-216

Mireau H, Lancelin D, Small I (1996) The same *Arabidopsis* gene encodes both cytosolic and mitochondrial alanyl-tRNA synthetases. Plant Cell 8:1027-1039

Oda K, Yamato K, Ohta E, Nakamura Y, Takemura M, Nozato N, Akashi K, Kanegae T, Ogura Y, Kohchi T, Ohyama K (1992a) Gene organization deduced from the complete sequence of liverwort *Marchantia polymorpha* mitochondrial DNA - A primitive form of plant mitochondrial genome. J. Mol. Biol. 223:1-7

Oda K, Yamato K, Ohta E, Nakamura Y, Takemura M, Nozato N, Akashi K, Ohyama K (1992b) Transfer RNA genes in the mitochondrial genome from a Liverwort, *Marchantia polymorpha*: The absence of chloroplast-like transfer RNAs. Nucleic Acids Res. 20:3773-3777

Pritchard AE, Seihlamer JJ, Mahalingam R, Sable CL, Venuti SE, Cummings DJ (1990) Nucleotide sequence of the mitochondrial genome of *Paramecium*. Nucleic Acids Res. 18:173-180

Rusconi CP, Cech TR (1996a) Mitochondrial import of only one of three nuclear-encoded glutamine tRNAs in *Tetrahymena thermophila*. EMBO J. 15:3286-3295

Rusconi CP, Cech TR (1996b) The anticodon is the signal sequence for mitochondrial import of glutamine tRNA in *Tetrahymena*. Genes and Dev. 10:2870-2880

Sangaré A, Weil JH, Grienenberger JM, Fauron C, Lonsdale D (1990) Localization and organization of tRNA genes on the mitochondrial genomes of fertile and male-sterile lines of maize. Mol. Gen. Genet. 223:224-232

Schneider A (1994) Import of RNA into mitochondria. Trends Cell. Biol. 4:282-286

Schneider A, Martin J, Agabian N (1994) A nuclear-encoded tRNA of *Trypanosoma brucei* is imported into mitochondria. Mol. Cell. Biol. 14:2317-2322

Simpson AM, Suyama Y, Dewes H, Campbell DA, Simpson L (1989) Kinetoplastid mitochondria contain functional tRNAs which are encoded in nuclear DNA and also contain small minicircle and maxicircle transcripts of unknown function. Nucleic Acids Res. 17:5427-5445

Small I, Maréchal-Drouard L, Masson J, Pelletier G, Cosset A, Weil JH, Dietrich A (1992) *In vivo* import of a normal or mutagenized heterologous transfer RNA into the mitochondria of transgenic plants: Towards novel ways of influencing mitochondrial gene expression ? EMBO J. 11:1291-1296

Smith HC, Sowden MP (1996) Base-modification mRNA editing through deamination - the good, the bad and the unregulated. Trends Genet. 12:418-424

Sprinzl M, Steegborn C, Hübel F, Steinberg S (1996) Compilation of tRNA sequences and sequences of tRNA genes. Nucleic Acids Res. 24:68-72

Suyama Y (1986) Two dimensional polyacrylamide gel electrophoresis analysis of *Tetrahymena* mitochondrial tRNA. Curr. Genet. 10:411-420

Tarassov IA, Entelis N, Martin R (1995a) Mitochondrial import of a cytoplasmic lysine-tRNA in yeast is mediated by cooperation of cytoplasmic and mitochondrial lysyl-tRNA synthetases. EMBO J. 14:3461-3471

Tarassov IA, Entelis N, Martin RP (1995b) An intact protein translocating machinery is required for mitochondrial import of a yeast cytoplasmic tRNA. J. Mol. Biol. 245:315-323

Tarassov IA, Entelis NS (1992) Mitochondrially imported cytoplasmic tRNA.(CUU) of *Saccharomyces cerevisiae* : *in vivo* and *in vitro* targeting systems. Nucleic Acids Res. 20:1277-1281

Unseld M, Marienfeld JR, Brandt P, Brennicke A (1997) The mitochondrial genome of *Arabidopsis thaliana* contains 57 genes in 366924 nucleotides. Nature Genet. 15:57-61

Weber-Lotfi F, Maréchal-Drouard L, Folkerts O, Hanson M, Grienenberger JM (1993) Localization of tRNA genes on the *Petunia hybrida* 3704 mitochondrial genome. Plant Mol. Biol. 21:403-407

Wolff G, Plante I, Lang BF, Kück U, Burger G (1994) Complete sequence of the mitochondrial DNA of the chlorophyte alga *Prototheca wickerhamii*. J. Mol. Biol. 237:75-86

mRNA Decay Machinery in Plants: Approaches and Potential Components

James P. Kastenmayer[1,2], Ambro van Hoof[1,3], Mark A. Johnson[1,4] and Pamela J. Green[1,2]

[1] Michigan State University Department of Energy Plant Research Laboratory, Michigan State University, East Lansing MI 48824-1312, USA
[2] Michigan State University Department of Biochemistry, East Lansing MI 48824 USA
[3] Current Address: University of Arizona/Howard Hughes Medical Institute Department of Molecular and Cellular Biology Tucson AZ 85721
[4] Michigan State University Department of Microbiology, East Lansing MI 48824 USA

Introduction

The abundance of a given mRNA is an important determinant of gene expression. This steady-state level is determined by a combination of the rates of synthesis and degradation of mRNA. Control at the level of message stability is of particular importance for genes whose expression is tightly regulated. Rapid degradation of transcripts in concert with transcriptional repression allows for the level of a transcript in the cytoplasm to be quickly diminished. This precise control of gene expression, afforded by rapid mRNA turnover, likely plays a significant role for plants, since as sessile organisms they must be able to quickly adapt to changes in the environment.

Studies of plant transcripts have identified sequence elements which are able to confer constitutive rapid degradation to reporter transcripts (see van Hoof and Green, 1997 for a recent review). These include multiple copies of the AUUUA sequence (Ohme-Takagi et al., 1993), an element implicated in the rapid turnover of several mammalian transcripts (Chen and Shyu, 1995), and a dimer of the DST sequence (Newman et al., 1993, Sullivan and Green, 1996). The DST sequence was originally

NATO ASI Series, Vol. H 104
Cellular Integration of Signalling Pathways in Plant Development
Edited by F. Lo Schiavo, R. L. Last, G. Morelli, and N. V. Raikhel

identified as a conserved element in the 3' untranslated region (UTR) of transcripts of the small auxin up RNAs (SAUR) genes of soybean (McClure et al., 1989). These sequences were shown to confer instability by demonstrating their ability to target stable reporter transcripts for rapid degradation (Ohme-Takagi et al., 1993; Newman et al., 1993). The 3' UTR of an *Arabidopsis SAUR* transcript, encoded by the *SAUR-AC1* gene, has also been shown to confer instability to reporter transcripts and therefore represents a naturally occurring instability element (Gil et al., 1996). With the identification of sequence elements which target transcripts for rapid decay in plants, the emphasis has shifted to elucidation of the mechanisms by which unstable transcripts are degraded.

Evidence for a potential difference between the major decay pathway in yeast and plants

Studies conducted during the past five years on transcripts of the yeast *Saccharomyces cerevisiae* have led to the identification of several mRNA degradation pathways (reviewed in Tharun and Parker, 1997). The pathway which appears to be responsible for the degradation of the majority of mRNAs is the "deadenylation dependent decay pathway". This pathway consists of several steps, and involves several different ribonucleases (Figure 1). The first step is rapid shortening of the poly(A) tail to an oligo(A) tail mediated by an as yet unidentified nuclease(s) (Decker and Parker, 1993). Transcripts with oligo(A) tails are then susceptible to a decapping enzyme, of which Dcp1p is a key component (Beelman and Parker, 1996). These decapped mRNAs are then degraded by a 5'-3' exonuclease, Xrn1p (Hsu and Stevens, 1993). The elucidation of this pathway has been greatly aided by two experimental strategies.

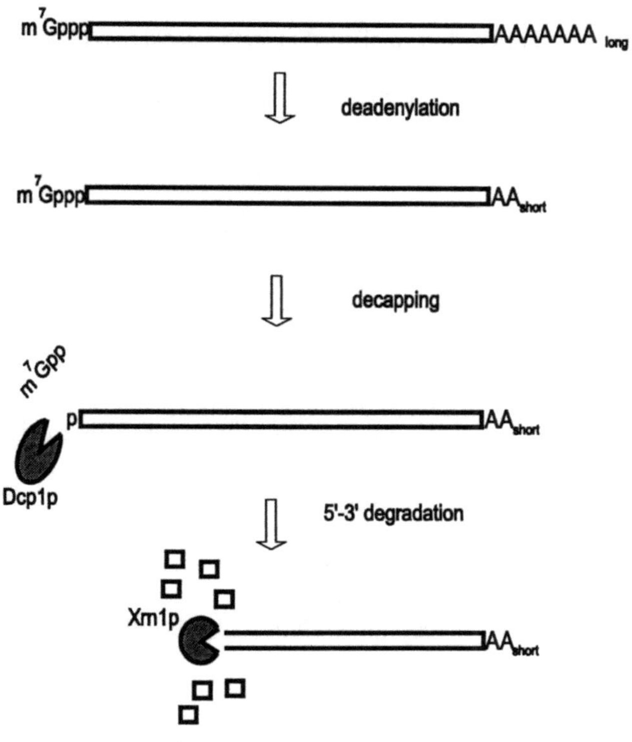

Figure 1. Deadenylation dependent decay pathway, the major pathway of mRNA turnover in yeast.

Inactivation of genes encoding enzymes of the turnover mechanism provided evidence of the roles of Xrn1p and Dcp1p (Hsu and Stevens, 1993; Beelman et al., 1996). The second strategy, which was particularly insightful, involved the insertion of a poly(G) tract into transcripts (Vreken and Raue, 1992; Decker and Parker, 1993). For transcripts degraded by the pathway in Figure 1, the poly(G) tract blocks the progression of Xrn1p resulting in the accumulation of an intermediate beginning at the poly(G) tract and extending to the 3' end of the deadenylated mRNA (Decker and Parker, 1993; Muhlrad et al., 1994, 1995). In studies following the degradation of these transcripts over time, it is also observed that the intermediates are not detected until the poly(A) tail has been shortened to an oligo(A) length (Decker and Parker 1993, Muhlrad et al., 1994, 1995). Insertion of strong secondary structures is therefore a powerful tool in elucidating mRNA decay mechanisms. It allows for several aspects of turnover to be studied simultaneously including poly(A) status, and the direction of action of exonucleases.

A similar strategy has been employed in the study of unstable plant reporter transcripts. A poly(G) tract of 25 guanosine residues has been inserted in the 3' UTR of reporter transcripts in plants (Figure 2). However, 3' degradation intermediates are not detected when these transcripts are produced in stably transformed tobacco cells grown in liquid culture, or in transgenic *Arabidopsis* (M. Johnson and P.J Green, unpublished results).

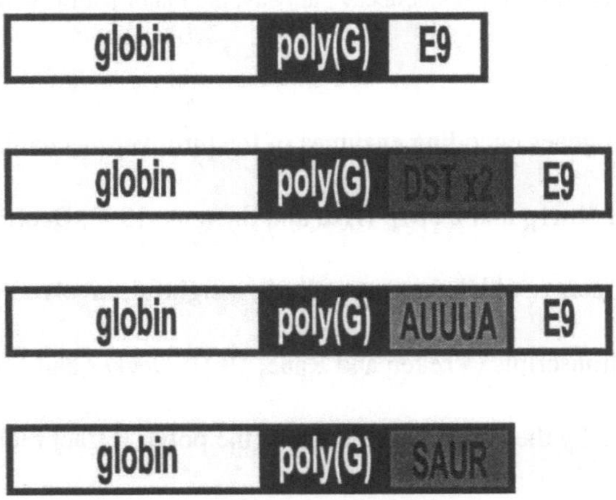

Figure 2. Reporter transcripts produced in tobacco cell culture and transgenic *Arabidopsis*

Similarly, several groups have found that the insertion of poly(G) tracts into transcripts produced in mammalian cells did not result in the accumulation of decay intermediates (L. Maquat, personal communication; A.-B. Shyu, personal communication; G.Goodall, personal communication). This indicates that there may be a difference between the major mRNA decay pathways in yeast and higher eukaryotes. Whether this difference is small or large remains to be determined as discussed below.

Approaches to elucidate mRNA decay mechanisms and machinery in plants

The identification of mRNA decay pathways and machinery is complicated in plants not only by the lack of intermediates following poly(G) insertion but also by the observation that most endogenous transcripts degrade without the accumulation of visible intermediates (a common observation in other eukaryotic systems). However, by identifying and characterizing putative plant mRNA decay enzymes we may not only pinpoint components of the decay machinery but also delineate features that are unique to plants. Because of the large scale sequencing efforts in *Arabidopsis* (e.g. Newman et al., 1994), putative mRNA decay enzymes can often be identified on the basis of sequence similarity.

One good example is Ath-DCP1, which is a possible homolog of the yeast decapping enzyme Dcp1p (Beelman et al, 1996; LaGraneur et al., submitted). Initially, three *Arabidopsis* ESTs were identified (181O21T7, 171N16T7, and ATTS1912) which all correspond to the same cDNA. DNA gel blots indicate that there is a single gene for Ath-DCP1 in *Arabidopsis*. The sequence of the complete coding region of the Ath-DCP1 cDNA is shown in Figure 3. Overall, the degree of similarity to the yeast enzyme is not high. Nevertheless, Ath-DCP1 does contain all of the residues found to be absolutely conserved among the yeast DCP1 and potential homologs from *P. faliparum, C. elegans, and H. sapiens* which argues that it is a bonafide homolog (bold in figure 3). Mutations in the yeast DCP1 that alter these invariant residues have been found to cause enzyme defects (S. Tharun and R. Parker, personal communication). Examining whether Ath-DCP1 has decapping activity *in vitro* or when expressed in yeast should now be straight forward.

10	20	30	40
MSQNGKIIPN	LDQNSTRLLN	LTVLQRI**D**PY	IEELLITAAH
50	60	70	80
VTFYEFNIEL	SQ**W**SRKDVE**G**	SLFVVK**R**STQ	PRFQFIVMNR
90	100	110	120
RNTDNLVENL	LGDFEYEVQG	PYLLY**R**NASQ	EVNGI**W**FYNK
130	140	150	160
RECEEVATLF	NRILSAYSKV	NQKPKASSSK	SEFEELEAKP
170	180	190	200
TMAVMDGPLE	PSSTARDAPD	DPAFVNFFSS	TMNLGNTASG
210	220	230	240
SASGPYQSSA	IPHQPHQPHQ	PHQPTIAPPV	AAAAPPQIQS
250	260	270	280
PPPLQSSSPL	MTLFDNNPEV	ISSNSNIHTD	LVTPSFFGPP
290	300	310	320
RMMAQPHLIP	GVSMPSAPPL	NPNNASHQQR	SYGTPVLQPF
330	340	350	360
PPPTPPPSLA	PAPTGPVISR	DKVKEALLSL	LQEDEFIDKI
370			
TRTLQNALQQ			

Figure 3. Deduced amino acid sequence of Ath-DCP1 cDNA (Accession number: AF007109). Bold letters indicate conserved residues between Ath-DCP1, and potential homologs from P. *faliparum, C. elegans, and H. sapiens.*

It is possible that most plant transcripts are decapped by Ath-DCP1 and then degraded in the 5' to 3' direction by an Xrn1p homolog, but we still must explain why the insertion of a poly(G) tract does not block the enzyme's progress. One simple reason for the lack of poly(G) 3' intermediates may be that the plant version of Xrn1p may progress right through poly(G) tracts. Alternatively, novel plant helicases may disrupt the secondary structure caused the by poly(G) tract so that the sequence is no longer able to block decay. If a plant XRN1 homolog can be identified, then it can be examined for blockage by poly(G) to help differentiate between these possibilities. Another hypothesis that could account for the data is that in plants the body of the mRNA is degraded by a pathway that is fundamentally different than the major (and minor) yeast pathway(s). For example, poly(G) intermediates would not be expected if endonucleases play the predominant role (Tanzer and Meagher, 1995). Elucidating such a novel mechanism would require efforts beyond the characterization of putative homologs.

One far-reaching approach is to design genetic strategies to isolate mutants of *Arabidopsis* that have defects in mRNA decay pathways. To this end we have incorporated copies of the DST sequence and the AUUUA repeat into selectable and screenable marker transcripts in transgenic *Arabidopsis* plants. By mutagenizing the transgenic plants and identifying M2 plants with elevated levels of these DST- or

AUUUA-containing transcripts, we hope to identify genetically components in the corresponding mRNA decay pathways. Mutations in RNA-degrading activities or RNA binding proteins that slow down decay would be expected to cause elevated accumulation of otherwise unstable transcripts. Once mutants are isolated, the isolation of the corresponding genes should be straight forward in *Arabidopsis* via map-based cloning. Thus far, the recognition of DST elements appears to be unique to plants, whereas AUUUA sequences are recognized in mammalian cells as well. Accordingly, the search for components of the DST- and AUUUA-mediated mRNA decay pathways by the aforementioned genetic approach has the potential to identify mRNA decay factors that are unique to plants or of broad significance.

Acknowledgments: This work was supported by grants from the NSF, DOE and USDA to PJG. We are grateful to Roy Parker for helpful discussions and for providing unpublished information about Dcp1p.

References

Beelman CA, Stevens A, Caponigro G, LaGrandeur TE, Hatfield L, Fortner DM, Parker R (1996) An essential component of the decapping enzyme required for normal rates of mRNA turnover. Nature **382**: 642-646

Chen C-YA, Shyu A-B (1995) AU-rich elements: characterization and importance in mRNA degradation. Trends Biochem. Sci. **20**: 465-470

Decker CJ, Parker R (1993) A turnover pathway for both stable and unstable mRNAs in yeast: evidence for a requirement for deadenylation. Genes Dev. **7**: 1632-1643

Gil P, Green PJ (1996) Multiple regions of the *Arabidopsis SAUR-AC1* gene control trascript abundance: the 3' untranslated region functions as an mRNA instability determinant. EMBO J. **15**: 1678-1686

McClure BA, Guilfoyle TJ (1989) Rapid redistribution of auxin-regulated RNAs during gravitropism. Science **243**: 91-93

Muhlrad D, Decker CJ, Parker R (1994) Deadenylation of the unstable mRNA encoded by the yeast *MFA2* gene leads to decapping followed by 5'-->3' digestion of the transcript. Genes Dev. **8**: 855-866

Muhlrad D, Decker CJ, Parker R (1995) Turnover mechanisms of the stable yeast PGK1 mRNA. Mol. Cell. Biol. **15**: 2145-2156

Newman TC, Ohme-Takagi M, Taylor CB, Green PJ (1993) DST sequences, highly conserved among plant *SAUR* genes, target reporter transcripts for rapid decay in tobacco. Plant Cell **5**: 701-714

Newman TC, de Bruijn F, Green PJ, Keegstra K, Kende H, McIntosh L, Ohlrogge J, Raikhel NV, Somerville SC, Tomashow M, Retzel E, Somerville CR (1994) Genes Galore: A summary of methods for accessing results from partial sequencing of anonymous Arabidopsis cDNA clones. Plant Physiology **106**: 1241-1255

Ohme-Takagi M, Taylor CB, Newman TC, Green PJ (1993) The effect of sequences with high AU content on mRNA stability in tobacco. Proc. Natl. Acad. Sci. USA **90**: 11811-11815

Shyu A-B, Belasco JG, Greenberg ME (1991) Two distinct destabilizing elements in the c-*fos* message trigger deadenylation as a first step in rapid mRNA decay. Genes Dev. **5**: 221-231

Sullivan ML, Green PJ (1993) Post-transcriptional regulation of nuclear-encoded genes in higher plants: the roles of mRNA stability and translation. Plant Mol. Biol. **23**: 1091-1104

Sullivan ML, Green PJ (1996) Mutational analysis of the DST element in tobacco cells and transgenic plants: Identification of residues critical for mRNA instability. RNA **2**: 308-315

Tanzer MM, Meagher RB (1995) Degradation of the soybean ribulose-1,5-bisphosphate carboxylase small-subunit mRNA, SRS4, initiates with endonucleolytic cleavage. Mol. Cell. Biol. **15**: 6641-6652

Tharun S and Parker R (1997) Mechanisms of mRNA Turnover in Eukaryotic Cells. *In* JB Harford, D Morris, eds, mRNA Metabolism & Post-Transcriptional Gene Regulation, Vol 17. Wiley-Liss Inc, pp 181-200

van Hoof A and Green, PJ Mechanisms of mRNA Turnover in Eukaryotic Cells. *In* JB Harford, D Morris, eds, mRNA Metabolism & Post-Transcriptional Gene Regualtion, Vol 17. Wiley-Liss Inc, pp 181-200

Molecular Cloning and Characterization of two cDNAs Encoding Enzymes Required for Secondary Cell Wall Biosynthesis in Maize.

Laura Civardi[1], Alain Murigneux[2], Patricia Tatout[2], Pere Puigdomènech[1] and Joan Rigau[1].

[1] Departament de Genètica Molecular, CID-CSIC, Jordi Girona 18-26, 08034 Barcelona, Spain.
[2] Biocem-Group Limagrain, 24 Avenue des Landais 63170 Aubière, France

ABSTRACT

Lignin is an essential component of vascular plants, being required for structural support and taking part in defense against pathogen attack. In fact, lignin is one of the most abundant polymers in plants second only to cellulose. In some cases, however, lignin represents an undesirable component of agronomically and industrially important plant species, for instance in the production of paper pulp or in forage plants where the lignin content interferes with digestibility in the animals. In recent years, several studies have been trying to elucidate the enzymes taking part in the lignification process in order to modify and reduce lignin content in plants (for a review see Campbell and Sederoff, 1996; Boudet and Grima-Pettenati, 1996). Among the enzymes involved in lignification, the cinnamoyl CoA reductase (CCR, EC 1.2.1.44) and the cinnamyl alcohol dehydrogenase (CAD, EC 1.1.1.195) have been described as key enzymes in this process. The enzyme CCR catalyzes the conversion of hydroxycinnamoyl esters to their corresponding aldehydes and the enzyme CAD catalyzes the conversion of these hydroxycinnamaldehydes to the corresponding alcohols (better known as monolignols) that are the monomeric precursors of lignin. CAD cDNAs have been cloned from several plants including *N.tabacum* (Knight *et al.*, 1992), *P.taeda* (O'Malley *et al.*, 1992), *A.cordata* (Hibino *et al.*, 1993), *E.gunnii* (Grima-Pettenati *et al.*, 1993), *P.abies* (Galliano *et al.*, 1993), *M.sativa* and *P.deltoides* (Van Doorsselaere *et al.*, 1995). Recently, for the first time, the cloning of a CCR cDNA from *E.gunnii* was reported (Lacombe *et al.*, 1997).

NATO ASI Series, Vol. H 104
Cellular Integration of Signalling Pathways in Plant Development
Edited by F. Lo Schiavo, R. L. Last, G. Morelli, and N. V. Raikhel

Our group is actively involved in characterizing at the molecular level the pathway that leads to lignin biosynthesis in maize. Maize is widely used as a forage crop in several European countries. Several studies have linked lignin content in forage grasses to reduced digestibility and dry matter intake in livestock. In fact, a number of genes have been linked to variations in lignin content in maize and sorghum and they have been introduced in breeding programs for forage maize. Among these genes it is worth mentioning the *brown midrib* genes (Barrière and Argillier, 1993; Barrière *et al.*, 1994). It has been shown that at least in one case (*bm3*) the *brown midrib* phenotype is due to a mutation occurring on the sequence coding for an enzyme of the lignin biosynthesis pathway, the caffeic acid O-methyltransferase (Vignols *et al.*, 1995). This result indicated the fundamental importance of acquiring further knowledge on the enzymes responsible for lignification in maize. These studies may enable us to modify and/or reduce lignin content in an important crop, or in related species, by selective control of the expression of one or more lignification genes (i.e. by antisense technology). As part of this project, here we report the cloning and the analysis of two maize cDNA clones encoding respectively the enzymes CAD and CCR.

KEYWORDS: Cinnamoyl CoA reductase, cinnamyl alcohol dehydrogenase, lignification, maize.

1 MATERIALS AND METHODS

1.1 cDNA library screening

A λZAPII cDNA library made from poly(A)$^+$RNA isolated from the non-meristematic region of 9-day-old maize roots, was screened. The CAD cDNA and the CCR cDNA from maize were cloned by using respectively a 1.1 kb *EcoRI* fragment of a CAD cDNA from sugarcane and a 1 kb *SmaI-EcoRV* fragment of a CCR cDNA, also from sugarcane, as probes (kind gifts from G. Selman, CIGB, La Habana, Cuba). Hybridization was carried out following standard conditions (Ausubel *et al.*, 1994).

1.2 Southern and northern analysis

Genomic DNA was isolated from maize (*Z.mays* W64A) leaves by the method of Dellaporta *et al.* (1983). The DNA was digested with restriction enzymes, fractionated on agarose gel and transferred to a nylon membrane. Total RNA was isolated from 9-day-old maize plants (Verwoerd *et al.*, 1989), separated on formaldehyde-agarose gel and transferred to a nylon membrane. The membranes were hybridized either to a 657 bp *SmaI-SpeI* fragment isolated from the CAD cDNA clone, or to an 810 bp *XhoI* fragment isolated from the CCR cDNA clone and containing the 3' untranslated region (see Fig. 2). Hybridization conditions were as described by Vignols *et al.* (1995).

1.3 RFLP mapping

RFLP mapping was done according to Murigneux *et al.* (1993b). The CAD cDNA probe was mapped on 114 F2 individual plants derived from a cross between two flint European lines (F2 x Co255). The CCR cDNA probe was mapped to the maize genome on a double haploid population of 71 lines (Murigneux *et al.*, 1993a) derived from the F1 cross DH5 x DH7. On both population have been mapped more than 100 RFLPs, including BNL (Brookhaven National Laboratory), UMC (University of Missouri) and LIM (Limagrain) probes. The mapping was done using the MAPMAKER software version 2.0.

2 RESULTS AND DISCUSSION

2.1 Cloning of a cDNA encoding the maize cinnamyl alcohol dehydrogenase

A clone corresponding to CAD cDNA and containing the full protein sequence was isolated by screening a maize root cDNA library. This clone of 1528 bp contains a 5' untranslated region of 127 bp and a 3' untranslated region of 296 bp. A unique open reading frame coding for a protein of 367 amino acids was found between nucleotide 128 and 1228. The maize CAD cDNA clone lacks a canonical consensus polyadenylation signal AATAAA, but it is possible to define some AT-rich and GT-rich regions that may be involved in mRNA 3' processing (Wu *et al.*, 1995).

A comparison analysis of the deduced amino acid sequence of the maize CAD with other CAD sequences (Fig. 1) revealed that, although this is the first cereal sequence reported, the CAD protein of maize is highly homologous to the other CAD sequences in plants. Indeed, it showed between 76 and 80% identity with the protein from other angiosperms and 67% identity with CADs cloned from gymnosperms. Interestingly, among the CAD sequences reported in the databases, there were four sequences

```
                                                                          Zn1 ...
Z.mays        MGSL-ASERKVVGWAARDATGHLSPYSYTLRNTGPEDVVVKVLYCGICHTDIHQAKNHLGASKYPMVPGHEVVGE  74
P.radiata     ....E-T.KT.T.Y....SS......T.N..KK.....I...I......S.LV.MR.EM.M.H...........I  74
P.taeda       ....E-..KT.T.Y....SS......T.N..KK.....I...I......S.LV.MR.EM.M.H...........I  74
P.abies       ....E-...T.T.Y....SS......T.....K.....I.R.I......S.LV.MH.EM.M.N...........V  74
E.botryoides  ....EK-..TTT.....PS.V...T.S........LYI...S.....S....I..D..M.H.............  74
E.gunnii      ....EK-..TTT.....PS.V...T.S........LYI...S..V..S....I..D..M.H.............  74
N.tabacum-1   ..G.EV-.KTTI.....PS.V...T...........E........L....L..V..D..M.N..L..........  74
N.tabacum-2   ....DV-.KSAI.....PS.L...T...........Q........L..S.L..V..D..M.N..L.........K  74
A.cordata     ....E.-...TT.....PS.V...T...E......FI.II............I..D...N.............  74
P.deltoides   ....ET-...I.....T.S...A..T.S..D......FI..IS..V........I..D..M.H.............  74
M.sativa      ...IE.A..TT..L..K.PS.I.T..T........D..YI.IH...V..S.L..I..D..M.N.............  75
A.thaliana    ...VE.G.K.AL.....PS.V.........S..AD..YI..IC...........I..D..M.N.............  75
consensus     **+++  *++ +*+**+*++*+*+**+*+**++*++*++*++++++++**+**+*++*++*++*+*+**+*+*******+  75

                                        Zn2
Z.mays        VVEVGPEVAKYGVGDVVGVGVIVGCCRECSPCKANVEQYCNKKIWSYNDVYTDGRPTQGGFASTMVVDQKFVVKI 149
P.radiata     .T.I.S..K.FK..EH....C...S..S.GN.NQSM....S.R..T....NH..T.........S.....M...R. 149
P.taeda       .T.I.S..K.FK..EH....C...S..S.GN.NQSM....S.R..T....NH..T.........S.....M...R. 149
P.abies       .T.I.S..K.FK..EH....C...S..S..N.NGSM....S.R..T....NH..T.........S.....M...R. 149
E.botryoides  .L...S..T..R...R..T.IV.....S.G..NSDQ..........N.........K.......GEI..G.R..... 148
E.gunnii      .L...S..T..R...R..T.IV.....S....NSDQ..........N.........K.......GEI..GER..... 149
N.tabacum-1   ......D.S.FK...T....LL..S..N.G...RDI.........NC........K.......KS........... 149
N.tabacum-2   .....AD.S.FK...T....LL..S..N.G...REI.........NC........K.......NS.....N..... 149
A.cordata     .....SD.T.FK...C..D.T.....KT.R....D..................K......SGH............ 149
P.deltoides   .....SD.TRFK..............S.KN.H...SEI.................K.......ES...H.....R. 149
M.sativa      .L...SN.TRFK..EI....LL....KS.RA.DSEI.................KI......EST..E........ 149
A.thaliana    .L...SD.S.FT..........V....GS.K..SSEL......R...........K.......D..I.N........ 149
consensus     *+*+*++*++++**+ **+*++** *++* +*  ++****+*+**++***++**++*****+ +++*+++****+* 150

                                                 NADP
Z.mays        PAGLAPEQAAPLLCAGVTVYSPLKHFGLTN-PGLRGGILGLGGVGHMGVKVAKAMGHHVTVISSSSKKRAEAMDH 223
P.radiata     .EN.PL..............F..M...AM.E-..KKC..................I...F.L.........D..KE...EV 223
P.taeda       .EN.PL..............F..M...AM.E-..KKC...........L...I...F.L.........D..KE...EV 223
P.abies       .EN.PL.................M....M.E-..KKC..................I...F.L.........D..KE..LEV 223
E.botryoides  .D..ES..D.-VM...........VR...KQ-S..................I.................D...T..LE. 222
E.gunnii      .D..ES......M............VR...KQ-S..................I.................D...T..LE. 223
N.tabacum-1   .E.M............I.......N...FKQ-S..................I.................N...Q..LE. 223
N.tabacum-2   .E.M............I.....FN...FNQ-S.F................I.................N...Q..LE. 223
A.cordata     .D.M....................T....KEIS..................L.................D..KE..I.. 224
P.deltoides   .D.MS..........L..............KQ-S..................I.................D...E...E. 223
M.sativa      .E......V..................S....K-T..........................L.........D..KK..LED 224
A.thaliana    .E.M.V....................S....-MAS..K..................I.................D..KE..IE. 224
consensus     *+++++**+*+++***+**+**+ +*++++ +*+++**********+***+***+*+*********+**++**+++ 225

Z.mays        LGADAYLVSSDAAAMAAAADSLDYIIDTVPVHHPLEPYLALLKLDGKLVLLGVIGEPLSFVSPMVMLGRKAITGS 298
P.radiata     .........K.TEK.ME..E.....M..I..A............TN....M...VP...H..T.LLI...RS.A.. 298
P.taeda       .........K.TEK.ME..E.....M..I..A............TN....M...VP...H..T.PLI...RS.A.. 298
P.abies       .........K..EK.QE..E.....M..I..A............TN....M...VP...H..T.LLI...RS.A.. 298
E.botryoides  ...........ENG.KE.T......F..I..V...................I.T...NA..Q.I..........S.... 297
E.gunnii      ...........ENG.KE.T......F..I..V...................I.T...NA..Q.I..........S.... 298
N.tabacum-1   ....D......TDK.QE.S..............G........S...I....I.M...NT..Q.I..........S.... 298
N.tabacum-2   ....D......TDK.QE................G....L..S...I....I.I...NT..Q.I..........S.... 298
A.cordata     ...........TQ.QE.................F........S.........I.M...NT..Q.I............... 299
P.deltoides   ....E......VES.QK...Q............V........S.........I.M...NA..Q..T.........S.... 298
M.sativa      ....N......TVG.QE................G........S...I....I.M...NT..Q..T.........S.... 299
A.thaliana    ....D.V....P.E.QRL...............F...D....C.........I.M...NT..Q..T.L.I....V.S.. 299
consensus     ****+*+**+* + *++++++****+**+**+***+**+*+**++*+**++***++ **+*++*+++****+++** 300

Z.mays        FIGSIDETAEVLQFCVDKGLTSQIEVVKMGYVNEALERLERNDVRYRFVVDVAGSNVEAEAAAADAASN 367
P.radiata     ....ME..Q.T.D..AE.KVS.M....GLD.I.T.M....K...............KLDN          357
P.taeda       ...GME..Q.T.D..AE.KVS.M....GLD.I.T.M....K...............ELDN          357
P.abies       ....ME..Q.T.D..AE.KVS.M....GLD.I.T.M...VK.............A..LDK          357
E.botryoides  ....MK..E.M.E..KE.........I..D...T......K................KLD          354
E.gunnii      ....MK..E.M.E..KE.........I..D...T......K.............V..KLD          356
N.tabacum-1   ....MK..E.M.D..KE..V.....I...D.I.T.M....K.............I..KLDQ         357
N.tabacum-2   ....MK..E.M.D..KE..V.....I...D.I.T.M....K...S..........KLDQ          357
A.cordata     ....MK..E.M.D..NE..I..T......D.I.T.F....K................KLDQ.T       360
P.deltoides   ....MK..E.M.E..KE..VA.M...I..D.I.T.F....K................KLIH         357
M.sativa      .V..VK..E.M.E.WKE.....M..I.T.D.I.K.F....K.............K..KF.E         358
A.thaliana    .....K..E...A..KE.....T..T..IDEL.I.F...RK.................LVE...TTTN  365
consensus     *+*+++**+*+*+*+++*+++* **++++++*+*++++*+****++***+*******++*+++
```

Fig. 1. Deduced amino acid sequence from the maize CAD clone aligned with the following CAD sequences: *A.cordata* GenBank AC: D13991, *A.thaliana* GenBank AC: Z31715, *E.gunnii* GenBank AC: X65631, *E.botryoides* Swiss-Prot AC: P50746, *M.sativa* GenBank AC: Z19573, *N.tabacum*-1 GenBank AC: X62343, *N.tabacum*-2 GenBank AC: X62344, *P.abies* GenBank AC: X72675, *P.deltoides* GenBank AC: Z19568, *P.radiata* GenBank AC: U62394 and *P.taeda* GenBank AC: Z37991. Dots indicate identical amino acid residues, dashes indicate gaps introduced to maximize the alignments. Conserved domains are in bold letters. Also included is a consensus line where * means that there are the same amino acids in all the sequences, + means that there are some conservative changes along the sequences, and a space means that there are significant changes on the protein at this position.

(*M.crystallinum* AC: U79770, *S.humilis* AC: L36456 and L36823 and *A.thaliana* AC: L37883) that showed much lower homology (approximately 50% identity at the amino acid level) to the other CAD genes. A more accurate sequence comparison indicated that these genes share higher similarity (between 62 and 66% amino acid identity) with a mannitol alcohol dehydrogenase (MTD) from *Apium graveolens* (Williamson *et al.*, 1995) and the Eli3 protein from *A. thaliana* that was recently shown to be a benzyl alcohol dehydrogenase (BAD) (Somssich *et al.*, 1996). It is possible that these sequences may be related to a different class of alcohol dehydrogenases (MTD or BAD) that are not involved in the lignin biosynthetic pathway. As determined in previously cloned CAD genes (Knight *et al.*, 1992; MacKay *et al.*, 1995), the maize CAD protein contains three highly conserved regions that are responsible for zinc atom binding and cofactor binding in several classes of alcohol dehydrogenase. Figure 1 shows the catalytic Zn1 binding consensus sequence (Zn1), the structural Zn2 consensus sequence (Zn2) and the NADP-binding site (NADP) identified in the CAD sequences analyzed so far.

Using the fragment *SmaI-SpeI* of the CAD cDNA (Fig. 2) a Southern blot analysis of genomic DNA from maize (Fig. 2A) clearly showed that only one hybridizing band was detected for each restriction enzyme digestion. These results indicate that in maize there is possibly only one CAD-encoding gene. An interesting result came from the RFLP mapping study. It revealed a very close linkage between the CAD probe and the *bm1* locus on the short arm of chromosome 5, between probes UMC1 and BNL7.71. The *brown midrib* phenotype in maize is correlated to a reduced and/or modified lignin content. In maize were described four independent *bm* mutants (*bm1* through *bm4*) (Neuffer *et al.*, 1968; Barrière and Argillier, 1993 and Barrière *et al.*, 1994), and recently our group demonstrated that the *bm3* mutants of maize are due to mutations in the gene encoding the caffeic acid O-methyltransferase (COMT) (Vignols *et al.*, 1995). Analogously, the RFLP mapping results reported here strongly suggest that the *bm1* mutant may be due to a mutation in the CAD gene. A northern blot with mRNAs of the four *bm* mutants of maize, which have a lower lignin content, did not show any change in the size of the band detected in any mutant and specifically in the *bm1* mutant (data not shown). The *bm1* mutant analyzed could be due to a punctual mutation in the sequence that converts the mRNA translated into a non-functional enzyme, but this point could not be proved definitively with the available results.

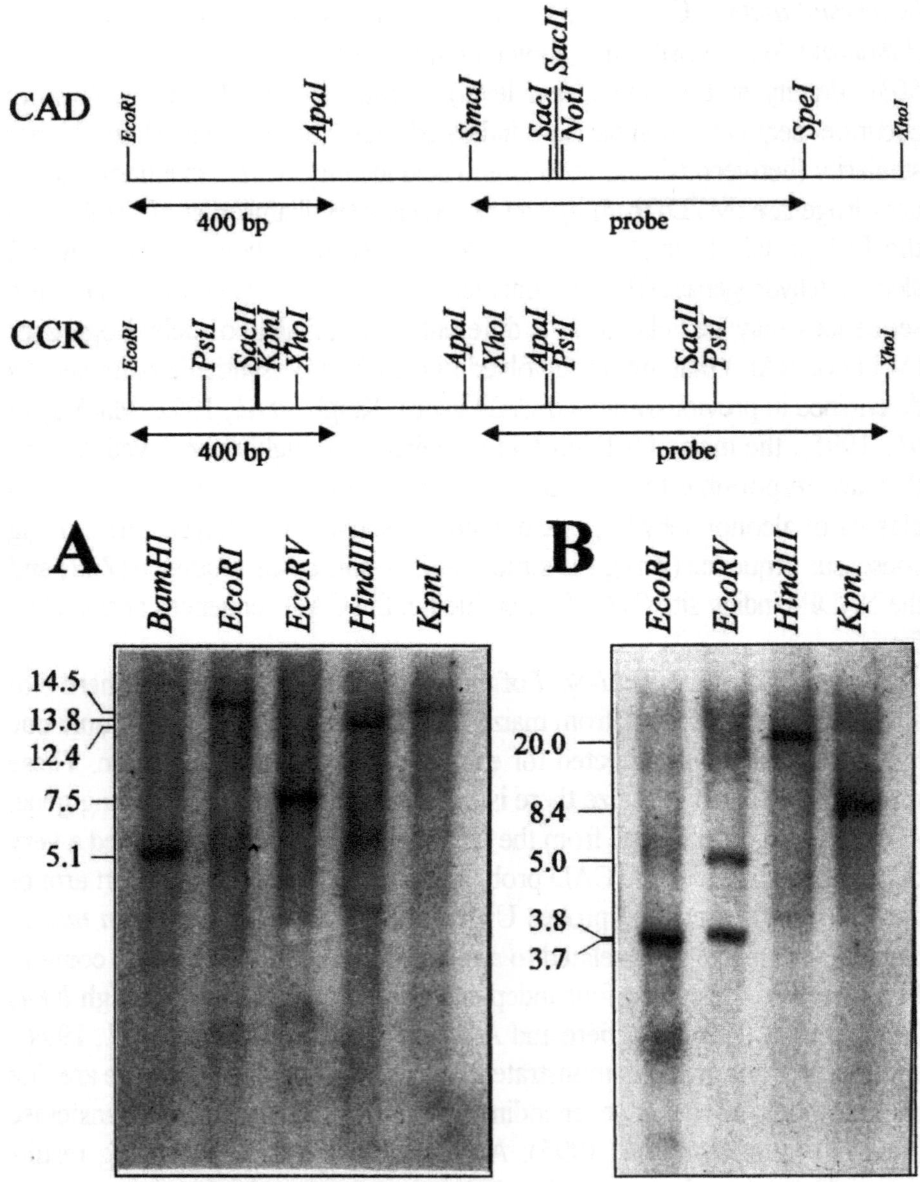

Fig. 2. Up: cDNA restriction map of the CAD and CCR cDNA clones from maize to show the probes used in Southern and northern analysis and their main restriction sites. Down: Southern blot analysis of maize genomic DNA. Ten μg of maize genomic DNA were loaded in each lane. The DNA was hybridized either to a ^{32}P-labeled CAD-specific (**A**) or to a ^{32}P-labeled CCR-specific probe (**B**). The EMBL accession numbers of the CAD and CCR cDNAs of maize here described are Y13733 and Y13734, respectively.

2.2 Cloning of a cDNA encoding the maize cinnamoyl-CoA reductase

Several positive clones were isolated from the maize root cDNA library by using the sugarcane CCR cDNA as a probe (see Materials and Methods). One near full-length clone of 1486 bp was sequenced and characterized. The clone presents a 5' untranslated region of 180 bp and a 3' untranslated region of 190 bp. A putative polyadenylation signal (AATAAA) is located at position 1444. An open reading frame coding for a protein of 371 amino acids is located between nucleotides 181 and 1293.

The CCR protein of maize shows great sequence similarity to other CCR sequences (Fig. 3). At the amino acid level the CCR cDNA has a sequence identity of between 70% (*N.tabacum*) and 85% (*F.arundinacea*). A recently published study on CCR from eucalyptus (Lacombe *et al.*, 1997), identified a conserved motif of 8 amino acids (KNWYCYGK) (I in Fig. 3) which was present in all the CCR sequences analyzed and was correlated to the catalytic site of the enzyme. Such sequence is present also in the CCR maize clone characterized in this study with the only difference that at position 177 the K residue was replaced by a R residue. In the N-terminal region there is another conserved motif involved in co-factor binding to the NADP (Lacombe *et al.*, 1997) (NADP in Fig. 3).

Using an *XhoI* fragment of the cDNA (Fig. 2) as a probe, a Southern blot analysis showed that under high stringency conditions (20 mM Na_2HPO_4, 0.1% SDS, 1 mM EDTA at 65°C), only one or two bands were detected by the CCR-specific probe (Fig. 2B). When the hybridization washes were performed at lower stringency, it was possible to detect some additional bands (data not shown), suggesting that in the maize genome are present one or a few sequences that cross-hybridize with the CCR sequence. At this time it is not possible to state whether these sequences represent another CCR gene or another class of reductases. The CCR probe maps on chromosome 1 between probes BNL7.08 and UMC128. BNL7.08 locus, mapped on chromosome 1, is about 10 cM below UMC58 often reported in maize maps (BNL7.08 maps also on chromosome 8). CCR is therefore about 15 cM below UMC58 and 10 cM above UMC128. The map position is at 50 cM away from the *bm2* mutant and therefore it is not linked to any known locus that could be related to lignin biosynthesis. However, it is important to notice that the hybridization pattern of the cDNA CCR probe showed more than one band for several restriction enzymes used. This clone may hybridize to another locus in the maize genome, but this was not determined during this study.

```
                                                          NADP
Z.mays         MTVVDAVVSSTDAGAPAAAATAVPAGNGQTVCVTGAAGYIASWLVKLLLEKGYTVKGTVRNPDDPKNAHLKALDG  75
F.arundinacea  .....------------...PQL.-.H.............G......R.......................   62
E.gunnii       .P.-..-----------------LP.S........G.F...I.....R...R...........G..RD.E.   57
P.trichocarpa  .P.-..--..-------------LS.Q...I....GG.F...M....D....R..A...A....S..RE.E.   59
N.tabacum      .P-S---..-----------------..I......G.F......I.......R........R..S..RE.ER  53
consensus      *++ +                     ****++****++*+***++**+**++*****+***+****+*+****+***  75

Z.mays         AAERLILCKADLLDYDAICRAVQGCQGVFHTASPVTDDPEQMVEPAVRGTEYVINAAAEAGTVRRVVFTSSIGAV 150
F.arundinacea  .TK...............A..E..H..............................D...............I 137
E.gunnii       .S...T.Y.G..M..GSLEE.IK..D..V..................I..KN..V.....-K............ 130
P.trichocarpa  .E...T.........ESLKEGI...D...............E......N..KN..I.....-K............ 133
N.tabacum      .K.T.T..R.....FQSLRE.IS..D.....R................I..KN..T.....-K............ 127
consensus      * ++*+*+++**+*++++ +++ **+**+**+*********+****** **++** ***+* +*************+ 150

                                       I
Z.mays         TMDPKRGPDVVVDESCWSDLEFCEKTRNWYCYGKAVAEHAAWETARRRGVDLVVVNPVLVVGPLLQATVNASIAH 224
F.arundinacea  ....N.......N..........K..K..........Q....A..K..I.................P.....A.. 211
E.gunnii       ....N.A-...............KS.K..........K...PEGKE......I.....L.....S.I....I. 204
P.trichocarpa  Y...NK.....I...........KN.K..........Q...DM.KEK.............L.....P......T. 207
N.tabacum      Y...N.D..K....T....PD..KN.K.......M...Q...DE..EK.....AI......L.....QN....VL. 202
consensus      +***++++*+*++*+*****++**+ *+*******+****+***+ ++++*+***++*****+***** ++***+ * 225

Z.mays         ILKYLDGSARTF-ANAVQAYVDVRDVADAHLRVFESPRASGRHLCAERVLHREDVVRILAKLFPEYPVPARCSDE 299
F.arundinacea  .........KKY-.....S........G..I....A.E...Y.........G...Q..S..L......T..... 286
E.gunnii       .....T...K.Y-..S.....H.K....L..VL.L.T.S...Y....S....G...E....F....N..TK.... 279
P.trichocarpa  .....T...K.Y-..S.....H......L..IL...T.S...Y..S.S....GE..E....F.....I.TK.... 282
N.tabacum      .H...T...K.YTS.SL....H......LR.IL.Y.T.S....Y....S....C...E....F.....I.TK...V 277
consensus      *+*** ***++++++*++*+**+*+***++*+++*++*++*+*****++*+*+*+**+*+*+*+++***+++**++***+ 300

Z.mays         VNPRKQPYKFSNQKLRDLGLQFRPVSQSLYDTVKNLQEKGHLPVLGERTTTEAADKDAPTAEMQQGGIAIRA 371
F.arundinacea  .........M.....Q......T..ND...E...S........L.PSKPEGLNGVTA                342
E.gunnii       ....VK...............E.T..K.C..E...S.........P-SPPEDSVRIQG               335
P.trichocarpa  K....................FE.T..K.C..E...S.........IPKQAAEESLKIQ               338
N.tabacum      TK..VK..........K....E.T..-.C..E...S.........IPTQ-KDEIIRIQSEKFRSS        338
consensus      ++**++***+*****+******+++*** ++**+****+*********+++      ++ ++++
```

Fig. 3. Deduced amino acid sequence from the maize CCR cDNA aligned with the following CCR sequences: *E.gunnii* Genbank AC: Z79566, *F.arundinacea*, *M.sativa* and *N.tabacum* CCR sequences are as reported by Boudet *et al* (1995). The signs are the same as in Figure 1.

2.3 Expression of CAD and CCR genes in maize

The expression pattern of the CAD and CCR genes in maize was determined by northern blot analysis (Fig. 4A and 4B). A transcript of approximately 1.5 kb was detected by both cDNA probes (the same probes used in Southern analysis, see Fig. 2), which is in close agreement with the size determined for the two full-length clones. For both genes the accumulation of mRNA followed the same pattern: the highest expression was present in the region of the plant corresponding to the elongating zone of the root (R1), while the expression seemed to decrease (especially in the case of the CAD gene) in the maturation zone of the root (R2). No expression was detected in the meristematic region of the root, corresponding to the root tip (T). In leaves (L) the mRNA was accumulated at significant levels, while little expression was detected in the mesocotyl (M).

The accumulation pattern shown by both the CAD and the CCR genes closely follows the pattern established for the COMT gene (Collazo *et al.*,

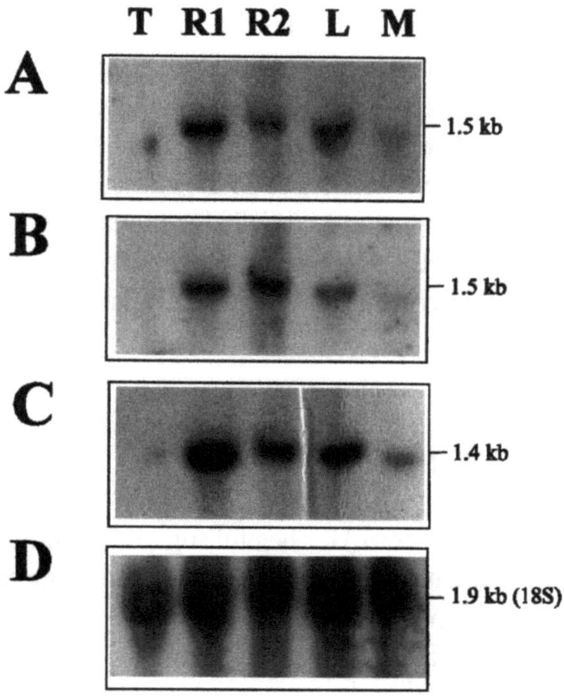

Fig. 4. Northern blot analysis of different maize RNA tissues. Total RNA was isolated from root tip (T), elongation zone of the root (R1), differentiating zone of the root (R2), leaf (L) and mesocotyl (M) of 9-day-old maize plants. Twenty μg of RNA were loaded in each lane and the blot was hybridized to a ^{32}P-labeled CAD-specific probe (**A**), to a ^{32}P-labeled CCR-specific probe (**B**), or to a ^{32}P-labeled COMT probe (**C**). Equal loading of RNA was determined by hybridizing the filter with a maize ribosomal DNA probe (**D**) (McMullen *et al.*, 1986).

1992; Fig. 3C). For this gene, subsequent *in situ* and promoter analyses did demonstrate that the COMT mRNA accumulated exclusively in cells undergoing active lignin synthesis in the non-meristematic region of the root and in the leaves (Capellades *et al.*, 1996). Analogous studies will determine the exact localization of the CAD and CCR transcripts in maize. Our results show that the genes coding for enzymes of the lignin biosynthesis pathway reported so far are coordinately expressed in the maize tissues undergoing lignification. This is in accordance with the specific differentiation of cells synthetizing secondary cell wall. It is also remarkable that two of the known maize mutations affecting lignin biosynthesis, *bm3* and *bm1*, occur on enzymes of the pathway, namely COMT and CAD. In the case of CCR no mutant has been described in maize. This fact may be due either to the fact that CCR appears to be encoded by more than one sequence, which is not the case for COMT or CAD, or that a null mutation on this gene may have highly deleterious effects on the plant. In fact, the allele of ***bm1*** here studied does not show any effect on the northern analysis of CAD, indicating that it may be the result of point mutations producing a partially functional enzyme instead of a null mutation. These results may be of interest for the production of transgenic plants with reduced lignin content using antisense technology based on the enzymes of the pathway.

ACKNOWLEDGMENTS

The authors thank P.Perez (Biocem) for helpful suggestions during the time of the project, B. Burr and University of Missouri for providing probes, and S. Tingey (Dupont) for the Mac Version of Mapmaker. The work has been carried out under grant BIO94/0734 from Plan Nacional de Investigación Científica y Técnica and within the framework of Centre de Referència de Biotecnologia de la Generalitat de Catalunya.

REFERENCES

Ausubel, F.M., Brent, R., Kingston, R.E., Moore, D.D., Seidman, J.G., Smith, J.A. and Struhl, K. eds. (1994). Current Protocols in Molecular Biology. Wiley Interscience. Ed. New York.

Barrière, Y., and Argillier, O. (1993). Brown-midrib genes of maize: a review. *Agronomie* **13**: 865-876.

Barrière, Y., Argillier, O., Chabbert, B., Tollier, M.T., and Monties, B. (1994). Breeding silage maize with brown-midrib genes. Freeding value and biochemical characteristics. *Agronomie* **14**: 11-21.

Boudet, A.M., Grima-Pettenati, J. And Goffner, D. (1995). DNA sequences coding for a cynnamoyl CoA reductase and their use in the regulation of plant lignin concentrations. Patent WO 95/27790 (PCT/FR95/00465).

Boudet, A.M. and Grima-Pettenati, J. (1996). Lignin genetic engineering. *Molecular Breeding* **2**: 25-39.

Campbell, M.M and Sederoff, R.R. (1996). Variation in Lignin Content and Composition. *Plant Physiol.* **110**: 3-13.

Capellades, M., Torres, M.A., Bastisch, I., Stiefel, V., Vignols, F., Bruce, W.B., Peterson, D., Puigdomènech, P. and Rigau, J. (1996). The maize caffeic acid O-methyltransferase gene promoter is active in transgenic tobacco and maize plant tissues. *Plant Mol. Biol.* **31**: 307-322.

Collazo, P., Montoliu, L. Puigdomènech, P. and Rigau, J. (1992). Structure and expression of the lignin O-methyltransferase gene from *Zea mays* L. *Plant. Mol. Biol.* **20**: 857-867.

Dellaporta, S.L., Wood, J. and Hicks, J.B. (1983). A Plant DNA Minipreparation: Version II. *Plant. Mol. Biol. Rep.* **1**: 19-21.

Galliano, H., Cabane, M., Eckerskorn, C., Lottspeich, F., Sandermann, H. and Ernst, D. (1993). Molecular cloning, sequence analysis and elicitor-/ozone-induced accumulation of cinnamyl alcohol dehydrogenase from Norway spruce (*Picea abies* L.). *Plant Mol. Biol.* **23**: 145-156.

Grima-Pettenati, J., Feuillet, C., Goffner, D., Borderies, G. and Boudet, A.M. (1993). Molecular cloning and expression of a *Eucaliptus gunnii* cDNA clone encoding cinnamyl alcohol dehydrogenase. *Plant Mol. Biol.* **21**: 1085-1095.

Hibino, T., Shibata, D., Chen, J.Q. and Higuchi, T. (1993). Cinnamyl Alcohol Dehydrogenase from *Aralia cordata*: Cloning of the cDNA and Expression of the Gene in Lignified Tissues. *Plant Cell Physiol.* **34**: 659-665.

Knight, M.E., Halpin, C. and Schuch, W. (1992). Identification and characterization of cDNA clones encoding cinnamyl alcohol dehydrogenase from tobacco. *Plant. Mol. Biol.* **19**: 793-801.

Lacombe, E., Hawkins, S., Van Doorsselaere, J., Piquemal, J., Goffner, D., Poeydomenge, O., Boudet, A.M. and Grima-Pettenati, J. (1997). Cinnamoyl CoA reductase, the first committed enzyme of the lignin branch biosynthetic pathway: cloning, expression and phylogenetic relationships. *Plant J.* **11**: 429-441.

MacKay, J.J., Liu, W., Whetten, R., Sederoff, R.R. and O'Malley, D.M. (1995). Genetic analysis of cinnamyl alcohol dehydrogenase in loblolly pine: single gene inheritance, molecular characterization and evolution. *Mol.Gen.Genet.* **247**: 537-545.

McMullen, M., Hunter, B., Phillips, R.L. and Rubenstein, I. (1986). The structure of the maize ribosomal DNA spacer region. *Nucleic Acids Res.* **14**: 4953-4968.

Murigneux, A., Barloy, D., Leroy, P. and Beckert, M. (1993a). Molecular and morphological evaluation of doubled haploid lines in maize. 1. Homogeneity within DH lines. *Theor. Appl. Genet.* **86**: 837-842.

Murigneux, A., Baud, S. and Beckert, M. (1993b). Molecular and morphological evaluation of doubled haploid lines in maize. 2. Comparison with single-seed descent lines. *Theor. Appl. Genet.* **87**: 278-287.

Neuffer, M.G., Jones, L. and Zuber, M.S. (1968) in: The Mutants of Maize, S. Matthias and H. Hamilton eds. (Madison, WI: Crop Science Society of America).

O'Malley, D.M., Porter, S. And Sederoff, R.R. (1992). Purification, Characterization, and Cloning of Cinnamyl Alcohol Dehydrogenase in Loblolly Pine (*Pinus taeda* L.). *Plant Physiol.* **998**: 1364-1371.

Somssich, I.E., Wernert, P., Kiedrowski, S. and Halbrock, K. (1996). *Arabidopsis thaliana* defense-related protein ELI3 is an aromatic alcohol:$NADP^+$ oxidoreductase. *Proc. Natl. Acad. Sci.* **93**: 14199-14206.

Van Doorsselaere, J., Baucher, M., Feuillet, C., Boudet, A.M., Van Montagu, M. and Inze, D. (1995). Isolation of cinnamyl alcohol dehydrogenase cDNAs from two important economic species: alfalfa and poplar. Demonstration of a high homology of the gene within angiosperms. *Plant Physiol. Biochem.* **33**: 105-109.

Verwoerd, T.C., Dekker, B.M.M. and Hoekema, A. (1989). A small-scale procedure for the rapid isolation of plant RNAs. *Nucleic Acids Res.* **17**: 2362.

Vignols, F., Rigau, J., Torres, M.A., Capellades, M. and Puigdomenech, P. (1995). The *brown midrib3* (*bm3*) Mutation in Maize Occurs in the Gene Encoding Caffeic Acid O-Methyltransferase. *Plant Cell* **7**: 407-416.

Williamson, J.D., Stoop,J.M., Massel, M.O., Conkling, M.A. and Pharr, D.M. (1995). Sequence analysis of a mannitol dehydrogenase cDNA from plants reveals a function for the pathogenesis-related protein ELI3. *Proc. Natl. Acad. Sci.* **92**: 7148-7152.

Wu, L., Takashi, U. And Messing, J. (1995). The formation of mRNA 3'-ends in plants. *Plant J.* **8**: 323-329.

Dissecting Light Repression of the Asparagine Synthetase gene (AS1) in Arabidopsis

Nora Ngai and Gloria Coruzzi

New York University,
Department of Biology,
Washington Square,
New York, N.Y. 10003

Key words: asparagine synthetase, light repression, Arabidopsis, transfactors

Introduction

In plants, asparagine is a key amino acid used to mobilize assimilated nitrogen from sources to sinks. Inorganic nitrogen assimilated initially into glutamine is converted into asparagine under certain metabolic conditions (see Figure 1). Biochemical, physiological and molecular studies suggest that asparagine biosynthesis in plants is a dynamic process that is tightly regulated throughout development and by environmental factors such as light (Sieciechowicz et al., 1988; Lam et al., 1994; Tsai and Coruzzi, 1990,1991). Physiological studies on pea and Arabidopsis have shown that when plants are dark-adapted, asparagine levels are elevated and asparagine becomes the predominant amino acid transported in the phloem (Urquhart and Joy, 1982; Schultz, 1994; Lam et al., 1995). Asparagine has a higher nitrogen to carbon ratio than glutamine, hence it is a more economical nitrogen transport compound in plants grown in the dark, when carbon skeletons are limiting. As such asparagine is the preferred nitrogen transport compound in dark-adapted plants (see Figure 1). By contrast, asparagine synthesis is repressed by light as evidenced by changes in amino acid levels and gene expression studies.

The asparagine synthetase genes in pea (AS1) and in Arabidopsis (ASN1) are each expressed at high levels in dark-grown or dark-adapted plants, and light represses their transcription (Tsai and Coruzzi, 1990; Lam et al., 1994). The transcriptional repression of AS genes in pea and Arabidopsis have been shown to be mediated at least in part by phytochrome. In Arabidopsis, the negative effect on ASN1 expression was shown to be also mediated through a light-induced increase in carbon metabolites. Lam et al. (1994) demonstrated that sucrose can mimic the effect of light by repressing ASN1 expression in the dark. This suggests that light may indirectly exert a negative effect on ASN1 expression by increasing the levels of carbon metabolites. In addition, repression of ASN1 by sucrose can be partially relieved by the addition of exogenous amino acids (glutamate, glutamine and

NATO ASI Series, Vol. H 104
Cellular Integration of Signalling Pathways in Plant Development
Edited by F. Lo Schiavo, R. L. Last, G. Morelli, and N. V. Raikhel

asparagine) (Lam et al., 1994), suggesting that it is the ratio of nitrogen and carbon metabolites that controls ASN1 gene expression. When levels of carbon metabolites are low relative to organic nitrogen, the ASN1 gene is expressed, leading to the biosynthesis of asparagine which stores and transports the excess nitrogen (Figure 1). Similar carbon repression of asparagine synthesis was also demonstrated in maize,

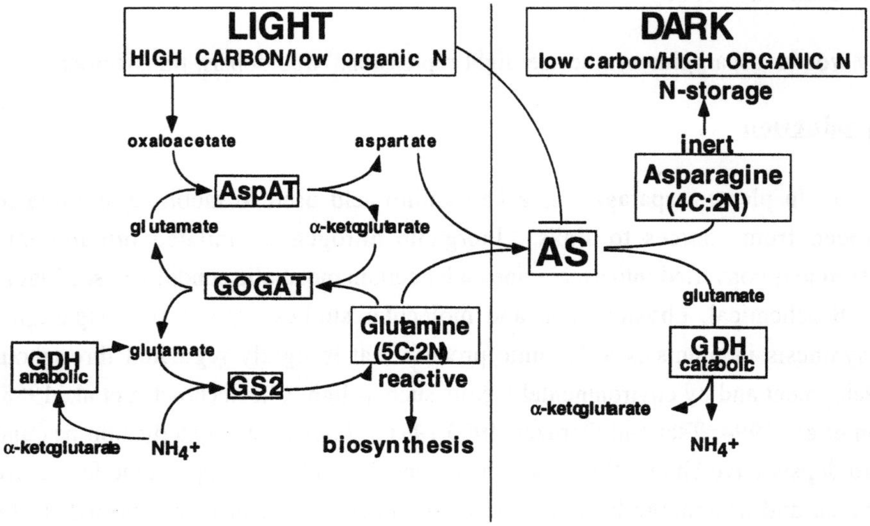

Figure 1. A model of the nitrogen assimilatory pathway in plants. Inorganic nitrogen in the form of nitrate is taken up by plants and reduced to ammonium (NH_4^+) by the sequential actions of nitrate and nitrite reductase (Lea, 1997). Under light conditions where plants have a high carbon to nitrogen metabolic ratio, ammonium is assimilated into glutamine via the combined actions of glutamine synthetase (GS2; E.C. 6.3.1.7) and glutamate synthase (GOGAT; E.C. 1.4.7.1). Glutamine and glutamate act as nitrogen donors for the synthesis of other amino acids and nucleic acids. Under dark conditions, there is a low carbon to nitrogen metabolic ratio in plants. Nitrogen assimilated into glutamine is converted to asparagine by asparagine synthetase (AS) where asparagine is synthesized for storage of nitrogen. (Abbreviations: AspAT, aspartate aminotransferase (E.C. 2.6.1.1) and GDH, glutamate dehydrogenase (E.C. 1.4.1.2)).

both at the level of asparagine accumulation (Oaks and Ross, 1984) and at the level of gene expression (Chevalier et al., 1996). In asparagus, AS gene expression increases in post-harvested spears when the sugar content is low (Davis and King, 1993). Thus, sucrose repression of asparagine synthesis occurs in a number of diverse species including dicots, legumes and monocots.

Light has been shown to exert reciprocal effects on AS and GS gene expression. In Arabidopsis, GLN2 (chloroplastic glutamine synthetase) expression is induced by light while ASN1 expression is repressed by light as measured at the mRNA level (Figure 2A) (Lam et al., 1995). This reciprocal regulation by light is reflected in the level of glutamine and asparagine measured in leaf extracts from light-grown vs. dark-adapted plants (Figure 2B) (Schultz, 1994; Lam et al., 1995). Glutamine levels are higher in light-grown plants and asparagine levels are higher in dark-adapted plants (Figure 2B). Studying the light-responsive elements involved in the light-regulation of GS and AS genes may provide insights into the mechanisms controlling light-regulated gene expression in plants (Tjaden et al., 1995; Neuhaus et al., 1997; Ngai et al., 1997) . The study of the pea AS1 promoter in Arabidopsis should enable the identification of cis-elements that are involved in the regulatory mechanism controlling light-activated transcriptional repression of AS1.

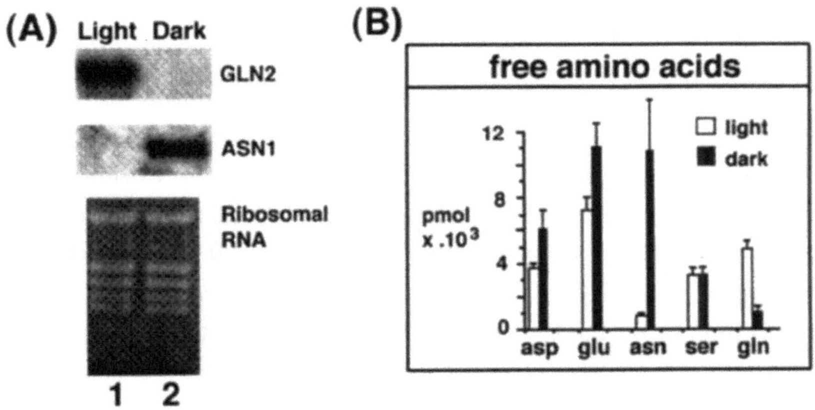

Figure 2. GLN2 and ASN1 are reciprocally controlled by light in Arabidopsis. (**A**) The level of GLN2 mRNA are highly expressed in light-grown plants (lane 1) and not in dark-adapted plants (lane 2). In contrast, the level of ASN1 mRNA are higher in dark-adapted plants (lane 2) and repressed in light-grown plants (lane 1) (Lam et al., 1995). (**B**) The glutamine and asparagine levels coincide with the changes in GLN2 and ASN1 mRNA levels in light vs dark-adapted Arabidopsis plants (Schultz, 1994; Lam et al., 1995).

Results and Discussion

Transgene expression is negatively regulated by light in transgenic Arabidopsis plants containing the AS1-GUS transgene

Deletion analysis of the AS1 promoter showed that as little as 88 bp of the promoter (-88 to +120) can confer light-repressed transcription to the GUS reporter gene in transgenic tobacco plants (Ngai et al, 1997). In addition, in a gain-of-function experiment, the -124 bp AS1 promoter was able to confer light-regulated expression to a 35S TATA element in transgenic tobacco (Ngai et al, 1997). To test whether the pea AS1 promoter can also confer light-regulated GUS expression in Arabidopsis, AS1-GUS containing Arabidopsis plants were created. A construct with the AS1 promoter from -148 to +120 fused in a transcriptional fusion to GUS was subcloned into the Ti binary vector PBI101.1 and introduced into *Agrobacterium tumefaciens* GV3101 (MP90) by electroporation. The exconjugants were used to transform the Columbia ecotype of Arabidopsis by vacuum infiltration (Bent et al., 1994). Seeds were collected and the resulting kanamycin resistant transgenic (T_0) plantlets were transferred to soil and grown to maturity.

T_1 seeds were collected from 7 independent transformants and AS1-GUS transgene expression was monitored histochemically and quantitatively. Transgenic AS1-GUS Arabidopsis seedlings displayed GUS activity in the vasculature tissue similar to results obtained in tobacco (Ngai et al., 1997) (data not shown). GUS activity was measured in extracts from approximately 200-300 Arabidopsis T_1 seedlings grown for 10-days in continuous light or dark on MSK media (50 ug/ml kanamycin) (Figure 3). For each independent transformant, the GUS activity was higher in dark-grown seedlings compared to light-grown (Figure 3). A nonparametric paired test between the GUS activity of light- vs. dark-grown seedlings showed their differences to be significant with a P value of 0.0156. The D:L ratio of independent transformants ranged from 1.5 (transgenic line 1) to 12.8 (transgenic line 13) (Figure 3). Thus the DNA elements contained within -148 bp of the pea AS1 promoter are sufficient to confer negative light regulation to a reporter gene in Arabidopsis.

Arabidopsis nuclear protein factors interact with putative repressor binding sites, Box B and Box C/C' present in -148 AS1 promoter

The shortest AS1 promoter fragment shown to confer light-activated transcriptional repression of AS1-GUS expression in transgenic tobacco plants was -88 bp (Ngai et al., 1997). Within the -88 bp AS1 promoter are DNA sequences that are homologous to sequences found in the promoters of two other pea genes that are transcriptionally repressed by light; pea AS2 (Tsai and Coruzzi, 1990) and pea phytochrome (Sato, 1988). Based on DNA homology, two short sequences in the

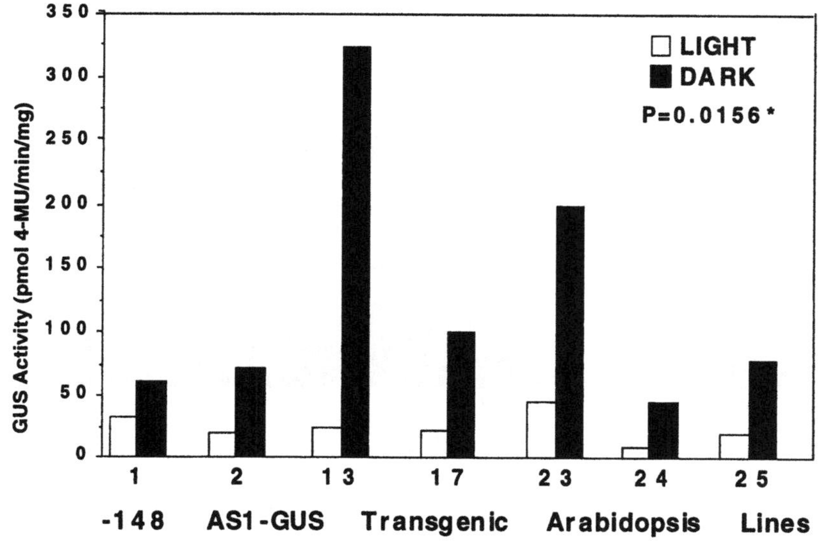

Figure 3. Transgenic Arabidopsis lines with the -148 AS1-GUS construct are repressed by light. Shown in the bar graph are the GUS activity of independent transgenic Arabidopsis lines 1, 2, 13, 17, 23, 24, and 25. The GUS activity was measured in extracts from 200-300 seedlings that were grown in the light (open bar) and in the dark (black bar) on MSK (50ug/ml kanamycin) for 10 days. The AS1-GUS expression of transgenic Arabidopsis plants are negatively regulated by light. The GUS activity in the light vs. dark were compared by nonparametric paired test by the statistical program Instat 2.0. * P values of < 0.05 are considered significant.

pea AS1 have been identified as possible candidates for negative light-regulatory elements: Box B (-61(AAACGACACCGTTT)-48) and Box C (-45(CTCCCAC)-39)/ C'(-88(TCCCGGTACACACTTCTT)-71) (Ngai et al., 1997) In addition, the Box C/C' sequence has homology to an inverse sequence of RE1 (-79(CCGCGCCC ATG)-70), a sequence shown to act as a repressor in the oat phytochrome gene which is also negatively regulated by light (Bruce et al., 1991).

In vitro gel-shift analysis was performed with nuclear extracts from Arabidopsis to identify whether these conserved DNA sequences (Box B and Box C/C') are involved in DNA-protein interaction (Ngai et al., 1997). The F1 and F2 DNA fragments of the AS1 promoter were used as labelled DNA probes. F1 encompasses the AS1 promoter from -73 to -33, while F2 encompasses the AS1 promoter from -88 to -48 (Figure 4A). Both F1 and F2 contain the Box B element.

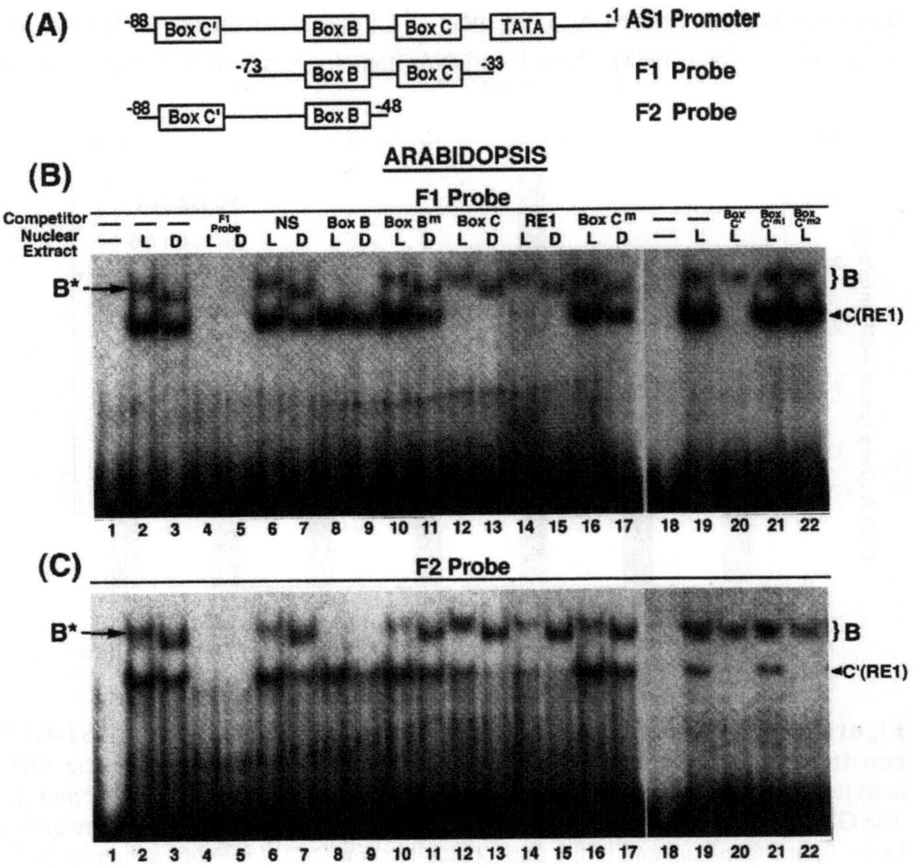

Figure 4. ***In vitro*** **gel-shift analysis with Arabidopsis nuclear extracts identifies DNA:protein complexes in the minimal functional 88 bp AS1 promoter element.** **(A)** A schematic presentation of the DNA fragments (F1 and F2) of the AS1 promoter used as a probe to detect DNA-protein bindings by gel-shift analysis. **(B)** Gel-shift analysis using the F1 (-73 to -33) as a probe. F1 has Box B and Box C. **(C)** Gel-shift analysis using the F2 (-48 to -88) as a probe. F2 has Box B and Box C'. For competition assays, DNA fragments were generated by annealing two complimentary oligos. Competitors are BoxB=AAA CGACACCGTTT, Box B^{m}=AAACGAAAAAGTTT, BoxC=AGCTCCCACCTTC, $BoxC^{m}$=AGCTCAAACCTTC, Box C'=TCCCGGTACACACTTCTT, $BoxC'^{m1}$= TCCCGGTAAAAACTTCTT, $BoxC'^{m2}$=TCAAGGTACACACTTCTT, and a dimer of RE1=CCGCGCCCATG (Bruce et al., 1991) (Bold letters identify mutated residues (C->A) in mutants). NS (non specific DNA fragment) is a gel purified 63 bp XbaI/SalI DNA fragment of Bluescript generated by PCR. Gel-shift analysis was performed with nuclear extracts from leaves of Arabidopsis grown in the light(L) or dark-adapted for 6 days(D).

F1 contains a Box C element while F2 contains a Box C' element. DNA-protein complexes were detected with nuclear extracts from leaves of Arabidopsis grown in the light (L) or dark-adapted for 6 days (D) (Figure 4B and C, lanes 2 and 3). Incubation with F1 and F2 probes showed similar shifts (Figure 4B and C). As determined by competition experiments, these shifts were found to be specific for the F1 and F2 probes. The shifts are competed with a 50-fold excess of unlabelled F1 or F2 DNA fragments (Figure 4B and C, lanes 4 and 5), but not by a 50-fold excess of non-specific (NS) DNA fragment (Figure 4B and C, lanes 6 and 7). To localize the specific DNA binding sequences within the F1 and F2 AS1 fragment, competition experiments with Box B, Box C and Box C' elements were performed. Both F1 and F2 have a DNA:protein interaction with DNA element Box B (upper band, Figure 4B and C). Competition experiments showed that the upper shifted band can be competed by an excess of unlabelled wild type Box B (Figure 4B and C, lanes 8 and 9), but not by an excess of unlabelled mutant Box B^{m} (Figure 4B and C, lanes 10 and 11) (see Figure 4 legend for sequences). The lower band was found to be a DNA:protein interaction with Box C or Box C'. Using F1 or F2 as a probe, the lower band can be competed by an excess of unlabelled Box C (Figure 4B and C, lanes 12 and 13), but not by an excess of mutant Box C^{m} (Figure 4B, lanes 16 and 17) (see Figure 4 legend for sequences). Similarly, the lower band can also be competed by Box C' (Figure 4B and C, lane 20) and not by an excess of mutant Box C'^{m1} (Figure 4B and C, lane 21). Competition with another Box C' mutant, Box C'^{m2} (see Figure 4 legend for sequence), can compete out the lower band detected with the F2 probe but not with the F1 probe (Figure 4B and C, lane 22). The cytosine nucleotides that have been changed to adenine in Box C'^{m2} (see Figure 4, legend) seemed to be involved in the DNA:protein binding of Box C but not in Box C'. The Box C or C' shift can each be competed specifically by the RE1, a putative repressor element previously defined in the oat phytochrome gene (Bruce et al., 1991) (Figure 4B and C, lanes 14 and 15). The nuclear protein factor that binds to either Box C, C', or RE1 could be a repressor binding protein which has multiple DNA binding sites. Alternatively, there could be a family of binding proteins that bind to Box C, C', or RE1 which may be involved in light-activated transcriptional repression.

Gel shift analysis showed two major DNA/protein complexes involving the Box B and Box C/C' elements contained within the -88 AS1 promoter (Ngai et al., 1997). The Box B and Box C/C' binding factors identified in tobacco are also present in extracts of both light or dark-grown/adapted Arabidopsis plants. However, in light-growth conditions, the protein(s) binding to Box B are modified (Ngai et al., 1997). That is the B complex is altered in mobility in extracts made from light-grown compared to dark-adapted Arabidopsis (see B* (arrow), Figure 4B and C, lane 2). This light-specific modification of the "B" shift suggests that there is a light-induced modification of the B-binding protein or a light-induced modification of a protein that binds to the B-binding protein. This light-induced modification of the B-binding protein may be involved in light-activated transcriptional repression of AS1. In addition, the nuclear protein that specifically binds to Box C or Box C' may also be involved in repression, as its binding is competed by RE1, a putative repressor element previously defined in the oat phytochrome gene (Bruce et al., 1991). How might B and C/C' DNA-protein complexes be involved in light regulation? The light-specific modification of the B complex may involve an additional factor as shown in the model (Figure 5, factor X). In this model, the B binding factor is a

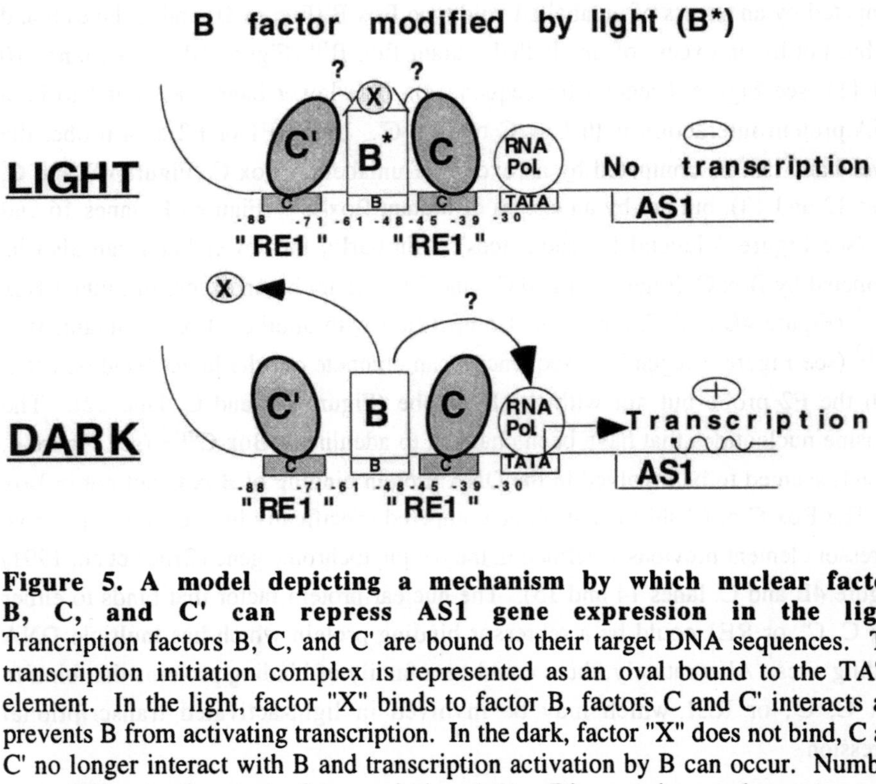

Figure 5. A model depicting a mechanism by which nuclear factors B, C, and C' can repress AS1 gene expression in the light. Trancription factors B, C, and C' are bound to their target DNA sequences. The transcription initiation complex is represented as an oval bound to the TATA element. In the light, factor "X" binds to factor B, factors C and C' interacts and prevents B from activating transcription. In the dark, factor "X" does not bind, C and C' no longer interact with B and transcription activation by B can occur. Numbers are relative to the start of transcription (+1). Diagram is not drawn to scale. RE1=negative element of oat phy A described in Bruce et al., 1991.

transcription activator, in which a light induced modification leads to an interaction directly or indirectly with repressors bound to Box C/C'. In the dark, this complex is abolished allowing the B transcription activator to function. This model supports a eukaryotic repression mechanism known as quenching (Johnson, 1995). This is only one of several models that can be proposed based on the data.

The two Box C elements (C and C') can compete with each other for factor binding. The nuclear protein that binds to the C or C' elements may be the same or it could be different. While one mutant of Box C' (Box C'^{m1}) affects binding to both Box C and C', the cytosine nucleotides mutated in Box C'^{m2} only affected factor binding with Box C and not with Box C'. These results suggest that there may be a preference of binding of the factor(s) to either Box C or Box C' site. As C and C' binding are each competed with the putative repressor element RE1 defined in the phytochrome (Bruce et al., 1991), they may be acting as repressor-binding DNA elements in the AS1 promoter. Neuhaus et al. (1997) demonstrated that co-injection of the AS1-GUS with a plasmid containing a tetramer of the C element of the AS1 promoter (TAGCTCCCACC) denoted as RE3 into tomato hypocotyl cells resulted in the inhibition of the down-regulation of the AS1-GUS expression in the light. In contrast, a tetramer of a mutated RE3 element (TAGCTC**GGT**CC, bold letters indicate mutated nucleotides) was not able to inhibit the down-regulation of the AS1-GUS expression by light. This suggests that a light-activated repressor is interacting with the RE3 element and that its removal (by competition) removed the repression of the AS1 promoter in the light.

The Box C/C' or RE3 elements were able to confer light-activated repression to a reporter gene in gain-of-function experiments (Neuhaus et al., 1997; Ngai et al., 1997). An RE3 tetramer placed between a 35S CaMV B domain (-343 to -90) and a minimal -46 35S CaMV TATA box was able to repress GUS expression specifically in the light (Neuhaus et al., 1997). A tetramer of the mutant RE3 however could not confer light-repression to the reporter gene (Neuhaus et al., 1997). Similarly, the -124 bp native AS1 promoter which has two C elements (Box C and Box C') placed upstream of a 35S CaMV (4XB4(-301 to -208) + A(-90 to +8))-GUS element was also able to confer light-activated transcriptional repression to a reporter gene (Ngai et al., 1997). In this construct, the Box C and C' elements were placed at a distance (approximately 416 bp) from the TATA (Ngai et al., 1997), unlike the RE3 element which was placed proximal to the TATA (Neuhaus et al., 1997). Therefore, the C elements (Box C/C' or RE3) appear to be able to function at a position that is distal (Ngai et al., 1997) or proximal (Neuhaus et al., 1997) to the TATA. These experiments demonstrate that the Box C/C'(RE3) elements are both necessary and sufficient to mediate light repression and can function in a position-independent site relative to the TATA in a heterologous context.

Conclusion

In Arabidopsis, the expression of the ASN1 gene was shown to be regulated by light in a negative manner (Lam et al., 1994). GUS expression in Arabidopsis plants containing the transgene -148 AS1-GUS were negatively regulated by light. Potential transcription regulatory elements, Box B, Box C, and Box C' located within the -148 bp AS1 promoter fragment had binding activity to Arabidopsis nuclear proteins by gel shift analysis. A light-specific DNA:protein interaction was detected with Box B. The nuclear factors that bind to Box C and C' elements of AS1 are competed by a putative repressor element "RE1" defined previously in the oat phytochrome gene (Bruce et al., 1991). In addition, the Box C/C' was shown to function in light-activated transcriptional repression of AS1-GUS in a position independent to the TATA (Neuhaus et al., 1997; Ngai et al., 1997). It is possible that the Box B and/or Box C/C' binding proteins may be repressors or accessory proteins that interact with putative repressor proteins. In a positive selection scheme using Arabidopsis, it should be possible to isolate mutants impaired in light repression of AS1. Consolidating biochemical with molecular and genetic approach will provide an insight to the mechanisms contolling light-activated transcriptional repression of AS1 in plants.

Acknowledgements

We would like to thank Ming-Hsiun Hsieh for advice and critical reading of the manuscript. This work was supported by DOE grant DEFG02-92ER20071.

References

Bent, A., Kunkel, B.N., Dahlbeck, D., Brown, K.L., Schmidt, R., Giraudat, J., Leung, J., Staskawicz, B.J. (1994) Science 265:183-396.

Bruce, W.B., Deng, X.-W. and Quail, P.H. (1991) A negatively acting DNA sequence elements mediates phytochrome-directed repression of phyA gene transcription. EMBO J. 10:3015-3024.

Chevaliar, C., Bourgeois, E., Pradet, A. and Raymond, P. (1996) Metabolic regulation of asparagine synthetase gene expression in maize (Zea mays L.) root tips. Plant Mol. Biol. 28:473-485.

Davis, K.M. and King, G.A. (1993) Isolation and characterization of a cDNA clone for a harvest induced asparagine synthetase from *Asparagus officinalis* L. Plant. Physiol. 102:1337-1340.

Johnson, A. (1995) The price of repression. Cell 81:655-658.

Lam, H.-M., Peng, S.S.-Y and Coruzzi, G.M. (1994) Metabolic regulation of the gene encoding glutamine-dependent asparagine synthetase in *Arabidopsis thaliana.* Plant Physiol. 106:1347-1357.

Lam, H.-M., Coschigano, K., Schultz, C., Melo-Oliveira, R., Tjaden, G., Oliveira, I., Ngai, N., Hsieh, M.-H. and Coruzzi, G. (1995) Use of *A. thaliana* mutants and genes to study amide amino acid biosynthesis. Plant Cell 7:887-898.

Neuhaus, G., Bowler, C., Hiratsuka, K., Yamagata, H. and Chua, N.-H. (1997) Phytochrome-regulated repression of gene expression requires calcium and cGMP. EMBO J. 10:2554-2564.

Ngai, N, Tsai, F.-Y and Coruzzi, G.M. (1997) Light-Induced Transcriptional Repression of the Pea AS1 Gene: Identification of Cis-elements and Transfactors. Plant J., in press.

Oaks, A. and Ross, D.W. (1984) Asparagine synthetase in Zea mays. Can. J. Bot. 62:68-73.

Sato, N. (1988) Nucleotide sequence and expression of the phytochrome gene in Pisum sativum: Differential regulation by light of multiple transcripts. Plant Mol. Biol. 11:697-710.

Schultz, C.J. (1994) A molecular and genetic dissection of the aspartate aminotransferase isoenzymes of *Arabidopsis thaliana*. PhD thesis. New York University, New York.

Sieciechowicz, K.A., Joy, K.W. and Ireland, R.J. (1988) The metabolism of asparagine in plants. Phytochemistry 27:663-671.

Tjaden, G., Edwards, J.W. and Coruzzi, G.M. (1995) *cis* elements and *trans*-acting factors affecting regulation of a non-photosynthetic light-regulated gene for chloroplast glutamine synthetase. Plant Physiol. 108:1109-1117.

Tsai, F.-Y. and Coruzzi, G.M. (1990) Dark-induced and organ-specific expression of two asparagine synthetase genes in *Pisum sativum*. EMBO J. 9:323-332.

Tsai, F.-Y. and Coruzzi, G.M. (1991) Light represses the transcription of asparagine synthetase genes in photosynthetic and non-photosynthetic organs of plants. Mol. Cell. Biol. 11:4966-4972.

Urquhart, A.A. and Joy, K.W. (1982) Transport, metabolism, and redistribution of Xylem-borne amino acids in developing pea shoots. Plant Physiol. 69:1226-1232.

Genetic Approaches to Understanding the Regulation of Tryptophan Biosynthesis

Katherine J. Denby and Robert L. Last

Boyce Thompson Institute for Plant Research and Section of Genetics and Development, Cornell University, Ithaca NY 14853-1801 USA.

Keywords. Tryptophan metabolism, *Pseudomonas syringae*, camalexin, signal transduction

1. The Role of the Tryptophan Biosynthetic Pathway

Little is known about how amino acid metabolism is regulated in higher plants despite our extensive knowledge of the same processes in prokaryotes and fungi (see reviews by Galili, 1995; Herrmann, 1995; Lam et al., 1995; Radwanski and Last, 1995). The physiological roles of different isozymes is not clear, regulatory elements in amino acid biosynthetic gene promoters have not been defined and, in contrast to microorganisms, nothing is known about factors interacting with these promoters to affect gene expression. However, genes have been cloned for all of the enzymes in the Arabidopsis tryptophan biosynthetic pathway, antibodies were produced against six out of the seven proteins, and mutants obtained or antisense plants constructed for six out of the seven proteins (Bender and Fink, 1995; Radwanski and Last, 1995; Li and Last, 1996). In addition, the induction of the tryptophan pathway in response to a variety of environmental conditions (as discussed below) provides an excellent opportunity to pursue genetic approaches to understanding the regulation of this important pathway.

Tryptophan biosynthesis is an interesting primary metabolic pathway to study because it leads not just to the synthesis of the amino acid tryptophan, but also to a variety of secondary metabolites. These include indole glucosinolates, monoterpenoid indole alkaloids, auxin, indole phytoalexins and anthranilate derived alkaloids, such as quinoline alkaloids (Larsen, 1981; Normanly et al., 1993; Tsuji et al., 1993; Kutchan, 1995). These compounds play diverse roles in plants, such as anti-herbivory, defence against microbial pathogens and regulation of growth and development. Secondary metabolites derived from the tryptophan pathway are likely to be required differentially in plants, both spatially and temporally, and some are highly inducible. When attempting to understand regulation of the tryptophan pathway, it is vital to know how primary metabolism is coordinated with secondary metabolism to meet changing requirements for tryptophan or its

NATO ASI Series, Vol. H 104
Cellular Integration of Signalling Pathways in Plant Development
Edited by F. Lo Schiavo, R. L. Last, G. Morelli, and N. V. Raikhel

precursors. This is also essential for attempts to increase secondary metabolite production for biotechnology.

2. Tryptophan Biosynthesis in Arabidopsis

The tryptophan biosynthetic pathway consists of 5 different enzymes, all of which are located in the chloroplasts (Zhao and Last, 1995). Anthranilate synthase, the first enzyme and committing step in the pathway, and tryptophan synthase, the last enzyme, each consist of two different subunits, alpha and beta (Last et al., 1991; Niyogi and Fink, 1992; Niyogi et al., 1993; Radwanski et al., 1995), whereas the other tryptophan biosynthetic enzymes are homomeric. Anthranilate synthase converts chorismate, derived from the shikimic acid pathway, into anthranilate, which is then converted into tryptophan via indole (see Figure 1). Many of the enzymes are encoded by multiple genes and are expressed at different levels. For example, anthranilate synthase alpha subunit is encoded by two genes, ASA1 and ASA2, and ASA1 is expressed at a ten-fold higher level than ASA2 (Niyogi and Fink, 1992).

Anthranilate synthase, the rate limiting enzyme, is allosterically regulated by the end product of the pathway, tryptophan. A feedback-resistant mutant was isolated that contains a mutation in the anthranilate synthase alpha subunit (Li and Last, 1996). This mutant accumulates three-fold the level of tryptophan found in wild-type plants. The level of the tryptophan biosynthetic proteins is not reduced in this mutant despite the excess accumulation of tryptophan. However, it appears that flux through the tryptophan pathway may also be controlled by regulating expression of the biosynthetic enzymes as ASA mRNA levels are increased following pathogen infection or wounding in Arabidopsis (Niyogi and Fink, 1992) and by a fungal elicitor in *Ruta graveolens* (Bohlmann et al., 1995).

3. The Tryptophan Enzymes Respond to a Variety of Environmental Stresses

3.1 Pathogenesis

Camalexin is the major known phytoalexin produced in Arabidopsis and is synthesized from an intermediate of the tryptophan pathway (Tsuji, et al., 1993). This phytoalexin accumulates to very high levels after infection with pathogens (Tsuji et al., 1992), which leads to the question of whether flux through the tryptophan pathway increases to compensate for the increased demand for intermediates. Indeed, after infiltration with a virulent strain of Pseudomonas (*Pseudomonas syringae pv. maculicola*) the tryptophan enzymes

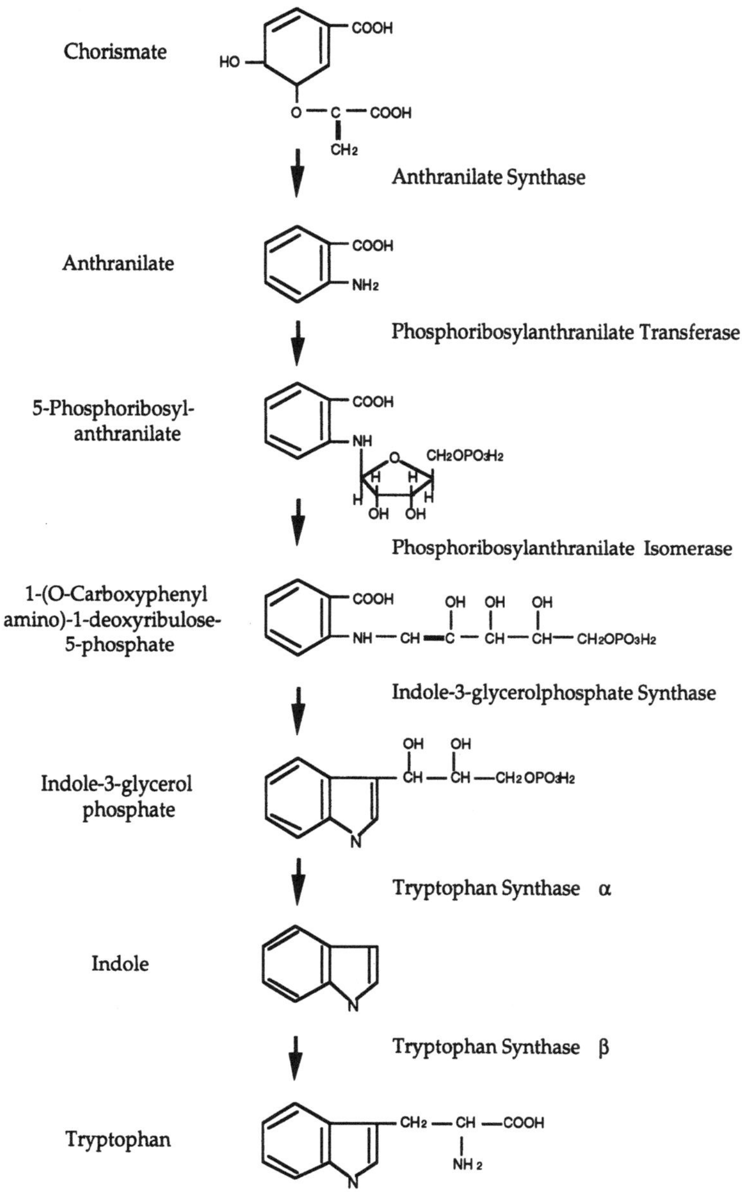

Fig. 1. Enzymes and metabolites of the tryptophan biosynthetic pathway.

are induced and increase approximately 3 fold by 2 days post-infiltration. Avirulent Pseudomonas and Xanthomonas also cause a similar induction of the tryptophan enzymes (Zhao and Last, 1996). Treatment of plants with abiotic elicitors, such as alpha amino butyric acid (unpublished results) and silver nitrate (Zhao and Last, 1996), also results in induction of the tryptophan enzymes. Camalexin accumulation and tryptophan protein induction appear to be tightly correlated: the greater the fold of induction of the tryptophan enzymes, the higher the level of camalexin that accumulates (Zhao and Last, 1996).

3.2 Oxidative Stress

Oxidative stress is another type of stress that causes induction of the tryptophan enzymes. Acifluorfen is a compound that inhibits haem and chlorophyll biosynthesis (Matringe et al., 1992) and is thought to cause increased production of free radicals and resulting oxidative stress. Arabidopsis plants treated with acifluorfen accumulate high levels of camalexin and show significant induction of the tryptophan enzymes (unpublished results). Exposure to ozone, an air pollutant well known to cause free radical production in plants (Kangasjärvi et al., 1994), induces PAT mRNA 6-8 fold in addition to various antioxidant enzyme mRNAs (Conklin and Last, 1995).

3.3 Amino Acid Starvation

In prokaryotes and fungi, the regulation of amino acid biosynthesis in response to amino acid starvation is well understood and different mechanisms operate in these two groups of organisms. In bacteria, the genes encoding enzymes of each amino acid biosynthetic pathway are physically grouped on the chromosome into operons. Each operon is specifically upregulated in response to a lack of the amino acid produced by that particular operon. For example, under conditions of tryptophan starvation, only the tryptophan operon will be induced (Yanofsky and Crawford, 1987). A very different system, that of general control, operates in yeast. Many genes encoding enzymes in different amino acid biosynthetic pathways are coregulated, and their expression is induced by starvation for any one of several amino acids (Hinnebusch, 1992).

In higher plants it appears that genes encoding the tryptophan biosynthetic enzymes are upregulated by starvation for various amino acids. However, the tryptophan pathway may be unique in this respect as this cross pathway control does not seem to extend to genes from other amino acid biosynthetic pathways. Guyer et al. (1995) recently demonstrated that histidine starvation in Arabidopsis seedlings caused induction of several amino acid biosynthetic genes in different biosynthetic pathways. In contrast, aromatic amino acid starvation (using the herbicide glyphosate) and branched-chain amino acid starvation (using primisulfuron) only led to induction of the two tryptophan enzyme genes tested and not of other amino acid biosynthetic genes assayed.

Similar results were obtained by C. Williams, J. Zhao and R. Last (unpublished results) demonstrating induction of the tryptophan enzymes in response to starvation for various amino acids. Treatment with glyphosate, which starves for aromatic amino acids by inhibiting EPSP synthase (Amrhein et al., 1980), and chlorsulfuron, which inhibits acetolactate synthase and starves for branched-chain amino acids (Ray, 1984), leads to a significant increase in the levels of the tryptophan enzyme mRNAs and proteins. Feeding lysine and threonine, which feedback inhibit aspartate kinase and hence starve for methionine (Bryan, 1980), also caused induction of the tryptophan pathway enzymes. This response appears not to be a general induction of amino acid biosynthetic genes as, for example, methionine synthase mRNA levels do not increase following these treatments (Williams, Zhao and Last, unpublished results).

Why is the tryptophan pathway responding to so many different stimuli (pathogens, abiotic elicitors and oxidative stress) in addition to amino acid starvation? It is possible that the tryptophan enzymes are responding to a more general stress signal rather than amino acid starvation *per se* and that biosynthesis of secondary metabolites is driving regulation of this primary metabolic pathway. Camalexin accumulates under all of the above conditions and there may be other secondary compounds, as yet uncharacterised, which are also highly inducible. The variety of secondary metabolites produced from tryptophan or its precursors, which may be differentially required by the plant, could explain why the tryptophan pathway responds to diverse stimuli. The tryptophan pathway may be responding to various signals to meet demands for secondary metabolite biosynthesis.

4. Genetic Approaches to Tryptophan Pathway Regulation

4.1 Analysis of Tryptophan Enzyme Regulation Using Previously Characterised Arabidopsis Mutants

This lab is pursuing a genetic approach to understanding how, and possibly why, the tryptophan pathway is regulated in response to these many different stimuli. Two complementary approaches have been taken. Firstly, induction of the tryptophan enzymes by pathogens and the various chemical treatments was investigated in mutants with known characterised defects in common signalling components. An effect on induction of the tryptophan enzymes in any of these mutants would suggest that the component that is defective has a role in the regulation of the tryptophan pathway.

Transgenic NahG plants express a bacterial salicylate hydroxylase that converts salicylic acid (SA) into catechol and prevents the accumulation of SA that occurs in response to pathogenesis (Delaney et al., 1994). The absence of SA accumulation prevents induction of a group of proteins, known as pathogenesis related (PR) proteins, which are normally induced following pathogen infection (Uknes et al., 1992; Lawton et al., 1995). As reported in Zhao and Last (1996) induction of the tryptophan enzymes and accumulation of camalexin is greatly attenuated in NahG transgenic plants. Salicylic acid would, therefore, appear to be a vital component of the signalling pathway regulating the tryptophan enzymes following pathogenesis. However, spraying of SA, which induces PR proteins (Delaney et al., 1995), had no effect on the levels of the tryptophan enzymes (Zhao and Last, 1996) indicating that SA is not sufficient for induction. Similarly, SA alone does not induce a barley thionin gene that is induced by phytopathogenic fungi and bacteria (Epple et al., 1995).

The Arabidopsis mutant *npr1* fails to induce PR proteins after pathogenesis (Cao et al., 1994). In addition, treatment with SA does not induce PR proteins in *npr1* plants, indicating that the NPR1 protein functions downstream of SA. Normal induction of the tryptophan enzymes and camalexin accumulation still occurs in this mutant (Zhao and Last, 1996) indicating that NPR1 is not required in the signal transduction pathway regulating the tryptophan pathway enzymes during pathogenesis.

The roles of ethylene and jasmonic acid in induction of the tryptophan enzymes following Pseudomonas infection have also been explored. Various mutants (*ein2, etr1-1 and etr 1-3*), which are insensitive to ethylene due to defects in two different components of the signal transduction pathway mediating a plant's response to ethylene (Roman et al., 1995), had no effect on induction of the tryptophan enzymes after Pseudomonas infection (unpublished results). Similarly, a triple fad mutant (*fad 3; fad 7; fad 8*), which cannot synthesise jasmonic acid (McConn and Browse, 1996), exhibited wild-type induction of the tryptophan enzymes after pathogenesis. These results are in contrast to the pathogen induction of an Arabidopsis defensin gene, *PDF1.2*. The induction of *PDF1.2* was almost completely blocked in both ethylene (*ein2*) and jasmonic acid (*coi1*) mutants (Penninckx et al., 1996).

The response of the tryptophan enzymes to amino acid starvation, oxidative stress and abiotic elicitors appears to be controlled somewhat differently than the induction in response to Pseudomonas infection. In NahG plants induction of the tryptophan enzymes and camalexin accumulation after the above chemical treatments is not reduced compared to wildtype, (Williams, Zhao and Last, unpublished), suggesting that SA is not playing an important role in the responses to these stress conditions. A model for signal transduction pathways regulating tryptophan enzyme induction is shown in Figure 2.

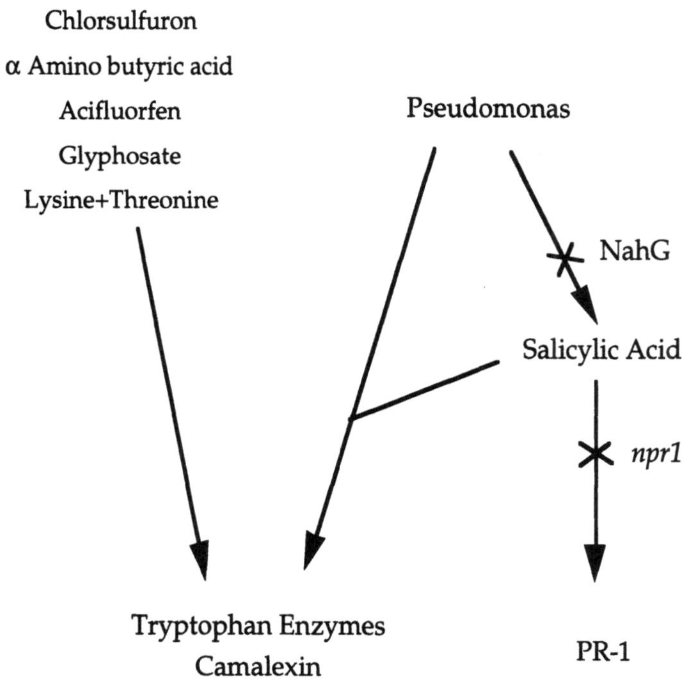

Fig.2. A model for the signal transduction pathways regulating induction of the tryptophan biosynthetic enzymes and accumulation of camalexin. The pathway regulating PR-1 protein induction is shown for comparison.

4.2 Isolation of Specific Tryptophan Pathway Regulatory Mutants

The second genetic approach is to isolate mutants with defects specifically in the regulation of the tryptophan enzymes. The tryptophan pathway appears to be regulated in a manner unique among amino acid biosynthetic pathways and identification of mutants will greatly facilitate dissection of the signal transduction pathways governing expression of the tryptophan genes. A transgenic line containing a fusion construct between the promoter of the PAT enzyme and the GUS coding region (Rose and Last, 1997) was mutagenised and used as the starting point for these mutant screens. Two classes of regulatory mutants have been isolated.

Initially, GUS staining of cotyledons from plate grown mutagenised plants was used to identify several constitutive mutants exhibiting higher expression of the tryptophan enzymes prior to any treatment. One advantage of this broad constitutive screen is that mutations may be identified in any of the signal transduction pathways outlined in Figure 2. Several mutants have been isolated that show constitutively increased levels of the tryptophan enzymes.

The second class of mutants that have been isolated show a specific defect in the response of the tryptophan enzymes to Pseudomonas infection. The same PAT-GUS fusion line was grown on soil and half of a leaf was infiltrated with a suspension of Pseudomonas bacteria. In wildtype plants, GUS staining increases in the infected area, reflecting induction of the endogenous tryptophan proteins. Mutants were identified that did not exhibit increased GUS staining in the area of the lesion. The types of mutants one could expect to identify in this screen include those defective in pathogen recognition or signalling between the pathogen and induction of the tryptophan enzymes. The secondary screen involved direct testing of the tryptophan enzyme levels through immunoblot analysis of infected leaves. Several mutants were identified that exhibited significantly lower levels of the tryptophan enzymes 2 days after Pseudomonas infection, compared to wildtype. These mutants may represent more than one genetic locus as camalexin accumulation is differentially affected. In some mutants camalexin still accumulated despite the lack of induction of the tryptophan enzymes. In others camalexin accumulation is also affected, hence, it is possible that these may be allelic to phytoalexin deficient *(pad)* mutants already isolated by Jane Glazebrook and co-workers (Glazebrook and Ausubel, 1994; Glazebrook et al., 1996; Glazebrook et al., 1997). The isolation of both types of mutants in which defects in induction of the tryptophan enzymes do or do not affect camalexin accumulation will facilitate dissection of differences and similarities in signalling mechanisms between camalexin and the tryptophan proteins, and aid in establishing epistatic relationships between the various mutants.

The isolation of mutants in which the tryptophan enzymes are not induced to wild-type levels following Pseudomonas infection also enables examination of the role of the tryptophan pathway during pathogenesis. Is the induction of the tryptophan pathway enzymes merely to accommodate camalexin biosynthesis or is tryptophan or other secondary metabolites produced from the pathway playing a yet to be determined role?

Acknowledgements

The authors wish to thank Chad Williams and Palitha Dharmawardhana for critical reading of the manuscript. This research was supported by National Institutes of Health Grant No. GM-43134.

Amrhein, H., Teus, B., Gehrke, P., and Steinrucken, H. C. (1980) The site of inhibition of the shikimate pathway by glyphosate. Plant Physiol. 66: 830-834.

Bender, J. and Fink, G.R. (1995) Epigenetic control of an endogenous gene family is revealed by a novel blue fluorescent mutant of Arabidopsis. Cell 83: 725-734

Bohlmann, J., De Luca, V., Eilert, U., and Martin, W. (1995) Purification and cDNA cloning of anthranilate synthase from *Ruta graveolens*; modes of expression and properties of native and recombinent enzymes. Plant J. 7: 491-501.

Bryan, J. K. (1980) Synthesis of the aspartate family and branched chain amino acids. *In* The Biochemistry of Plants, 5, B. J. Miflin, eds Academic Press New York, pp. 403-452.

Cao, H., Bowling, S. A., Gordon, S., and Dong, X. (1994) Characterization of an Arabidopsis mutant that is nonresponsive to inducers of systemic acquired resistance. Plant Cell 6: 1583-1592.

Conklin, P. L., and Last, R. L. (1995) Differential accumulation of antioxidant mRNAs in *Arabidopsis thaliana* exposed to ozone. Plant Physiol. 109: 203-212.

Delaney, T. P., Friedrich, L., and Ryals, J. A. (1995) An *Arabidopsis* signal transduction mutant defective in chemically and biologically induced resistance. Proc. Natl. Acad. Sci. USA 92: 6602-6605.

Delaney, T. P., Uknes, S., Vernooij, B., Friedrich, L., Weymann, K., Negrotto, D., Gaffney, T., Gut Rella, M., Kessmann, H., Ward, E., and Ryals, J. (1994) A central role of salicylic acid in plant disease resistance. Science (Washington D C) 266: 1247-1250.

Epple, P., Apel, K., and Bohlmann, H. (1995) An Arabidopsis thionin gene is inducible via a signal transduction pathway different from that for pathogenesis-related proteins. Plant Physiol 109: 813-820.

Galili, G. (1995) Regulation of lysine and threonine synthesis. Plant Cell 7: 899-906.

Glazebrook, J., and Ausubel, F. M. (1994) Isolation of phytoalexin-deficient mutants of *Arabidopsis thaliana* and characterization of their interactions with bacterial pathogens. Proc. Natl. Acad. of Sci. USA 91: 8955-8959.

Glazebrook, J., Rogers, E. E., and Ausubel, F. M. (1996) Isolation of Arabidopsis mutants with enhanced disease susceptibility by direct screening. Genetics 143: 973-982.

Glazebrook, J., Zook, M., Mert, F., Kagan, I., Rogers, E. E., Crute, I. R., Holub, E. B., Hammerschmidt, R., and Ausubel, F. M. (1997) Phytoalexin-deficient mutants of Arabidopsis reveal that *PAD4* encodes a regulatory factor and that four *PAD* genes contribute to downy mildew resistance. Genetics 146: 381-392.

Guyer, D., Patton, D., and Ward, E. (1995) Evidence for cross-pathway regulation of metabolic gene expression in plants. Proc. Natl. Acad. Sci. USA 92: 4997-5000.

Herrmann, K. M. (1995) The Shikimate Pathway: Early Steps in the Biosynthesis of Aromatic Compounds. Plant Cell 7: 907-919.

Hinnebusch, A. G. (1992) General and pathway-specific regulatory mechanisms controlling the synthesis of amino acid biosynthetic enzymes in *Saccharomyces cerevisiae In* The Molecular and Cellular Biology of the Yeast *Saccharomyces*, 2, E. W. Jones, J. R. Pringle and J. B. Broach, eds Cold Spring Harbor Laboratory Press Cold Spring Harbor, NY, pp. 321-414.

Kangasjärvi, J., Talvinen, J., Utriainen, M., and Karjalainen, R. (1994) Plant defence systems induced by ozone. Plant Cell Env. 17: 783-794.

Kutchan, T. M. (1995) Alkaloid Biosynthesis - The Basis for Metabolic Engineering of Medicinal Plants. Plant Cell 7: 1059-1070.

Lam, H. M., Coschigano, K., Schultz, C., Melo Oliveira, R., Tjaden, G., Oliveira, I., Ngai, N., Hsieh, M. H., and Coruzzi, G. (1995) Use of Arabidopsis mutants and genes to study amide amino acid biosynthesis. Plant Cell 7: 887-898.

Larsen, P. O. (1981) Glucosinolates. *In* The Biochemistry of Plants, 7, E. E. Conn, eds Academic Press New York, pp. 501-525.

Last, R. L., Bissinger, P. H., Mahoney, D. J., Radwanski, E. R., and Fink, G. R. (1991) Tryptophan mutants in *Arabidopsis*: the consequences of duplicated tryptophan synthase ß genes. Plant Cell 3: 345-358.

Lawton, K., Weymann, K., Friedrich, L., Vernooij, B., Uknes, S., and Ryals, J. (1995) Systemic acquired resistance in Arabidopsis requires salicylic acid but not ethylene. Mol. Plant Microbe Interact. 8: 863-870.

Li, J., and Last, R. L. (1996) The *Arabidopsis thaliana trp 5* mutant has a feedback-resistant anthranilate synthase and elevated soluble tryptophan. Plant Physiol. 110: 51-59.

Matringe, M., Camadro, J. M., Block, M. A., Joyard, J., Scalla, R., Labbe, P., and Douce, R. (1992) Localization within chloroplasts of protoporphyrinogen oxidase, the target enzyme for diphenyl ether-like herbicides. J. Biol. Chem. 267: 4646-4651.

McConn, M., and Browse, J. (1996) The critical requirement for linolenic acid is pollen development, not photosynthesis, in an Arabidopsis mutant. Plant Cell 8: 403-416.

Niyogi, K. K., and Fink, G. R. (1992) Two anthranilate synthase genes in Arabidopsis: Defense-related regulation of the tryptophan pathway. Plant Cell 4: 721-733.

Niyogi, K. K., Last, R. L., Fink, G. R., and Keith, B. (1993) Suppressors of *trp1* fluorescence identify a new Arabidopsis gene, *TRP4*, encoding the anthranilate synthase beta subunit. Plant Cell 5: 1011-1027.

Normanly, J., Cohen, J. D., and Fink, G. R. (1993) *Arabidopsis thaliana* auxotrophs reveal a tryptophan-independent biosynthetic pathway for indole-3-acetic acid. Proc. Natl. Acad. Sci. USA 90: 10355-10359.

Penninckx, I. A. M., Eggermont, K., Terras, F., Thomma, B., De Samblanx, G. W., Buchala, A., Metraux, J. P., Manners, J. M., and Broekaert, W. F. (1996) Pathogen-induced systemic activation of a plant defensin gene in Arabidopsis follows a salicylic acid-independent pathway. Plant Cell 8: 2309-2323.

Radwanski, E. R., and Last, R. L. (1995) Tryptophan biosynthesis and metabolism: Biochemical and molecular genetics. Plant Cell 7: 921-934.

Radwanski, E. R., Zhao, J., and Last, R. L. (1995) *Arabidopsis thaliana* tryptophan synthase alpha: Gene cloning and expression analysis. Molec. Gen. Genet. 248: 657-667.

Ray, T. B. (1984) Site of Action of Chlorsulfuron. Inhibition of Valine and Isoleucine Biosynthesis in Plants. Plant Physiol. 75: 827-831.

Roman, G., Lubarsky, B., Kieber, J. J., Rothenberg, M., and Ecker, J. R. (1995) Genetic Analysis of Ethylene Signal Transduction in *Arabidopsis thaliana*: Five Novel Mutant Loci Integrated into a Stress Response Pathway. Genetics 139: 1393-1409.

Rose, A. B., and Last, R. L. (1997) Introns act post-transcriptionally to increase expression of the *Arabidopsis thaliana* tryptophan pathway gene *PAT1*. Plant J. 11: 455-464.

Tsuji, J., Jackson, E. P., Gage, D. A., Hammerschmidt, R., and Somerville, S. C. (1992) Phytoalexin accumulation in *Arabidopsis thaliana* during the hypersensitive reaction to *Pseudomonas syringae* pv. *syringae*.. Plant Physiol. 98: 1304-1309.

Tsuji, J., Zook, M., Somerville, S. C., Last, R. L., and Hammerschmidt, R. (1993) Evidence that tryptophan is not a direct biosynthetic intermediate of camalexin in *Arabidopsis thaliana*. Physiol. Molec. Plant Path. 43: 221-229.

Uknes, S., Mauch Mani, B., Moyer, M., Potter, S., Williams, S., Dincher, S., Chandler, D., Slusarenko, A., Ward, E., and Ryals, J. (1992) Acquired resistance in Arabidopsis. Plant Cell 4: 645-656.

Yanofsky, C., and Crawford, I. P. (1987) The tryptophan operon *In Escherichia coli* and *Salmonella typhimurium*: Cellular and Molecular Biology, 2, F. C. Neidhardt, J. L. Ingraham, K. B. Low, B. Magasanik, M. Schaechter and H. E. Umbarger, eds American Society for Microbiology Washington, pp. 1453-1472.

Zhao, J., and Last, R. L. (1995) Immunological characterisation and chloroplast import of the tryptophan biosynthetic enzymes of the flowering plant *Arabidopsis thaliana*. J. Biol. Chem. 270: 6081-6087.

Zhao, J., and Last, R. L. (1996) Coordinate regulation of the tryptophan biosynthetic pathway and indolic phytoalexin accumulation in Arabidopsis. Plant Cell 8: 2235-2244.

Indole Alkaloid Biosynthesis in *Catharanthus roseus*: The Establishment of a Model System

Vincenzo De Luca, Benoit St-Pierre, Felipe Vazquez Flota and Pierre Laflamme

Institut de Recherches en Biologie Végétale, Département de Sciences Biologiques, Université de Montréal, 4101 rue Sherbrooke est, Montréal, Québec, Canada, H1X 2B2

Abstract. Plants possess the inherent ability to produce unique natural product chemistries as a result of differences that have been selected for throughout evolution. Selection through gene mutation and subsequent adaptation of metabolic pathways create new secondary products. The biosynthesis and accumulation of many secondary products, however, remain under remarkable control of the biotic and abiotic environments. The readiness of secondary metabolic pathways to sense the environment appears to be intimitely related to their ability to evolve and is represented by the existence of several hundred thousand plant-based secondary metabolites. *Catharanthus roseus* which produces nearly 200 different monoterpenoid indole alkaloids, displays a cell- tissue- development- and environment-specific control over their biosynthesis and accumulation. This chapter describes the experimental approaches which have helped to understand various aspects of indole alkaloid biosynthesis in *Catharanthus* and which make it a good model system to study its regulation.

Keywords. Monoterpenoid indole alkaloid biosynthesis, *Catharanthus roseus*, vindoline, compartmentation, developmental and environmental regulation

1 Introduction

Plants possess the unique ability to convert the basic ingredients of life into substances of small molecular weight which are known as secondary metabolites or natural products. The term secondary or non-essential has been utilized to differentiate these products of plant metabolism from primary or essential

NATO ASI Series, Vol. H 104
Cellular Integration of Signalling Pathways in Plant Development
Edited by F. Lo Schiavo, R. L. Last, G. Morelli, and N. V. Raikhel

metabolites which are absolutely required for life. It is commonly believed, however, that secondary metabolites are essential in allowing plants to cope and to interact with their biotic and abiotic environments. In this context, human cultures, which have always been fascinated by plant secondary metabolites, have used them as stimulants, depressants, and hallucinogens, as well as to heal and hunt animal as well as human prey (Mann, 1994).

More recently, many thousands of natural products have been identified by chemists who were preoccupied with the identification of active principles, their structural elucidation and their chemical as well as biochemical synthesis. These chemical investigations have evolved into the modern fields of organic chemistry, pharmacology and biochemistry and have led to the characterization of the major classes of plant natural products as well as to the elucidation of some of their biochemical pathways (Mann, 1994).

The availability of structural as well as biochemical information about the biosynthesis of plant natural products are now leading to the enzymatic and molecular characterization of these pathways. The typical experimental approaches used in these studies involve knowledge of protein function, use of this knowledge to purify the protein to homogeneity, use of homogeneous protein to clone the corresponding gene and the subsequent characterization of the gene. These strategies have been adopted since genetic approaches to clone genes using well-characterized model systems such as those of *Arabidopsis thaliana* have not been possible. In this context, genetic approaches using the *Arabidopsis* model system are useful for isolating genes generally occurring in plants, but it has been speculated that this strategy alone may not be very helpful for cloning genes which are invoved in species-specific secondary pathways.

2 The vinca alkaloids and the *Catharanthus roseus* model system

Catharanthus roseus (L.) G. Don, commonly known as the Madagascar periwinkle belongs to the Apocynaceae family. The secondary metabolites characteristic of this family are the monoterpenoid indole alkaloids which are derived from the shikimate and the mevalonate pathways. The biosynthesis of the indole moiety requires tryptamine which is derived from the decarboxylation of tryptophan by the enzyme tryptophan decarboxylase (TDC). The biosynthesis of the terpenoid moiety requires secologanin which is derived from geraniol via a series of enzymatic conversions (reviewed in De Luca, 1993 and in Meijer *et. al.*, 1993a). The first committed step leading to the monoterpenoid indole alkaloids involves strictosidine synthase which catalyses the stereospecific condensation of tryptamine and secologanin to yield 3-α(S)-strictosidine. Among the several thousand indole alkaloids which have been characterized and which are derived from this central intermediate, several have been found to be valuable agents for the treatment of hypertension and for a number of antineoplastic ailments (Farnsworth, 1975). In particular, vinblastine and vincristine from *Catharanthus roseus* are valuable chemotherapeutic agents currently used in the treatment of a number of cancerous diseases (Fig 1).

The therapeutic value of vinblastine and vincristine, their low abundance in the plant combined with their high cost and their complex chemical structures have prompted extensive efforts to find affordable chemical synthetic routes (Langlois *et. al.*, 1976, Kutney *et. al.*, 1976). In addition, significant efforts have been made to screen for vinblastine producing cell and organ cultures (De Luca and Kurz, 1988; Van der Heijden *et. al.*, 1989). These studies provided excellent training opportunities in natural product chemistry and cell culture for students and postdoctoral fellows, but the remarkable technical advances realized in the area of total synthesis of dimeric indole alkaloids and in the culture of *Catharanthus* cells did not lead to commercially relevant products. Vinblastine and vincristine continue to be extracted and purified from field grown *Catharanthus roseus* plants.

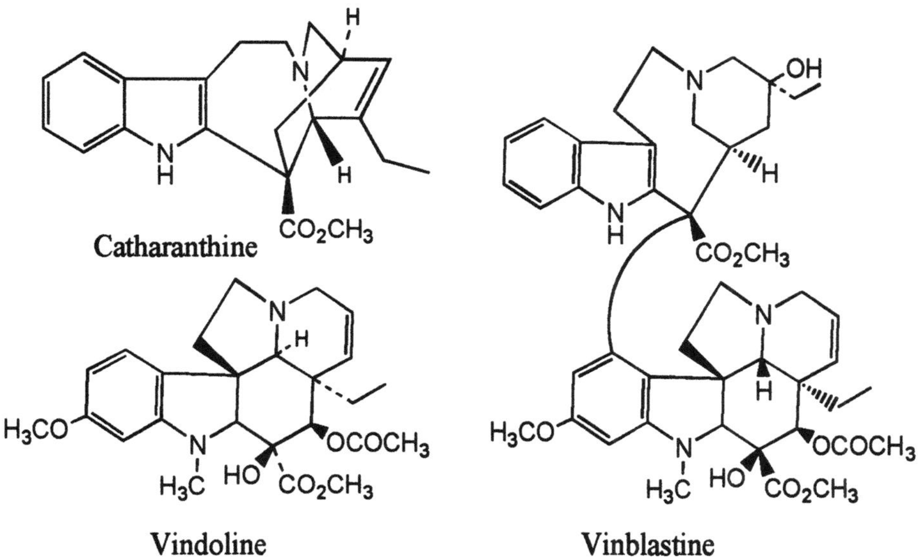

Figure 1: The dimeric indole alkaloids, vinblastine and vincristine, are derived from Catharanthine and Vindoline monomers.

2.1 Cell-, Tissue-, and Development-Specific Control of Indole Alkaloid Biosynthesis

The efforts to produce cell cultures which accumulate commercially interesting secondary metabolites have not lived up to the original promises. In the context of *Catharanthus roseus* cell lines or organ cultures developed in different laboratories around the world, very similar results were achieved. Invariably, similar successes were reported for the selection of cell lines (De Luca and Kurz, 1988; Van der Heiden *et. al.*, 1989) and of organized root cultures which accumulated complex indole alkaloids such as ajmalicine, catharanthine and tabersonine. A few reports also suggested that some root culture lines (O'Keefe *et. al.*, 1997 and references therein) accumulate traces of vindoline, the second component to catharanthine required in the assembly of vinblastine and vincristine (Fig 1). The capability of producing significant levels of vindoline, however, only reappeared in shooty teratomas (O'Keefe *et. al.*, 1997) or with the regeneration of shoots from callus cultures (Constabel *et. al.*, 1982).

The importance of cell-, tissue-, and development-specific controls in the biosynthesis of secondary metabolites is clearly illustrated in several studies performed with *Catharanthus roseus.*

2.1.1 The Biosynthesis of Vindoline in *Catharanthus roseus*

The experiments of Constabel *et. al.*, (1982) suggested that the ability to synthesize and accumulate vindoline coincided with the presence of shoots and suggested that vindoline biosynthesis is restricted to the leaf. In contrast, the ability to make catharanthine and 16-methoxytabersonine was shared between roots and shoots, and this may explain why root-like cell cultures can accumulate these compounds after appropriate treatments (reviewed in De Luca and Kurz, 1988; Van der Heiden *et. al.*, 1989).

The use of germinating seedlings of *Catharanthus roseus* growing in the light or in its absence proved to be of great importance in elucidating how vindoline is made in the intact plant and in dissecting the unique cell- tissue- and development-specific events involved in activating the vindoline pathway (Fig 2). If *Catharanthus* seeds are germinated and grown in the absence of light, they accumulate high levels of tabersonine (Fig 2B, lane a) as well as small amounts of 4 other intermediates beyond tabersonine. The growth of etiolated seedlings in the presence of light stimulated the quantitative, large scale turnover of tabersonine as well as the intermediates into vindoline (Fig 2B, Balsevich *et. al.*, 1986) and suggested that tabersonine is converted into vindoline by the sequence of reactions proposed in Fig 2A. Tabersonine (Fig 2B, Lane a) is hydroxylated (Fig 2A, reaction 1), 16-hydroxytabersonine (Fig 2B, lane b) is *O*-methylated (Fig 2A, reaction 2), 16-methoxytabersonine (Fig 2B, lane c) is hydrated across the double bond at position 2,3 (Fig 2A, reaction 3), 16-methoxy-2,3-dihydro-3-hydroxytabersonine (Fig 2B, lane d) is *N*-methylated (Fig (2A, reaction 4), 16-methoxy-2,3-dihydro-3-hydroxy-*N*-methyltabersonine (Fig 2B, lane e) is hydroxylated (Fig 2A, reaction 5) and deacetylvindoline (Fig 2B lane f) is *O*-acetylated (Fig 2A, reaction 6) to yield vindoline. The pattern of accumulation of indole alkaloid intermediates in etiolated versus light-grown seedlings also suggested that each of the first 2 hydroxylations (Fig 2B, lanes b and d) clearly catalyse rate-limiting steps in the conversion of tabersonine into vindoline. Furthermore, hydration of the 2,3 double bond may be particularly rate limiting as the methoxytabersonine levels only drop slightly in light grown seedlings (Fig 2B, lane c). Another valid interpretation, however, may be that the remaining methoxytabersonine in light-grown seedlings represents the root-specific pool of this alkaloid which is not available for further transformation into vindoline.

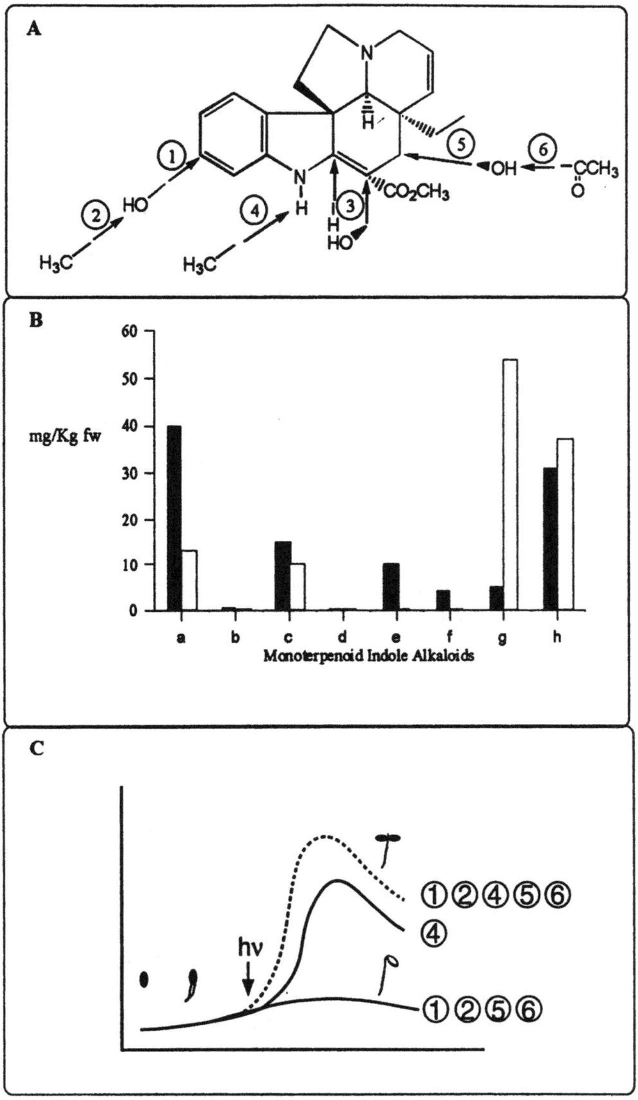

Figure 2: A) Sequence of reactions numbered 1 to 6 which convert tabersonine into vindoline. The generic names use the numbering system for the aspidospermidine alkaloids described in Chemical Abstracts (Collective Substance Index vol 106-115, 12CS, 1987-1991). B) The concentrations of tabersonine (lane a), 16-hydroxytabersonine (lane b), 16-methoxytabersonine (lane c), 16-methoxy-2,3-dihydro-3-hydroxytabersonine (lane d), 16-methoxy-2,3-dihydro-3-hydroxy-N-methyltabersonine (lane e), deacetylvindoline (lane f), vindoline (lane g) and catharanthine (lane h) in etiolated (black bars) or in light treated seedlings (white bars) This figure was modified from Balsevich *et. al.*, 1986. C) Appearance of relative enzyme activities during seedling development for reactions numbered 1 to 6 described in panel A and in Figure 3. The solid lines represent enzyme activities appearing during etiolated growth, whereas the dotted line represents enzyme activities occurring after etiolated seedlings were treated with light (*hv*). This figure was modified from St-Pierre *et. al.*, 1995 and De Luca *et. al.*, 1993.

The results obtained for vindoline accumulation (Fig 2B) suggest that the pathway leading to tabersonine production is fully active in etiolated seedlings. In addition the pathway for catharanthine biosynthesis is also functional in dark-grown seedlings. This is clearly illustrated in Fig 2B, lane h, which shows that etiolated seedlings accumulate almost the same levels of catharanthine as do seedlings grown in the presence of light. A lack of light requirement for catharanthine and tabersonine biosynthesis also partially explains why plant roots, stems and leaves as well as cell and root cultures are capable of making these indole alkaloids but not vindoline (reviewed in De Luca and Kurz, 1988). The role of light in activating vindoline accumulation was also studied at the enzyme level. Enzymes which catalyse steps 1,2,4, 5 and 6 (Fig 2A) were assayed at different developmental stages in etiolated and light-grown seedlings (Fig 2C). The results obtained showed that all these activities except for the N-methyltransferase are found at very low levels throughout etiolated seedling development whereas light treatment activates all 5 enzyme activities (Fig 2C) to coincide with the quantitative conversion of indole alkaloid intermediates into vindoline (Fig 2B). Further studies with roots, hypocotyls and cotyledons in seedlings and roots, stems and leaves in plants showed that the majority of vindoline biosynthetic activities (reactions 4, 5 and 6) occurred in leaves and cotyledons, with much lower activities occurring in other above-ground plant parts, but none of these activities were found in roots.

2.1.2 Subcellular Compartmentation of the Pathway for Vindoline Biosynthesis

The tissues of *Catharanthus roseus* may contain several hundred different indole alkaloids some of which may be very toxic. It was suggested by the histochemical studies of Yoder and Mahlberg (1976) that indole alkaloids may only be found in specialized parenchyma cells and in laticifers which may be useful in sequestering these toxic compounds from normal metabolic processes. In addition biochemical studies have shown that the pathway for the biosynthesis of indole alkaloids occurs in several sub-cellular compartments of the cell, including the cytosol, endoplasmic reticulum, the vacuole and the chloroplast (De Luca *et. al.*, 1992, Meijer *et. al.*, 1993). The few enzymatic steps which have been studied have shown that the geraniol hydroxylase and tabersonine-16 hydroxylase (Fig 2A, reaction 1) are reactions which probably occur on the external face of the endoplasmic reticulum. Strictosidine synthase is a glycosylated protein which has been immunologically localized in vacuoles of *Catharanthus* roots (McKnight *et. al.*, 1991). The *N*-methyltransferase (Fig 2A, reaction 4) involved in the late stages of vindoline biosynthesis is bound to chloroplast thylakoids whereas the hydroxylase and *O*-acetyltransferase which catalyse the last 2 steps in vindoline biosynthesis are both cytosolic proteins.

2.1.2 Subcellular Compartmentation of the Pathway for Vindoline Biosynthesis

The tissues of *Catharanthus roseus* may contain several hundred different indole alkaloids some of which may be very toxic. It was suggested by the histochemical studies of Yoder and Mahlberg (1976) that indole alkaloids may only be found in specialized parenchyma cells and in laticifers which may be useful in sequestering these toxic compounds from normal metabolic processes. In addition biochemical studies have shown that the pathway for the biosynthesis of indole alkaloids occurs in several sub-cellular compartments of the cell, including the cytosol, endoplasmic reticulum, the vacuole and the chloroplast (De Luca *et. al.*, 1992, Meijer *et. al.*, 1993). The few enzymatic steps which have been studied have shown that the geraniol hydroxylase and tabersonine-16 hydroxylase (Fig 2A, reaction 1) are reactions which probably occur on the external face of the endoplasmic reticulum. Strictosidine synthase is a glycosylated protein which has been immunologically localized in vacuoles of *Catharanthus* roots (McKnight *et. al.*, 1991). The *N*-methyltransferase (Fig 2A, reaction 4) involved in the late stages of vindoline biosynthesis is bound to chloroplast thylakoids whereas the hydroxylase and *O*-acetyltransferase which catalyse the last 2 steps in vindoline biosynthesis are both cytosolic proteins.

2.2 The Enzymes Involved in the Late Stages of Vindoline Biosynthesis.

The conversion of tabersonine to vindoline involves 6 enzymatic steps, several of which have been investigated in our laboratory. The isolation and characterization of intermediates which was described in section 2.1.2, provided alkaloid substrate information for the design of enzyme assays. The enzyme reactions together with their requirements are described in Figure 3.

2.2.1 Tabersonine-16-Hydroxylase

Tabersonine-16-hydroxylase is a cytochrome P450 dependent monooxygenase which requires NADPH, O_2 and tabersonine for enzyme activity (Fig. 3). Enzyme activity was detected in crude protein extracts of leaves and in light grown seedlings. After partial purification by linear sucrose density gradient centrifugation, the enzyme was biochemically localized to the endoplasmic reticulum (St. Pierre and De Luca, 1995). The enzyme was inhibited by carbon monoxide, clotrimazole, miconazole and by cytochrome c. Several putative cytochrome P450 genes have been cloned from *Catharanthus roseus*, (Hotz *et. al.*, 1995; Meijer *et. al.*, 1993), but none have yet to be identified as authentic tabersonine-16-hydroxylase clones.

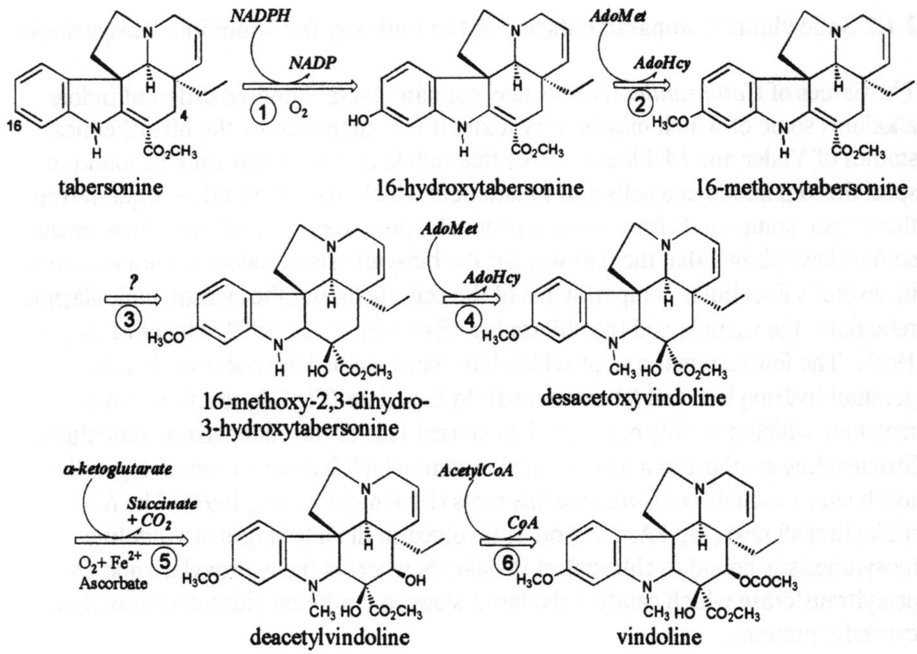

Figure 3: Enzymology of the reactions numbered 1 to 6 corresponding to Figure 2A which convert tabersonine into vindoline.

2.2.2 *S*-Adenosyl-L-Methionine-16-Hydroxytabersonine-*O*-Methyltransferase

This 16-*O*-methyltransferase (Fig. 3) has been detected in cell cultures of *Catharanthus roseus* (Fahn *et. al.*,1985) as well as in plant extracts. The enzyme requires *S*-Adenosyl-L-Methionine as a methyl donor and appears to show high specificity for 16-hydroxytabersonine. However this enzyme remains to be further characterized.

2.2.3 *N*-methyltransferase

This enzyme with a unique compartmentation in chloroplast thylakoids (De Luca and Cutler, 1987), also requires *S*-Adenosyl-L-Methionine as a methyl donor (Fig. 3) and appears to show high specificity for its substrate 16-methoxy-2,3-dihydro-3-hydroxytabersonine (De Luca *et. al.*, 1987) . The enzyme has been partially purified by ultracentrifugation and by solubilization of the pellet with CHAPS detergent (Dethier and De Luca, 1993). The solubilized enzyme showed an apparent Mr of 60,000 Da as determined by sucrose density gradient centrifugation. Although the detergent solubilized enzyme showed a slightly altered indole alkaloid substrate specificity, enzyme activity was absolutely dependent on a reduced 2,3 double bond, confirming its status as the third to last step in vindoline biosynthesis. The reason for the thylakoid localisation of this enzyme reamains to be elucidated.

2.2.4 Deacetoxylvindoline 4-hydroxylase

In contrast to membrane associated tabersonine-16-hydroxylase (Fig. 3) which is a cytochrome P450 monooxygenase, the second to last step in vindoline biosynthesis is catalysed by a soluble and probably cytosolic 2-oxoglutarate-dependent dioxygenase (DeCarolis *et. al.*, 1990). The enzyme was purified to apparent homogeneity using a combination of chromatography procedures including a key α-ketoglutarate-sepharose affinity chromatography technique (DeCarolis and De Luca, 1993). The M_r of the native and denatured enzyme was 45 and 44.7 kDa, respectively which suggests that this dioxygenase exists as a monomeric protein. The purified protein which was isolated from young leaves, occurs as 3 charged isoforms with pI values of 4.6, 4.7, and 4.8. The purified enzyme catalyses the addition of O_2 to desacetoxyvindoline and to α-ketoglutarate resulting in the formation of deacetylvindoline, succinate and CO_2. The release of $^{14}CO_2$ was measured and verified experimentally in enzyme assays using (1-^{14}C)α-ketoglutarate as a cosubstrate (DeCarolis *et. al.*, 1990). The purified enzyme showed an absolute requirement for desacetoxyvindoline, O_2, α-ketoglutarate and its activity was significantly enhanced by addition of ascorbate and ferrous ions to the reaction mixture.

The purification protocol was used to produce sufficient protein for microsequencing and three peptides were successfully sequenced. One of the three peptides (ELISEENPPIYK) which showed significant sequence homology to amino acids 304-316 (ELLNQDNPPLYK) of hyoscyamine-6-hydroxylase isolated from *Hyoscyamus niger* (Matsuda *et. al.*, 1991) was used to make a degenerate oligonucleotide based on this peptide in order to screen a *Catharanthus roseus* seedling cDNA library. Using this approach three clones were isolated and the open reading frame on each clone encoded a putative protein containing the sequence of the three peptides previously obtained by microsequencing (Vazquez-Flota *et. al.*, 1997). Southern blot analysis indicated that the hydroxylase is present as a single copy gene in a diploid genome. Isolation, sequencing and analyses of genomic clones identified the presence of a 205 bp and a 1720 bp intron which also occur in corresponding stretches of amino acids which are highly conserved in three other characterized plant dioxygenase genes. This information together with further analysis showed that 2 of our cDNA clones in fact represented dimorphic alleles whose sequences differed only in their 3'-untranslated region. These cDNA clones encoded a 401 amino acid protein with a M_r of 45.5 kDa which is virtually identical with the molecular weights obtained for enzyme purified from leaves (DeCarolis and De Luca, 1993).

The identity of this hydroxylase was completed when it was functionally expressed in *E. coli*. Enzyme assays with bacterial extracts expressing this protein showed high levels of desacetoxyvindoline-4-hydroxylase activity. Removal of O_2, desacetoxyvindoline, α-ketoglutarate, ascorbate or ferrous ions completely abolished the activity of the recombinant protein.

The amino acid sequence of this protein showed a high degree of similarity to a growing family of dioxygneses including anthocyanidin synthase (Davis, 1993), hyoscyamine 6β-hydroxylase (Matsuda *et. al.*, 1991), flavonol synthase (Holton *et. al.*, 1993), flavanone 3β-hydroxylase (Britsch *et. al.*, 1993), ethylene forming enzyme (Holdsworth *et. al.*, 1987), gibberellin 20-oxidase (Phillips *et. al.*, 1992) and isopenicillin *N*-synthase (Ramon *et. al.*, 1987). The sequence comparisons revealed that although amino acid similarities to other dioxygenases extended throughout the protein, areas of greater homology were observed in the C-terminal domain of the enzyme and several highly conserved amino acid residues could be identified which characterized the binding sites for ferrous ions.

Preliminary studies on the expression of desacetoxyvindoline-4-hydroxylase in different plant tissues provided a strong correlation between the levels of detectable mRNA and enzyme activities. The highest levels of hydroxylase transcripts and enzyme activities were found in leaves, whereas these were much less abundant in stems. Neither transcripts nor enzyme activities were detected in roots, thus confirming the restricted expression of this late step in vindoline biosynthesis as documented in earlier studies (DeCarolis *et. al.*, 1990). Enzymatic studies with developing seedlings had shown that while etiolated seedlings had trace levels of desacetoxyvindoline-4-hydroxylase activity, light treatment rapidly activated both enzyme activity and vindoline accumulation (Fig 2). The measurement of hydroxylase transcripts and immunologically reactive hydroxylase antigen produced quite different results. Etiolated seedlings started accumulating important amounts of hydroxylase transcript and protein, but showed little or no enzyme activity at different stages of growth. In contrast, light treatment of seedlings caused a rapid increase of enzyme activity but only caused a slight increase in both hydroxylase transcript and in protein compared to etiolated controls. These results suggest that in seedlings the activating effect of light may occur at a point downstream of transcription which remains to be elucidated (Vazquez-Flota *et. al.*, 1997).

2.2.5 Deacetylvindoline 4-*O*-acetyltransferase

The last step in vindoline biosynthesis is catalysed by an Acetyl CoA dependent 4-*O*-acetyltransferase (DAT) (Fig 3). The enzyme was purified to homogeneity by classical chromatography methods and by CoA affinity chromatography to yield a 33/21 kDa (Powers *et. al.*, 1990) or 26/20kDa (Fahn *et. al.*, 1990) heterodimeric protein. The two purification procedures which were slightly different both suggested that DAT was a heterodimer, but the differences in molecular weight were difficult to explain. The characterization of a heterodimeric *O*-acetyltransferase which catalyses the last step in cephalosporin biosynthesis (Matsuda *et. al.*, 1992) in *Acremonium chrysogenium*, suggested that the heterodimeric nature of DAT might be relevant to enzyme activity. Additional studies with rat choline-*O*-acetyltransferase (Wu *et. al.*, 1995) showed that proteolytic processing of this protein activates enzyme activity. Molecular characterization of the gene from *Acremonium chrysogenium* showed that the heterodimeric subunits were derived from a single gene product and that proteolysis must be involved in their formation.

The purified DAT heterodimer was used to make polyclonal antibodies in rabbit and to sequence of a number of oligopeptides derived from each subunit. The latter involved purification of subunits by sodium dodecyl sulphate polyacrylamide gel electrophoresis (SDS-PAGE), transfer of the subunits from the SDS-PAGE to nitrocellulose membranes, tryptic digest of the individual subunits, purification of oligopeptides by high performance liquid chromatography and microsequencing of selected oligopeptides.

Several approaches were used to attempt to clone DAT from *Catharanthus roseus* cDNA and genomic libraries. The first approach involved the screening of a cDNA expression library produced from light-induced etiolated seedlings. This approach was not successfull since antibody screening yielded numerous false positive clones. Later studies revealed that the DAT gene was not present in this particular cDNA library and that a cDNA expression library made from young leaf tissue would have been more suitable to clone DAT.

The second approach involved screening the cDNA and genomic libraries with degenerate oligonucleotide probes based on one of our numerous oligopeptide sequences. This approach had succeeded in the first attempt to isolate 3 deacetoxyvindoline 4-hydroxylase cDNA clones (Vazquez-Flota *et. al.*, 1997). Several degenerate probes based on one of the DAT peptide sequences were used to screen both cDNA and genomic libraries. All attempts to clone DAT by this approach resulted in the isolation of numerous cDNA and genomic clones which after extensive and time-consuming analyses proved to be false positives. Interestingly, later studies revealed that the hydroxylase was much more abundant than the *O*-acetyltransferase in the cDNA library produced from light-induced seedlings.

The third approach involved the screening of the cDNA and genomic libraries with standard polymerase chain reaction (PCR) approaches. Once again, numerous false positive clones were obtained and much time was spent in identifying these as false positives. All the problems alluded to in the use of this technique were encountered. Since many of the oligonucleotides used for PCR had a considerable level of degeneracy, they amplified false positives rather than authentic products. It was also not known what order in the DAT protein sequence the oligopeptides would fall and thus the experimental design required each pair of oligonucleotides to be produced in both orientations. In addition, since it was not known what size of product to expect and several PCR products of different sizes were produced, it was not easy to identify authentic fragments.

Many of these problems were solved when it was decided to use a protocol developed by Lingner *et. al.*, (1991) that uses PCR to amplify a single peptide from a protein sequence. This approach involves the design of degenerate oligonucleotides corresponding to the NH_2- and COOH- terminal ends of a peptide and leads to the production of a PCR product of defined size. Peptide DAT21p54

(**EFDISNFLDIDAYLSDSWC**) from the small subunit was used for this purpose to synthesize 2 pairs degenerate oligonucleotides based on **EFDIS** and **SDSWC** (St Pierre *et. al.*, 1997). The size of the PCR product expected was 73 bp (Fig 4) and was obtained with only one of 4 possible combinations of degenerate oligonucleotides. The PCR product which contained appropriate restriction sites, was purified by native polyacrylamide gel electrophoresis and was cloned into an appropriate vector. After sequencing this fragment, it was concluded that it encoded the complete sequence of peptide DAT21-54. The unique DNA sequence encoded the internal amino acid sequence of peptide DAT21-54 (**NFLDIDAYL**) and it was used to design specific primers to clone the DAT gene by 3'RACE (rapid amplification of cDNA ends, Frohman *et. al.*, 1988). The resulting 3'RACE product encoded a putative open reading frame encoding 77 amino acids which included another tryptic fragment of the small subunit of DAT.

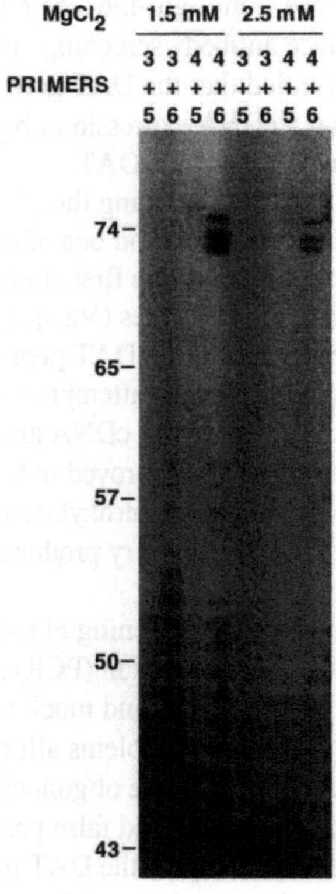

Figure 4: PCR amplification using of a 73 bp fragment corresponding to the nucleotide sequence encoding peptide DAT21-54. Primers
3: 5'-GCGAATTC**GA(G/A)TT(C/T)GA(C/T)AT(C/T/A)TC**
4: 5'-GCGAATTC**GA(G/A)TT(C/T)GA(C/T)AT(C/T/A)AG**,
5 **C(G/A)CT(G/A)(T/A)(G/C)IACCAC(G/A)**CCTAGGCG-5',
6: **G(I/A)CT(G/A)(T/A)(G/C)IACCAC(G/A)**CCTAGGCG-5' were used in the combinations indicated and only primer pair 4 + 6 produced a 73 bp fragment.. Note the use of EcoR1 arms which later enabled the cloning of the PCR fragment.

Since DAT was not represented in the seedling cDNA library, the rest of the DAT sequence was obtained by screening a genomic library using the 3'RACE product as a probe. One clone (gDAT-6) contained a putative open reading frame encoding a protein with 439 amino acids which included 9 out of 10 tryptic peptide sequences obtained from both subunits of DAT. The predicted molecular mass was 49,890 Da which was in good agreement with Mr of 54 and 45 kDa estimated by SDS-PAGE and by gel filtration chromatography, respectively (Powers *et. al.*, 1990). The clone was expressed with an N-terminal extension of 6 histidines (HIS_6-DAT) in *E. coli* to facilitate purification to homogeneity by Ni-NTA affinity chromatography (QIAgen, Chatsworth, CA). Enzyme assays with bacterial extracts expressing this protein showed high levels of DAT activity. Removal of deacetylvindoline or acetyl-CoA completely abolished the activity of the recombinant protein. When purified HIS_6-DAT protein as well as crude *C. roseus* leaf protein extracts were analysed by Western immunoblot after SDS-PAGE, single bands with Mr 50 kDa or greater were visualized and suggested that the heterodimer isolated during enzyme purification was probably an artifact of purification. Further studies are required to determine the basis of dimer formation.

Nucleic and amino acid sequence analysis of DAT revealed significant sequence similarity to nine plant genes with no known biochemical function and placed DAT in the class of microbial chloramphenicol-*O*-acetyltransferases (Shaw and Leslie, 1991). These results indicate that the plant genes may also be *O*-acetyltransferases whose substrate specificities need to be elucidated and that many more *O*-acetyltransferases related to DAT await to be discovered in plants.

Preliminary studies on the expression of DAT in different plant tissues show the highest levels of DAT transcripts and enzyme activities in leaves, whereas these were much less abundant in stems. As was found for the hydroxylase, neither transcripts nor enzyme activities were detected in roots.. Enzymatic studies with developing seedlings showed that the low levels of DAT activity were increased rapidly by light treatment (Fig 2). The measurement of DAT transcripts and immunologically reactive DAT antigen showed that in contrast to hydroxylase, only light treated seedlings accumulated DAT transcript and protein, strictly correlating with increases of enzyme activity. These results reveal some differences in the regulation of the last two steps in vindoline biosynthesis which need to be studied more fully (St-Pierre *et. al.*, 1997)

3 Conclusions

Most of the enzymes involved in the conversion of tabersonine into vindoline have been characterized. The genes which are responsible for the last 2 steps in vindoline biosynthesis have been cloned and are partially described in this chapter. The hydroxylase responsible for the 4th to last step in vindoline biosynthesis remains to be elucidated. This reaction as well as the other tho hydroxylations are probably key rate limiting steps in the sequence of conversions of tabersonine into vindoline since similar levels of the immediate precursor (Fig 2B, lane c) are present in etiolated and in light treated seedlings. Developing this enzyme assay will be useful to understand why part of the pathway of vindoline biosyntheis occurs in the chloroplast.

The molecular tools to study tryptophan decarboxylase (De Luca et al, 1989) and strictosidine synthase (McKnight et al, 1990) which catalyse common reactions leading to the biosynthesis of all the indole alkaloids occurring in *C. roseus* can now be coupled to molecular studies with deacetoxyvindoline 4-hydroxylase and deacetylvindoline 4-*O*-acetyltransferase to differentiate between catharanthine and vindoline biosynthesis. Can etiolated seedlings be made to produce vindoline under appropriate induction other than light? The hydroxylase and acetyltransferase probes can be used to study the effects on the vindoline pathway of inducers in dark grown seedlings and of inhibitors in light grown seedlings. These tools can be further used to dissect the signal transduction cascade which activates/derepresses the biosynthesis of vindoline and can be used to ask why vindoline biosynthesis is so regulated compared to catharanthine biosynthesis. One question that also needs to be studied is the requirement for light-, development- and tissue-specific expression of vindoline biosynthesis in relation to the toxicity of the dimeric alkaloids likely to be produced. Does the production of vindoline also trigger the accumulation of toxic dimeric alkaloids and are specific structures or compartments required for this? The molecular tools which are now available can identify through *in situ* hybridization and immunological studies whether vindoline biosynthesis occurs in the same cell or whether several cell-types are involved. In addition, the role of laticifers and idioblasts (Cutler and Mersey, 1986) in vindoline biosynthesis or in its sequestration can be studied. Successful completion of these types of studies may provide insight into the relationships between the need for biochemical and morphological differentiation in relation to activation of secondary product accumulation. This understanding may then lead to further elaborate the basic rules followed by plants to manufacture secondary metabolites in order to control their production.

Acknowledgements

This work was supported by the Natural Sciences and Engineering Research Council and Le Fonds poir la Formation de Chercheurs et l'Aide à la Recherche . F.V-F. was supported by scholarships from the National Council for Science and Technology (CONACyT, Mexico) and from Les Bourses d'Excellence se la Faculté des Études Supérieurs de L'Université de Montréal. P.L. was supported from Les Bourses d'Excellence se la Faculté des Études Supérieurs de L'Université de Montréal.

References

_Balsevich J, De Luca V, Kurz WGW (1986) Altered alkaloid pattern in dark grown seedlings of *Catharanthus roseus*. The isolation and characterization of 4-desacetoxyvindoline: A novel indole alkaloid and proposed precursor of vindoline. Heterocycles 24: 2415-2421.

_Bristsch L, Dedio J, Saedler H, Forkmann G (1993) Molecular characterization of flavanone 3ß-hydroxylases. Consensus sequence, comparison with related enzyme and the role of conserved histidine residues. Eur J Biochem 217: 745-754.
_Constabel F, Gaudet LaPraire P, Kurz WGW, Kutney JP (1982) Alkaloid production in *Catharanthus roseus* cell cultures XII. Biosynthetic capacity of callus from original explants and regenerated shoots Plant Cell Rep 1: 139-142.
_Cutler, A., Mersey, B (1986) Differential distribution of specific indole alkaloids in leaves of *Catharanthus roseus.* Can J Bot 64: 1039-1045.
_Davis KM (1993) A Malus cDNA with homology to *Anthirrhinun candida* and *Zea A2* genes. Plant Physiol 103: 1015.
_De Carolis E, Chan F, Balsevich J, De Luca V (1990) Isolation and characterization of a 2-oxoglutarate dependent dioxygenase involved in the second-to-last step in vindoline biosynthesis. Plant Physiol 94: 1323-1329.
_De Carolis and V. De Luca (1993) Purification to homogeneity and characterization of a 2-oxoglutarate dependent dioxygenase involved in vindoline biosynthesis in *Catharanthus roseus.* J. Biol. Chem. 268: 5504-5511.
_De Luca V, Cutler AJ (1987) Subcellular localization of enzyme involved in indole alkaloid biosynthesis in *Catharanthus roseus.* Plant Physiol 85: 1099-1102.
_De Luca V, Balsevich J, Tyler RT, Kurz WGW (1987) Characterization of a novel *N*-methyltransferase (NMT) from *Catharanthus roseus.* Plant Cell Rep 6: 458-461.
_De Luca V, Marineau C, Brisson N (1989) Molecular cloning and analysis of cDNA encoding a plant tryptophan decarboxylase: comparison with animal dopa decarboxylase. Proc Nat Acad Sci USA 86:2582-2586.
_De Luca V (1993) Indole Alkaloid Biosynthesis. *In* Methods in Plant Biochemistry. "Enzymes of Secondary Metabolism" (P. Lea ed.) 9: 345-368 Academic Press.
_De Luca V, R. Aerts, S. Chavadej, E. De Carolis and A.M. Alarco (1993) The biosynthesis of monoterpenoid indole alkaloids in *Catharanthus roseus. In* Current Topics in Plant Physiology, Proceedings of the Seventh Annual Penn. State Symposium, "Biosynthesis of Amino Acids", (H.E. Flores and B.K. Singh, eds). 6: 275-284.
_De Luca V, W.G.W. Kurz (1988) Monoterpenoid Indole Alkaloids. *In* Cell Culture and Somatic Cell Genetics of Plants (F. Constabel and I Vasil, eds.) 5: 385-401, Academic Press.
_Dethier M. and V. De Luca (1993) Partial purification of an *N*-methyltransferase which catalyses the third to last step in vindoline biosynthesis in *Catharanthus roseus.* Phytochemistry 32: 673-678.
_Farnsworth NR (1975) The *Catharanthus* alkaloids (WI Taylor and NR Farnsworth eds) M.Dekker, New York.
_Fahn W, Lausermair E, Deus-Neumann B, Stöckigt J (1985) The late enzymes in vindoline biosynthesis. *S*-adenosyl-L-Methionine:11-*O*-demethyl-17-*O*-deacetylvindoline 11-*O*-methyltransferase, and unspecific acetylesterase. Plant Cell Rep 4:337-340.
_Fahn W, Stöckigt J (1990) Purification of acetyl-CoA:17-*O*-deacetyl 17-*O*-acetyltransferase from *Catharanthus roseus* leaves. Plant Cell Rep 8: 613-616.

_Frohman MA, Dush MK, Martin GR (1988) Rapid production of full length cDNAs from rare transcripts: Amplification using a single gene-specific oligonucleotide primer. Proc Nat. Acad Sci USA 85: 8998-9002.
_Holdsworth MJ, Bird CR, Ray J, Schuch W, Grierson D (1987) Structure and expression of an ethylene-related mRNA from tomato. Nuc Acid Res 15: 731-739.
_Holton TA, Brugleira F, Tanaka Y (1993) Cloning and expression of flavonol synthase from *Petunia hybrida*. Plant J. 4: 1003-1010.
_Hotz M, Schröder G, Schröder J (1995) Cinnamate 4-hydroxylase from *Catharanthus roseus*, and a strategy for the functional expression of plant cytochrome P450 proteins as translational fusions with P450 reductase in *Escherichia coli* FEBS Letters 374: 345-350.
_Kutney JP, Hibino T, Jahngen E, Okutani T, Ratcliffe AH, Treasurywala AM, Wunderly S (1976) Total synthesis of indole and dihydroindole alkaloids. IX. Studies on the synthesis of bisindole alkaloids in the vinblastine-vincristine series. The biogenetic approach. (1976) Helv Chim Acta 59: 2858-2882.
_Langlois N., Gueritte F., Langlois Y., Potier P (1976) Application of a modification of the Polonovski reaction to the synthesis of vinblastine-type alkaloids. J Am Chem Soc 98:7017-7024.
_Lingner J, Kellermann J, Keller W: (1991) Cloning and expression of the essential gene for poly(A) polymerase from *S. cerevisiae*. Nature 354: 496-498.
_Mann J (1994) Chemical aspects of biosynthesis, Oxford University Press.
_Matsuda J, Okabe S, Hashimoto T, Yamada Y (1991)Molecular cloning of hyoscyamine 6ß-hydroxylase, a 2-oxoglutarate-dependent dioxygenase from cultured roots of *Hyoscyamus niger*. J Biol Chem 226: 9460-9464.
_Matsuda, A., Sugiura, A., Matsuyama, K., Matsumoto, H., Ichikawa, S., Komatsu, K (1992) Molecular cloning of acetyl CoA: deacetylcephalosporin C *O*-acetyltransferase cDNA from *Acremonium chrysogenium* : sequence and expression of catalytic activity in yeast. Biochem. Biophys. Res. Comm. 182: 995-1001.
_McKnight TD, Roessner CA, Devagupta R, Scott AI, Nessler Cl (1990) Nucleotide sequence of a cDNA encoding the vacuolar protein strictosidine synthase from *Catharanthus roseus*. Nucl Acids Res 18:4939.
_McKnight TD, Bergey DR, Burnett RJ, Nessler CL (1991) Expression of enzymatically active and correctly targeted strictosidine synthase in transgenic tobacco plants. Planta 185: 148-152.
_Meijer AM, Verpoorte R, Hoge JH (1993a) Regulation of enzymes and genes involved in terpenoid indole alkaloid biosyntheis in *Catharanthus roseus*. J. Plant Res. Special Issue 3:145-164.
_Meijer AM, Sour E, Verpoorte R, Hoge JH (1993b) Isolation of cytochrome P450 cDNA clones from the higher plant *Catharanthus roseus* by a PCR strategy. Plant Mol Biol 22: 379-383.
_O'Keefe BR, Mahady GB, Gills JJ, Beecher CWW (1997) Stable vindoline production in transformed cell cultures of *Catharanthus roseus*. J Nat Prod 60:261-264.

_Phillips A, Ward DA, Ukness S, Appleford NEJ, Lange T, Huttly AK, Gaskin, P, Graebe JE, Hedden P (1995) Isolation and expression of three gibberellin 20-oxidase cDNA clones from Arabidopsis. Plant Physiol 108: 1049-1057.
_Power R, Kurz WGW, De Luca V (1990) Purification and characterization of acetylcoenzyme A:deacetylvindoline 4-*O*-acetyltransferase from *Catharanthus roseus*. Arch Biochem Biophys 279:370-376.
_Ramon D, Carramolino L, Patiño C, Sánchez F, Peñalva MA (1987) Cloning and characterization of the isopenicillin *N* synthase gene mediating the formation of the β-lactam ring in *Aspergillus nidulans*. Gene 57: 171-181.
_Shaw WV, Leslie AGW (1991) Chloramphenicol acetyltransferase. Ann Rev Biophys Chem 20:363-386.
_St-Pierre B, De Luca V (1995) A cytochrome P450 monooxygenase catalyzes the first step in the conversion of tabersonine to vindoline in *Catharanthus roseus*. Plant Physiol 109: 131-139.
_St-Pierre B, Laflamme P, Alarco AM, De Luca V (1997) Molecular cloning and characterization of DAT from *Catharanthus roseus*. Submitted for publication.
_Van der Heijden R, Verpoorte R, Ten Hoopen HJG (1989) Cell and tissue cultures of *Catharanthus roseus* (L.) G. Don: a literature survey. Plant Cell Tissue Org Culture 18:231-280.
_Vazquez-Flota F, De Carolis E, Alarco AM, De Luca V (1997) Molecular cloning and characterization of desacetoxyvindoline-4-hydroxylase. A 2-oxoglutarate dependent-dioxygenase involved in the biosynthesis of vindoline in *Catharanthus roseus* (L.) G. Don. Plant Mol Biol 34:935-948.
_Wu D, Ahmed SN, Lian W, Hersh LB (1995) Activation of rat choline acetyltransferase by limited proteolysis. J. Biol. Chem 270: 19395-19401.
_Yoder LR, Mahlberg PG: (1976) Reactions of alkaloid and histochemical indicators in laticifers and specialized parenchyma of *Catharanthus roseus*. Am J Bot 63: 1167-1173.

Surface Control of Cell Polarity, the Plane of Cell Division, and Cell Fate in *Fucus* Embryos

Ralph S. Quatrano

Department of Biology, University of North Carolina, Chapel Hill, NC 27599-3280, USA

Keywords. cell polarity, *Fucus*, secretion, microfilaments, asymmetric division, cell wall

1 Introduction

When spherical zygotes of the brown alga *Fucus* are subjected to a gradient of light during the first several hours after fertilization, a bud or rhizoid is formed on the shaded side of the zygote at 14 hours. This polar growth is followed by an asymmetric cell division at 24 hours, with the cell plate always oriented perpendicular to the light axis. The resulting embryo is comprised of two unequal cells; the smaller, more elongated rhizoid cell emerges from the shaded region of the embryo, while the larger, more rounded thallus cell is on the lighted side. These cells differ not only in shape but in organellar content and molecular composition, both in the cytoplasm and cell wall. It is clear that within the first 24 hours after fertilization, external gradients imposed on the zygote can establish a polarity which results in the differentiation of the first two cells of the embryo and sets the developmental axis for the whole organism (Fowler & Quatrano 1995; Quatrano & Shaw 1997).

There are several properties of this system which allow one to experimentally approach the question of how molecular asymmetries are established in reference to external, spatial cues. Eggs of the Fucales are large (75-100 μm in diameter), apolar, lack a large central vacuole, and along with sperm, are released from mature plants under laboratory conditions. Fertilization occurs in a defined sea water medium, with the resulting diploid zygote adhering to the substratum within a few hours. This attachment allows for easy changing of solutions and for orientation of populations of zygotes to various external gradients. Most importantly, the polarity of the zygote is not irreversibly established until about 10 hours after fertilization. During the time between 4 and 10 hours, the direction of the light can be changed several times, with the resulting polar growth of the rhizoid and orientation of the plane of division always determined by the last light gradient. Not only is the egg apolar, but the polarity is labile and experimentally

NATO ASI Series, Vol. H 104
Cellular Integration of Signalling Pathways in Plant Development
Edited by F. Lo Schiavo, R. L. Last, G. Morelli, and N. V. Raikhel

determined. This polar development occurs synchronously for several cell divisions in the absence of surrounding cells. Hence, the process by which light perception is converted into a polar axis can be directly studied (Goodner & Quatrano 1993; Kropf 1994).

It appears that these polar processes in *Fucus* embryos are common not only to algae but to many different organisms, including higher plants. For example, the pattern of cell division in *Fucus* is quite similar to embryogenesis in *Arabidopsis*, suggesting that analogous morphogenetic processes may be operating in these embryos. Polar cell divisions in plants occur not only during early embryogenesis but also during microsporogenesis and throughout vegetative development, for example, in the development of stomata. The mechanism by which a cell identifies cortical sites for the directed transport of cytoplasmic material to support polar growth and/or cell plate formation is easier to study in *Fucus* than in most of these higher plant systems (Fowler & Quatrano 1997).

Much of our recently published work on the establishment of cell polarity in *Fucus* zygotes has focused on these questions:
1). how are external gradients translated into cortical asymmetries, and what are the nature of the molecules that are localized?
2). what are the processes required to stabilize the initial polar axis established by the light gradient, i.e. to fix the molecular asymmetries so that external gradients can no longer influence their redistribution?
3). what is the nature of the positional information set by the light gradient that orients the first cell division?

2 Establishment of a Polar Cell

2.1 What is the nature of the cortical asymmetries that are localized in response to a light gradient?

We have developed cytological markers to monitor the orientation of the polar axis, which include probes for vesicle targeting, actin, free calcium and dihydropyridine (DHP) receptors (Kropf et al 1989; Shaw & Quatrano 1996a; Shaw & Quatrano 1996b). Hence, we can temporally and spatially monitor the localization of cortical sites relative to the direction(s) of unilateral light imposed during the first 10 hours after fertilization. For example, we used fluorescently-labeled dihydropyridine (FL-DHP) to vitally stain *Fucus* zygotes during the establishment of cell polarity (Shaw & Quatrano 1996a). Distribution of FL-DHP in the plasma membrane is initially symmetrical before becoming asymmetrical in response to a unilateral light gradient, i.e. accumulating at the shaded side of the zygote and predicting the site of polar outgrowth of the rhizoid. The asymmetric distribution of FL-DHP receptors coincides temporally and spatially with an increase in intracellular calcium, as measured by microinjected calcium green dextran. Based on the site, timing, photo-reversibility, and actin dependence of the asymmetric localization of FL-DHP receptors, we concluded that FL-DHP is a vital probe for monitoring the initial orientation of the polar axis. Furthermore, we proposed that FL-DHP receptors correspond to ion channels that are directed to the future site of polar growth to create the changes in local calcium concentration required for setting the initial polar axis (Shaw & Quatrano 1996a). FL-DHP can now be used to literally visualize the orientation of the polar axis in a large population of developing zygotes, as well as to assay for potential intermediates in the response pathway; from a unilateral light gradient to the first observable cellular asymmetry in the plasma membrane. Hence, external gradients are translated into a cortical asymmetry by the actin-dependent translocation of existing plasma membrane molecules, forming a target site for subsequent vesicle transport to support the polar outgrowth of the rhizoid.

2.2 What are the processes required to stabilize the plasma membrane asymmetries established by the light gradient ?

We have also utilized toluidine blue O as a probe to follow Golgi vesicle accumulation and polar secretion into the cell wall at the site of future polar growth (Shaw & Quatrano 1996b). The deposition of Golgi-derived material (GDM) into the cell wall at this target site is temporally coincident with the time at which a change of light direction cannot redirect the target site for vesicle secretion. Does preventing secretion block polar axis fixation? We treated zygotes continually with brefeldin-A (BFA) to selectively interrupt secretion of GDM into the cell wall, to determine if this secretion is required to fix the polar axis. We showed that in the presence of BFA, the stabilization or fixation of the light-induced polar axis in *Fucus* zygotes is prevented, as well as polar growth of the rhizoid. However, BFA appears not to interfere with the initial orientation of the axis, establishment of the target site (i.e. DHP receptors, F-actin) or in the timing of the first cell division. Hence, based on the timing of GDM deposition into the wall, and the inhibition of polar axis fixation in the presence of BFA, we proposed that local secretion of GDM transforms the target site into the fixed site which will direct polar growth of the rhizoid (Shaw & Quatrano 1996b).

Evidence for the role of polarized secretion in axis fixation further refines our model of the axis stabilizing complex (ASC), a hypothetical structural unit which serves to stabilize molecular asymmetries in the plasma membrane at the target site for vesicle deposition (Quatrano 1990). Based on the requirement for an intact cell wall and actin cytoskeleton to stabilize the polar axis (Kropf et al 1989; Kropf et al 1988), we proposed that the ASC was composed of a cell wall component(s) which interacts with the actin cytoskeleton through a putative transmembrane receptor (Kropf et al 1988). This model is similar to that of focal adhesions in mammalian cells, i.e. a complex that involves an extracellular matrix component (e.g.) vitronectin and its associated sulfated heparin binding to a receptor peptide in the transmembrane protein (e.g. integrin) which is linked to the actin cytoskeleton through actin-binding proteins (e.g. vinculin) (Burridge & Chrzanowska-Wodnicka 1996). Based on the BFA results, we now believe that the requirements for a cell wall and F-actin to fix a polar axis are directly related to the BFA-sensitive secretory event (Shaw & Quatrano 1996b). Our ASC model would now add a step in the assembly of the ASC that would include the insertion of molecules into the cell wall which could provide the extracellular ligand(s) for the postulated ASC transmembrane receptor (Fowler & Quatrano 1995).

What genes might be involved in the transport and secretion of the Golgi-derived vesicles directed to the fixed site? The temporal and spatial processes of the orientation of the polar axis, fixation, and expression in *Fucus* are similar to the stages of bud-site selection, polarity establishment, and bud growth during the asymmetric division of budding yeast (Chant & Stowers 1995). Polarized growth of the budding yeast employs selective targeting of proteins to a patch on the cell surface at the bud site. Protein delivery is mediated by translocation of vesicles along cytoskeletal filaments to the bud site and appears to be regulated by GTPase cycles that specify docking of vesicles with the targeting patch. Ras-related GTPases act as molecular switches, cycling between an inactive, GDP-bound state, and an active, GTP-bound state that signals downstream molecules (Chant & Stowers 1995). Members of the rac/rho/CDC42 family of GTPases regulate the actin cytoskeleton at localized sites at the plasma membrane. For example, the yeast CDC42 gene is required for polarity both in budding, and in the growth response to mating pheromone, processes that require localization of actin to the site of polarized growth and are analogous to Fucus zygote development. The rab GTPase family, which includes the rab/YPT1/SEC4 genes, regulates vesicle transport and fusion in yeast and mammals (Roemer et al 1996; Drubin & Nelson 1996). Genes localized in the cytoplasmic regions of polar bud growth in yeast are found in *Fucus* (e.g. rac-like GTPases) and their role in the process of polar growth in *Fucus* is now being investigated.

2.3 What is the nature of the positional information set by the light gradient that orients the first cell division?

We noticed that the orientation of the first division plane in zygotes continuously incubated in BFA was random with respect to the light gradient. Hence, the resulting division was symmetric rather than asymmetric and raised the interesting possibility that the cell wall was providing some type of positional cues for orienting the plane of cell division.

These recent studies also showed that the site or sites of polar growth (i.e. the polar axis) was always properly oriented to the light gradient (i.e. on the shaded side), independent of the position of the division plane. We proposed that the establishment of zygotic cell polarity, and not the position of the first division plane, is critical for the formation of the initial embryonic pattern in *Fucus* (Shaw & Quatrano 1996b). Our data in *Fucus* is also consistent with the hypothesis that the factor(s) in the cell walls of *Fucus* embryos responsible for rhizoid differentiation are deposited simultaneously with and/or are the same positional signal for polar growth identified in our experiments (Berger et al 1994; Brownlee & Berger 1995). The role of targeted secretion from the Golgi, resulting in an asymmetry of the macromolecules present in the zygotic cell wall of *Fucus*, provides an exciting paradigm to apply to other plant systems for the study of extracellular positional information during development.

As a start to identify proteins that are localized to this domain, we have cloned a gene expressed in *Fucus* zygotes whose protein contains several regions of basic amino acids. These regions are similar to the heparin binding domain of human vitronectin and are recognized by an antibody to this domain. A segment of the recombinant *Fucus* protein containing one of these domains can bind the heparin-like, sulfated polysaccharide (F2) which is localized exclusively in the tip of the emerging rhizoid.

3 Summary

External gradients such as unilateral light are perceived and translated by apolar zygotes of *Fucus* into a cortical asymmetry by the actin-dependent translocation of existing plasma membrane molecules (DHP receptors) to the shaded side. This corresponds to the orientation of the polar axis, and with a vital marker (FL-DHP) to visualize this process, we can now assay for "upstream" events in the response pathway, i.e. intermediates in the pathway from a unilateral light gradient to the first observable cellular asymmetry in the plasma membrane. In addition to the DHP receptors, F-actin and an accumulation of free calcium are other components localized in this region of the cell cortex. These components (among others) form a "target site" for Golgi vesicle targeting and secretion. We now know that secretion of these vesicles into the cell wall are essential to fix the polar axis and to complete the ASC. Furthermore, this secretion appears to provide positional information required to orient the first cell division plane.

Finally, since the pattern of cell divisions in the *Fucus* embryo is similar to that of *Arabidopsis* (Brownlee & Berger 1995), observations in this system may be relevant to several embryo mutants. For example, the *fass* mutation in *Arabidopsis* has abnormal planes of early cell division, yet the apical-basal body plan is maintained (Torres-Ruiz & Jürgens 1994), a situation analogous to that seen in *Fucus* embryos blocked in targeted secretion (see above). The *knolle* mutant exhibits a radial pattern defect with abnormal cell divisions (e.g. lacking cross walls, abnormal number of spindles, etc.) and cell enlargement during embryogenesis (Lukowitz et al 1996). A similar mutation has been described in pea, i.e. cytokinesis-defective, which can be mimicked by caffeine treatment of wild type pea seedlings which results in dividing cells failing to generate a cell plate (Liu et al 1995). The KNOLLE protein has homology to syntaxins, a protein family involved in vesicular trafficking, and has led to the proposal that the KNOLLE might act "as a cell-plate specific syntaxin facilitating the transport of vesicles to or their fusion at the plane of cell division". The gene product of another embryo mutant in *Arabidopsis*, GNOM, has been cloned and shares a domain homologous to the yeast secretory protein Sec7p (Shevell et al 1994). These mutations of early plant embryogenesis are pointing to the importance of vesicle movement and secretion in cell plate formation and the control of the pattern of early cell divisions, as we have demonstrated during *Fucus* embryogenesis.

4 References

Berger, F., Taylor, A., Brownlee, C. 1994. Cell fate determination by the cell wall in early *Fucus* development. *Science* 263:1421-1423

Brownlee, C., Berger, F. 1995. Extracellular matrix and pattern in plant embryos: on the lookout for developmental information. *Trends in Genetics* 11:344-348

Burridge, K., Chrzanowska-Wodnicka, M. 1996. Focal adhesions, contractility, and signaling. *Annu. Rev. Cell Develop. Biol.* 12:463-519

Chant, J., Stowers, L. 1995. GTPase cascades choreographing cellular behavior: movement, morphogenesis, and more. *Cell* 81:1-4

Drubin, D. G., Nelson, W. J. 1996. Origins of cell polarity. *Cell* 84:335-344

Fowler, J. E., Quatrano, R. S. 1995. Cell polarity, asymmetric division, and cell fate determination in brown algal zygotes. *Sem. Develop. Biol.* 6:347-358

Fowler, J. E., Quatrano, R. S. 1997. Plant cell morphogenesis: Plasma membrane interactions with the cytoskeleton and cell wall. *Annu. Rev. Cell Develop. Biol.* 13:697-743

Goodner, B., Quatrano, R. S. 1993. *Fucus* embryogenensis: A model to study the establishment of polarity. *Plant Cell* 5:1471-1481

Kropf, D. L. 1994. Cytoskeletal control of cell polarity in a plant zygote. *Develop. Biol.* 165:361-71

Kropf, D. L., Berge, S. K., Quatrano, R. S. 1989. Actin localization during *Fucus* embryogenesis. *Plant Cell* 1:191-200

Kropf, D. L., Kloareg, B., Quatrano, R. S. 1988. Cell wall is required for fixation of the embryonic axis in *Fucus* zygotes. *Science* 239:187-90

Liu, C.-M., Johnson, S., Wang, T. L. 1995. *cyd*, a mutant of pea that alters embryo morphology is defective in cytokinesis. *Develop. Genet.* 16:321-331

Lukowitz, W., Mayer, U., Jurgens, G. 1996. Cytokinesis in the Arabidopsis embryo involves the syntaxin-related KNOLLE. *Cell* 84:61-71

Quatrano, R. S. 1990. Polar axis fixation and cytoplasmic localization in *Fucus*. In *Genetics of Pattern Formation and Growth Control*. pp. 31-46 Wiley-Liss, Inc.

Quatrano, R. S., Shaw, S. L. 1997. Role of the cell wall in the determination of cell polarity and the plane of cell division in *Fucus* embryos. *Trends in Plant Science* 2:15-21

Roemer, T., Vallier, L. G., Snyder, M. 1996. Selection of polarized growth sites in yeast. *Trends in Cell Biology* 6:434-441

Shaw, S. L., Quatrano, R. S. 1996a. Polar localization of a dihydropyridine receptor on living *Fucus* zygotes. *J. Cell Sci.* 109:335-342

Shaw, S. L., Quatrano, R. S. 1996b. The role of targeted secretion in the establishment of cell polarity and orientation of the division plane in *Fucus* zygotes. *Development* 122: 2623-2630

Shevell, D. E., Leu, W.-M., Gillmor, C. S., Xia, G., Feldmann, K. A., Chua, N.-H. 1994. EMB30 is essential for normal cell division, cell expansion, and cell adhesion in Arabidopsis and encodes a protein that has similarity to Sec7. *Cell* 77:1051-1062

Torres-Ruiz, R. A., Jürgens, G. 1994. Mutations in the *FASS* gene uncouple pattern formation and morphogenesis in *Arabidopsis* development. *Development* 120:2967-2978

INDUCTION AND PROGRESSION OF SOMATIC EMBRYOGENESIS

Fiorella Lo Schiavo

Department of Biology,University of Padua,Via G. Colombo 3, 35131 Padova,Italy

Somatic embryogenesis is a cloning process well known in plants. It is the process by which somatic cells develop through the stages of embryogeny to give identical whole plants without genetic fusion.

Although a number of specialized examples of somatic embryogenesis have been reported to occur also in vivo, the process is best known as a pathway for regeneration in vitro.

1 INDUCTION PROCESS

The first phenomenon we should like to understand is that of *induction.*

We have analyzed a few systems where embryogenesis occurs in a direct or in an indirect way. *Direct* occurs when embryogenic cells develop directly from explant cells, *indirect* when a number of unorganized, non-embryonic mitotic cycles are interposed between differentiated explant tissue and recognizable embryonic structure. The examples will be taken from carrot which has been and still is the model system (Terzi and Lo Schiavo 1990), but reference will also be made to Arabidopsis. Young seedlings of carrot can be induced directly or indirectly under the influence of different hormonal stimuli.

In the presence of ABA carrot seedlings stop growing and this arrest is manifested in all portions, most notably in the leaves primordia.

NATO ASI Series, Vol. H 104
Cellular Integration of Signalling Pathways in Plant Development
Edited by F. Lo Schiavo, R. L. Last, G. Morelli, and N. V. Raikhel

At the histological level lack of cell extension is clearly noticeable. Epidermal cells are rich in cytoplasm, ER and show small vacuoles. Differentiation of the leaves is blocked which, probably, remain in an undifferentiated state. At t=20 days, somatic embryos begin to appear first on leaves primordia and cotyledons and then on roots and hypocotyls.

These embryos originate in the protoderm which undergoes an intense proliferating activity. The division planes are irregular and generate several layers of small, cytoplasmic-rich cells. In a few spots those divisions give rise to globular structures clearly isolated from the surrounding cells; from these isolated structures somatic embryos develop. Hence, at least in this case, direct embryogenesis starts only from epidermal cells, no dedifferentiation is seen, but a block of cell extension seems to be the main initial event for the induction of somatic embryos.

If we expose carrot young seedlings in culture medium supplemented with auxin (usually 2-4D), 5-7 days later the epidermal layer and the parenchyma cells enlarge and come off without dividing, procambium and provascular cells proliferate. Procambium cells enlarge and detach from the stelar cylinder into the culture medium. These suspension cells proliferate and in 10-20 days give rise to an embryogenic cell line (Guzzo et al. 1994; Guzzo et al. 1995). In fact, in about 20 days after explantation, a particular cell type appears, very small (10-20 μm) which after a series of regular divisions generates what Halperin (Halperin 1966) called proembryogenic masses (PEM). An interesting observation is that one asymmetric division is essential for PEM formation.

Another example we want to make comes from Arabidopsis. Arabidopsis can regenerate via somatic embryogenesis not only of the direct type, but also of the indirect type, through the formation of an embryogenic cell culture (Pillon et al.1996).

As starting material we used zygotic embryos of stages between heart-shaped and bent torpedoes. These embryos ,when exposed to auxin (particularly 2-4D), give rise to direct embryogenesis. It is reported in the literature that zygotic embryos in which the growth of the main axis is weak or suppressed are more likely to give rise to somatic embryos from superficial cells (Hu ans Sussex 1971; Maheswaran and

Williams 1986; Petrova and Williams 1986). In general the zygotic embryo, submitted to various types of stress reacts by forming somatic embryos or embryogenic callus.

If the somatic embryos obtained from zygotic Arabidopsis embryos are exposed to high auxin concentration, embryos dedifferentiate in callus, and globular-heart and torpedo stages can be induced by proliferating cells via indirect embryogenesis.

Our results show that in Arabidopsis, zygotic and somatic embryos behave differentially in vitro.

So, in general, in *direct embryogenesis* embryos or plantlets fail to mature normally and give rise to successive cycles of embryos.

In other words, from the initial explant we get somatic embryos that in turn generate secondary embryos; this happens recurrently in a continous way and a special procedure is needed to interrupt the process.

In *indirect embryogenesis* the embryogenic cell line proliferates and generates continuously pro-embryogenic masses. If PEM's are placed in the proper conditions they will develop into mature embryos and plantlets, otherwise they will degenerate.

In both cases, in spite of the different mechanism of induction, the production of (initial) embryos can go on for long periods, the difference being the stage at which the embryos are produced : initial in the case of indirect and final in the case of direct embryogenesis.

2 DEVELOPMENTAL MUTANTS : AUXIN-RESISTANT LINES

During induction of the indirect type in carrot embryogenic cell cultures originate from which developmental mutants with extreme phenotype can be obtained and easily handled.

Among different types of developmental mutants auxin-resistant (auxin-R) lines have been isolated..

All our auxin-R lines result impaired in development. This result does not come as a surprise because auxin acts at every level of indirect somatic embryogenesis.

At early stages (Lo Schiavo 1995) auxin is instrumental in the acquisition of totipotency and at high concentration induces asymmetric divisions (Dudits et al. 1991; Guzzo et al 1994), the first step in PEM formation .

At intermediate stages, there is a change in IAA metabolism : IAA synthesis re-starts via indole (Michalczuk et al. 1992) and polar transport is established.

At late stages, IAA intervenes in the acquisition of a proper polarity (Schiavone and Cooke 1987), of a proper embryo axis and is involved in the correct morphogenetic development of root and cotyledons (Okada et al 1991; Liu et al. 1993).

Our attempt has been at first to identify which interaction with auxin was altered in our mutants, whether uptake or metabolism, perception or signal transduction. This was done on the basis of physiological parameters and a correlation with the type of block we observe in embryogenesis.

Let's consider some of them :

4w 89 : is characterized by a high level of internal unbound IAA and a low level of bound IAA that can justfy a reduction in IAA and 2-4D uptake and in an overinduction of IAA efflux.

This line is capable of making the initial stage of embryogenesis but then the globular form degenerates probably because the free level of IAA is too high and it stays high because of a defect in conjugation.

4w151 : makes embryos that develop more or less normally up to heart stage (but they show multiple cotyledons, an indication that something is wrong with IAA transport) and torpedoes where they are blocked. Interestingly enough, the cotyledonary half of the embryo is blocked, whereas the root keeps on extending as if it is preparing for germination. Keeping in mind that the root is capable of sensing low concentrations of auxin and considering that the various parameters : influx, efflux, conjugation are not dramatically affected, we interpret this mutant as one of perception.

4w13 : cannot make embryos, not even the initial stages. The internal level of free IAA and its conjugates is very high, influx carrier is normal, efflux carrier apparently absent. We think it cannot embryogenize because the internal level of IAA cannot go down.

4w111 : cannot make early embryos because of an altered hormonal balance; it is resistant to IAA, 2-4D, GA and ABA, i.e. all those we tested. Influx of IAA is low, 2-4D is normal. Efflux is high in spite of the low internal level of IAA.
It could be a mutation in signal transduction or it could be due to a multiple resistance protein such as described for lipophilic substances in bacteria and animal cells.

In *4w77* : early embryonal stages look normal, but then alterations in cotyledons are seen and on sectioning one sees multiple axes.
It looks like a mutation of transport : efflux looks normal but the influx carrier is altered showing reduced uptake for IAA and 2-4D and not showing competition for IAA and 2-4D.

From *RNA differential display* experiments, performed on these mutants, we isolated a certain number of clones. 8 of them have been sequenced. Two out of 8 are homologous to known dehydrogenases (ADH and NADH), the others show no homologies.

3 PROTEIN SECRETION-EMBRYO DEVELOPMENT

In recent years a certain amount of work has been dedicated in our laboratory, as well as in others', on the proteins secreted during development . These studies established a strict relationship between secretion and embryogenesis summarised in the above statements:

- Embryos are characterized by a specific set of extracellular proteins (deVries et al. 1988);

- A correct type of secretion is necessary during the developmental process (Lo Schiavo et al. 1990);
- If the pattern of secreted proteins is altered, this has consequences on the embryogenic process (Lo Schiavo et al. 1986; Baldan et al.1995);
-This developmental block, in some cases at least, can be corrected by addition to the embryo culture of conditioned medium or specific extracellular proteins (Lo Schiavo et al.1990; Cordewener et al. 1991; De Jong et al. 1992).

Another outcome has been the identification of a critical step in embryogenesis : conversion of globular to heart i.e. acquisition of polarity. This step is controlled by secreted proteins as shown by cell mutants impaired in embryogenesis, or by the effect of drugs known to interfere with secretion.
I will briefly mention some of the evidence that relates secretion to embryo development.

3.1 EFFECTS OF BREFELDIN A ON SOMATIC EMBRYOS

A set of experiments was carried out with BFA that is known to inhibit protein secretion (Capitanio et al. in press).
When this drug is added at 7,5 μg/ml for 48 hrs to an embryogenic cell culture the process is dramatically affected. Late forms are rare or lacking. Most of the forms either remain as globular or poly-embryos are formed.
In order to follow the timing of BFA action on embryo development, embryogenesis was synchronized and globular and heart-shaped embryos were purified and then submitted to a 48 hrs 7,5μg BFA treatment. The results were scored 16 days later.
The effect of BFA on globular embryos is dramatic, and at the time of scoring, relatively few hearts and torpedoes are seen and those are not really normal and show abnormality in the cotyledonary primordia.
Instead heart-shaped embryos showed only a slowing-down of the morphogenetic process but no blocked embryos nor polyembryos were seen .

At the cytological level, the treated embryos showed structural alterations : both ER and GA were heavily affected. Golgi stacks were almost completely disassembled and clusters of vesicles were formed .

When globular embryos are incubated for 48 hrs into conditioned medium containg BFA, the BFA-induced block of embryogenesis is not seen. But at the microscopic level, the late embryos obtained, showed normal roots and abnormal cotyledons.

In this case, it is interesting that complementation by conditioned medium was not complete: polarity was acquired normally, no polyembryos were formed, the embryonal root developed normally but cotyledonary primordia still maintained a typical abnormality, meaning that this kind of alteration cannot be rescued by extracellular proteins.

3.2 TS11 EMB-MUTANT

The characterization of ts11 isolated and characterized by us and Sacco de Vries' laboratory, has also some bearing on the relationship : *proper secretion-proper development.*

Ts11 is unable to form somatic embryos at the non-permissive temperature of 32°C, but the block can be overcome by addition to the medium of a 32kDa acidic endochitinase (De Jong et al. 1992).

Some data on secretion in ts11 are briefly summarized (Baldan et al. in press). A cytohistological analysis of the blocked forms reveals that the endomembrane system is altered. The endoplasmic reticulum (ER) forms dilated cysternae and vesicles that accumulate electron dense material . We observe points of continuity between ER and the plasma membrane.

The ultrastructural organization of the cell wall is also altered in ts11 embryos at 32°C. The outer wall of the protodermal cells shows a loose fibrillar texture, holes and discontinuities are evident and the surface is frayed .

Suspension-growing ts11 cells show quantitative differences in terms of protein secretion with respect to wt cells. Ts11 released a lower amount of protein in the medium during the culture cycle. Protein synthesis and secretion are compared by pulse and chase experiments on 4 days old culture. Radioactivity recovered from cell homogenates as TCA-precipitable material is almost identical for the two cell lines, but the pattern of newly synthesized proteins released into the medium is different.

To acquire more information, we analysed the location of individual proteins such as binding protein BiP (an ER-resident molecular chaperon) and glutamine synthase (GS), a cytosolic protein. Neither of the two is normally expected to be secreted. Wt cells do not secrete BiP at any time of the culture cycle, whereas ts11 cells accumulate a progressively increasing amount of BiP in the medium.

Quantitation of the western blot indicate that, at 7 days of culture, about 1,5% of total BiP is located extracellularly. Western blot performed using antibodies raised against glutamine synthase showed that also this cytosolic protein was released by ts11 but not by wt, about O.8% of the total GS is located extracellularly.

By comparing preparations of vesicles from wt and ts11 we notice in ts11 a reduced presence of coated vesicles, both clathrin and non-clathrin coated, revealing several alterations along the secretory patway.

However in order to identify the primary lesion in the secretory pathway altered in ts11, we should have a better knowledge of the plant secretory pathway. Very little is known at present and we started to isolate some components of non-clathrin coated vesicles. So far we managed to produce antibodies and to get a partial purification of a VAMP-like protein, whose nature has confirmed by the specific interaction with the tetanus toxin and which we hope to sequence and eventually clone.

When we add an acidic endochitinase to the culture medium, ts11 embryo development block is removed (De Jong et al. 1992). At the morphological level chitinase-rescued cultures show embryos close to normal. At the ultrastructural level, endomembrane alterations were reduced with both ER, cisternae and GA stacks having a regular appearance.

Unfortunatly it is difficult to have enough endochitinase to perform a complementation experiment on secretion. However we see that wt conditioned medium prevents BIP secretion .

A puzzling question is the complementation of ts11 defect by endochitinase. Chitin is generally thought to be absent from higher plants, although oligomers of N-acetylglucosamine may be present in secondary cell walls. Chitinase cutting are of such oligomers can release signal molecules active in embryogenesis.

Of course the isolation of novel signal molecules can help in the understanding of this mechanism.

On the other hand we are investigating on the bifunctional nature of this enzyme. Some chitinases possess an active site for phospholipase (PLA) in the middle of the chitin binding domain. So that our working hypothesis at present is that in the absence of chitin the enzyme works as a phospholipase and generates signal molecules necessary for embryo development.

In pathogenic condition (plant-pathogen interactions) the phospholipase active site is covered by chitin and cannot generate the necessary signals. Then the cells cannot complete their developmental programme and probably go in apoptosis. According to this working hypothesis ts11 produces less of the chitinase (De Jong et al. 1995) with lipase activity and the embryos degenerate.

Preliminary data show that PLA added to the medium can complement the secretory defect of ts11 and that chitin added to wt makes a sort of phenocopy.

I would like to conclude by pointing out that whatever the role of chitinase, it cannot be just coincidence that in several plant systems, mutants or phenocopies blocked in somatic or zygotic embryogenesis (Shevell et al. 1994; Lukowitz et al.1996), show alterations in secretory pathway and that these alterations severely affect development.

REFERENCES

Baldan B., Frattini C., Guzzo F., Branca C., Terzi M.,Mariani P, Lo Schiavo F. (1995) A stage-specific block is produced in carrot somatic embryos by 1,2-benzoxazole-3-acetic acid. Plant Sci 108:85-92

Baldan B., Guzzo F., Filippini F. Gasparian M., Lo Schiavo F., Vitale A., de Vries S.C. Mariani P., Terzi M. (1997) The secretory nature of the lesion of carrot cell variant ts11, rescuable by endochitinase. Planta in press

Capitanio G., Baldan B., Filippini F., Terzi M., Lo Schiavo F., Mariani P. (1997) Morphogenetic effects of Brefeldin A on embryogenic cell cultures of *Daucus carota* L. Planta in press

Cordewener J., Booij H., van der Zandt H., van Engelen F., van Kammen A., de Vries S.C. (1991) Tunicamycin-inhibited carrot somatic embryogenesis can be restored by secreted cationic peroxidase isoenzymes. Planta 184:478-486

De Jong A.J., Cordewener J. lo Schiavo F. Terzi M. Vandekerckhove J., van Kammen A., de Vries S.C. (1992) A carrot somatic embryo mutant is rescued by chitinase. Plant Cell 4:425-476

De Vries S.C., Booij H., Janssens R., Vogels R., Saris L., Lo Schiavo F., Terzi M., van Kammen A. (1988) Carrot somatic embryogenesis depends on the phytohormone-controlled presence of correctly glycosylated extracellular proteins. Genes &Dev. 2:462-476

Dudits D., Bogre L., Gyorgyey J(1991) Molecular and cellular approaches to the analysis of plant embryo development from somatic cells in vitro. J.Cell Sci. 99:475-484

Guzzo F., Baldan B. Mariani P., Lo Schiavo F., Terzi M. (1994) Studies on the origin of totipotent cells in explantes od Daucus carota L. J.Exp.Bot. 45:1427-1432

Guzzo F., Baldan B., Levi M.,Sparvoli E., Lo Schiavo F., Terzi M., Mariani P. (1995) Early cellular events during induction of carrot explants with 2,4D. Protoplasma 185:28-36

Halperin W (1966) Alternative morphogenetic events in cell suspensions. Am. J. Bot. 53:443-453

Hu C.Y., Sussex I.M. (1971) In vitro development of embryoids on cotyledons of *Ilex aquifolium*. Phytomorphology 21:103-107

Liu C.-m., Xu Z.-h, Chua N.-H. (1993(Auxin polar transport is essential for the establishment of bilateral symmetry during early plant embryogenesis. Plant Cell 5:621-630

Lo Schiavo F., Quesada-Allue L.A., Sung Z.R. (1986) Tunicamycin affects embryogenesis but not proliferation of carrot. Plant Science 44:65-73

Lo Schiavo F.,Giuliano G., de Vries S.C., Genga A. Bollini R., Pitto L., Nuti-Ronchi V., Cozzani F., Terzi M. (1990) A carrot cell variant temperature-sensitive for embryogenesis reveals a defect in the glycosylation of extracellular proteins. Mol. Gen. Genet. 223:385-393

Lo Schiavo F. (1995) Early events in embryogenesis. In: Somatic embryogenesis and sybthetic seeds I (Y.P.S. Bajaj editor) Springer Verlag, Berlin pp 20-29

Lukowitz W., Mayer U., Jurgens G. (1996) Cytokinesis in the Arabidopsis embryo involves the syntaxin-related KNOLLE gene product. Cell 84:61-71

Maheswaran G., Williams E.G. (1986) Clonal propagation of *Trifolium pratense*, *Trifolium resupinatum* and *Trifolium subterraneum* by direct somatic embryogenesis on cultered immature embryos. Plant Cell Rep. 3:165-168

Michalczuk L., Cooke T.J., Cohen JD (1992) Auxin levels at different stages of carrot somatic embryogenesis. Pphytochemistry 41:1097-1103

OkadaK., Ueda J., Komaki M.K., Bell C.J., Shimura Y. (1991) Requirement of the auxin polar transport system in early stages of Arabidopsis floral bud formation. Plant Cell 3:677-684

Petrova A. Williams E.G. (1986) Direct somatic embryogenesis from immature zygotic embryos of flax (*Linum usitatissimum* L.). Plant Physiol. 126:155-161

Pillon E., Terzi M., Baldan B. Mariani P. Lo schiavo F. (1996) A protocol for obtaining embryogenic cell lines from Arabidopsis. Plant Journal 9:573-577

Schiavone F.M., Cooke T.J. (1987) Unusual patterns of somatic embryogenesis in domesticated carrot: developmental effects of exogenous auxin transport inhibitors. Cell Differ. 21:53-62

Shevell D.E., Leu W.M., Gillmore C.S., Xia G., Feldman K.A., Chua N.H. (1994) EMB30 is essential for normal cell division, cell expansion and cell adhesion in Arabidopsis and encodes a protein that has similarity to Sec7. Cell 77:1051-1062

Terzi M. Lo Schiavo F. (1990)Somatic Embryogenesis. In : Bhojwani SS (ed) Plant tissue culture: application and limitations. Elsevier, Amsterdam, pp 54-66

Identification of genes expressed during *Arabidopsis thaliana* embryogenesis using enhancer trap and gene trap *Ds*-transposons

Casper W. Vroemen, Nicole Aarts, Paul M.J. In der Rieden, Ab van Kammen and Sacco C. de Vries

Department of Molecular Biology, Wageningen Agricultural University, Dreijenlaan 3, 6703 HA Wageningen, The Netherlands

Abstract. The technique of enhancer trap and gene trap mutagenesis has been exploited to identify new molecular markers for specific cell-types, tissues and regions in the *Arabidopsis thaliana* embryo and seedling. Screening of a population of 373 independent gene trap and 431 enhancer trap lines revealed that 25% of the gene trap insertions, and 81% of the enhancer trap insertions displayed *GUS* expression patterns in the embryo, seedling, silique, seed coat, or flower. A total of 39 lines expressed the *GUS* gene in the embryo. Except for one, all of these also displayed *GUS* expression at other stages of development. The insertion lines with specific *GUS* expression patterns in the embryo provide valuable markers for establishment of cell fate or position in embryo mutant backgrounds. Genomic DNA flanking the insertions was amplified by TAIL-PCR, and found to contain transcribed regions of a gene in all gene trap insertions, and in about a quarter of the enhancer trap insertions. Thus, enhancer trap and gene trap mutagenesis allow isolation of genes expressed during *Arabidopsis* embryogenesis based on expression pattern.

Keywords. Gene trap, enhancer trap, Arabidopsis, embryogenesis

1 Introduction

A fundamental question in developmental biology concerns the molecular mechanisms underlying the establishment of polarity and body pattern. In plants, the stereotyped body organization of the seedling is laid down during embryogenesis, and may be viewed as the super-imposition of two patterns, one along the apical-basal or longitudinal axis, and one along the radial axis (Mayer *et al.*, 1991). *Arabidopsis* provides an excellent model system for the genetic dissection of pattern formation during embryogenesis, since, as in other crucifers, the cell division pattern is largely invariant. Also, numerous mutations affecting the body organization of the embryo have been described (reviewed by Mordhorst *et al.*, 1997). A serious problem in the analysis of embryo pattern formation is the shortage of molecular markers for specific cells and regions in the embryo (Jürgens, 1995). Such markers are important, because it is often difficult to establish cell-identity in mutant embryo backgrounds (Devic *et al.*, 1996; Vroemen *et al.*, 1996; Yadegari *et al.*, 1994). Molecular markers for specific cells or regions in the developing *Arabidopsis* embryo identified so far, include the *AtLTP1* gene, expressed in the embryo protoderm (Thoma *et al.*, 1994;

NATO ASI Series, Vol. H 104
Cellular Integration of Signalling Pathways in Plant Development
Edited by F. Lo Schiavo, R. L. Last, G. Morelli, and N. V. Raikhel

Vroemen *et al.*, 1996; Yadegari *et al.*, 1994), the *SCARECROW* gene, expressed in the endodermal cell lineage (Di Laurenzio *et al.*, 1996), the *STM* gene (Long *et al.*, 1996) and the *CLV1* gene (Clark *et al.*, 1997), expressed in the presumptive shoot apical meristem, and the *ATML1* gene, which is expressed in all cells of the embryo proper until the eight-cell stage, in the protoderm from the sixteen cell-stage until the late heart-stage, and in the L1 layer of the shoot apical meristem in the mature embryo (Lu *et al.*, 1996). GUS markers for regions in the *Arabidopsis* embryo, such as the root tip (*POLARIS*), cotyledons and shoot and root apices (*EXORDIUM*), and root cap (*COLUMELLA*) have recently been used to investigate mechanisms involved in establishing polar organization in *Arabidopsis* embryos and seedlings (Topping *et al.*, 1994; Topping and Lindsey, 1997).

As part of a strategy to identify molecular markers for specific cell-types and regions in the *Arabidopsis* embryo, we have undertaken a gene / enhancer trap insertional mutagenesis screen, using the *Ac* / *Ds*-transposon based system described by Sundaresan *et al.* (1995). The gene and enhancer trap elements carry a *GUS* reporter gene that can respond to *cis*-acting transcriptional signals at the site of integration. A particularly useful aspect of this system is that it allows the identification of genes not only by mutant phenotype, but as well by their expression pattern. Many genes that have no visible phenotype upon disruption, because they are functionally redundant or their mutant phenotype is only visible under certain conditions and would be missed in screens for mutant phenotypes (Goebl and Petes, 1986), may be identified by expression pattern in gene trap and enhancer trap screens. Moreover, gene trap and enhancer trap mutagenesis can identify genes that are essential in both the development of the early embryo and later development. The function of such genes in later development can be obscured by an early lethal phenotype (Mlodzik *et al.*, 1990; Springer *et al.*, 1995).

Here we describe the use of gene and enhancer trapping to identify genomic sequences that direct gene expression in the *Arabidopsis* embryo. The collection of insertions we have obtained provides a set of molecular markers for specific cell-types, tissues, organs and regions in the developing embryo, and can also be used to clone the corresponding genes.

1.1 Design of the gene and enhancer trap elements

The two-element transposon system used in this study employs *Ac* starter lines, homozygous for an immobilized *Ac* element, that are crossed to one of two different *Ds* starter lines, *DsG* or *DsE*, homozygous for a non-autonomous gene trap or enhancer trap *Ds* transposon, respectively. This system was developed by Sundaresan *et al.* (1995). Figure 1 outlines schematically the *Ac*, *DsG*, and *DsE* T-DNA vectors and transposons. The *Ac* element contains a CaMV 35S promoter-*Ac* transposase fusion that causes high frequencies of *Ds* excision *in trans* (Swinburne *et al.*, 1992), and is "wings-clipped", meaning that it cannot transpose because it lacks one of the *Ac* termini.

The gene trap element *DsG* is designed to detect expression of a chromosomal gene when inserted within the transcribed region. For this purpose, the *DsG* element contains a promoterless *GUS* gene, whose expression relies on transcription from the tagged chromosomal gene (Friedrich and Soriano, 1991; Gossler *et al.*, 1989; Kerr *et*

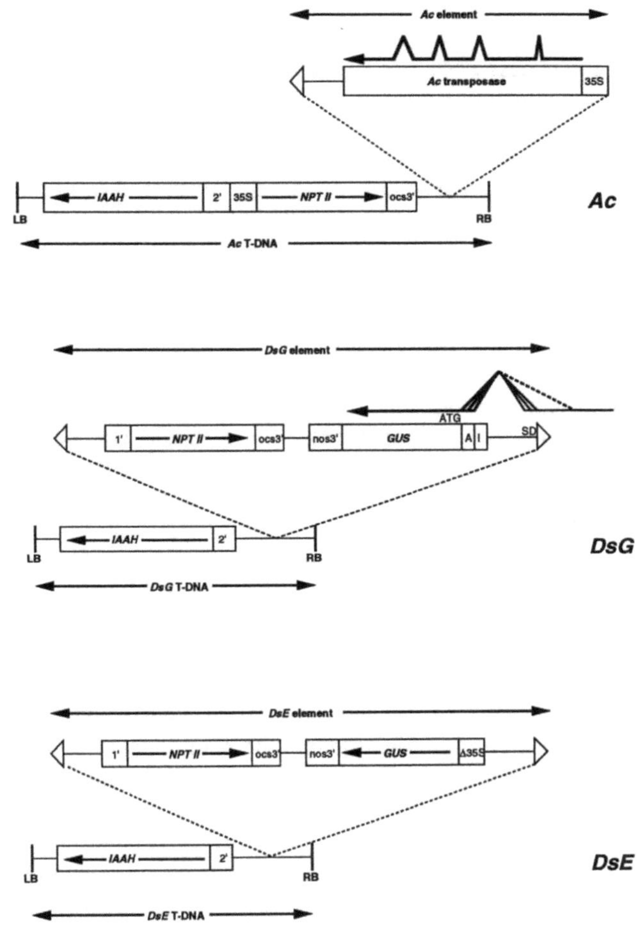

Fig. 1. T-DNA vectors and transposon constructs used in this study.
Ac : T-DNA carrying immobilized *Ac* element.
DsG : T-DNA carrying gene trap element *DsG*.
DsE : T-DNA carrying enhancer trap element *DsE*.

5' (drawn left) and 3' (drawn right) borders of the *Ds* element are represented as open triangles. (35S) CaMV 35S promoter; (*IAAH*) indole acetic acid hydrolase gene conferring sensitivity to NAM; (*NPTII*) neomycin phosphotransferase gene, conferring resistance to kanamycin; (1' and 2') 1' and 2' T-DNA promoter, respectively; (ocs3') octopine synthase terminator; (nos3') nopaline synthase terminator; (LB and RB) left border and right border sequences, respectively, of the T-DNA; (*GUS*) β-glucuronidase gene; (ATG) ATG-startcodon of *GUS* gene; (A) triple splice acceptor; (I) fourth intron of *Arabidopsis* G-protein gene *GPA1*; (SD) multiple splice donor sites at 3' end of *Ds*-element, covering all three reading frames; (Δ35S) -1 to -46 bp region of CaMV 35S promoter. Broken and dotted arrows in *DsG* represent splicing if insertion is into an intron or exon, respectively.
This figure has been adapted from Sundaresan *et al.* (1995).

al., 1989). Upstream of the *GUS* ATG startcodon, an intron of the *Arabidopsis GPA1* gene (Ma *et al.*, 1990) and a synthetic oligonucleotide containing two additional consensus splice acceptors have been fused, to provide for a splice acceptor in every reading frame. If *DsG* should insert into an intron, with the *GUS* gene in the same orientation as the tagged gene, as shown schematically in Figure 2A, splicing occurs from the splice donor of the chromosomal intron to the splice acceptors upstream of the *GUS* gene (Nussaume *et al.*, 1995). By contrast, if inserted in an exon in the correct orientation (Figure 2B), multiple splice donor sites, naturally existing at the 3' end of the *Ds* element and covering all possible reading frames (Wessler *et al.*, 1987), are exploited. The sequence between these splice donor sites and the splice acceptors just upstream of the *GUS* gene can be spliced out from the transcript, resulting in a fusion of the endogenous transcript and the *Ds* borne *GUS* transcript. The presence of three splice acceptor sequences, covering all three reading frames, in combination with multiple splice donors, also in each possible reading frame, ensures that for each possible reading frame the *DsG* element could insert into, at least one combination of splice donor and acceptor sequences generates an in-frame fusion between the endogenous and the *GUS* RNA (Nussaume *et al.*, 1995). As is clear from Figure 2C, no functional *GUS* fusion transcript is formed if the *DsG* element inserts into a gene in an orientation opposite to that of the tagged gene, which is expected to occur in half of the insertions.

With the *DsE* element, expression of the *GUS* reporter gene is dependent on *DsE* insertion near to chromosomal enhancer sequences. As shown in Figure 1, *DsE* contains a *GUS* gene fused to a minimal -1 to -46 bp CaMV 35S promoter, which is not active in the absence of enhancer sequences (Benfey *et al.*, 1989). When the *DsE* element inserts in the proximity of a chromosomal gene, within or outside of the coding region, *GUS* gene expression can be activated by a neighboring chromosomal enhancer (Bellen *et al.*, 1989; Bier *et al.*, 1989; Klimyuk *et al.*, 1995; O'Kane and Gehring, 1987; Sundaresan *et al.*, 1995; Wilson *et al.*, 1989). Since enhancers are known to act in an orientation-independent manner, insertions in either orientation may result in *GUS* expression (Figures 2D and 2E). In addition to the *GUS* reporter gene, both types of *Ds* elements carry a *NPTII* gene, which confers resistance to kanamycin as a selection marker.

1.2 Selection for plants carrying transposed *Ds* elements

A limitation of the *Ac* / *Ds* transposable element system for transposon tagging is the preferential transposition to sites that are closely linked to the donor locus (Bancroft and Dean, 1993; Belzile and Yoder, 1992; Dooner and Belachew, 1989; Greenblatt, 1984; Jones *et al.*, 1990; Keller *et al.*, 1993; Osborne *et al.*, 1991). In the selection scheme used in this study, local transpositions are eliminated, because donor-T-DNA located enhancers could directly cause *GUS* expression in lines carrying a *Ds* insertion close to the donor T-DNA (Klimyuk *et al.*, 1995). Moreover, by selecting against local transpositions, a more or less random distribution of *Ds* insertions throughout the genome can be obtained.

Figure 3 outlines the selection scheme used to generate lines carrying a stable *Ds* insertion at a location unlinked, or loosely linked, to the donor T-DNA. These lines are referred to as transposants (Bellen *et al.*, 1989), and their selection is

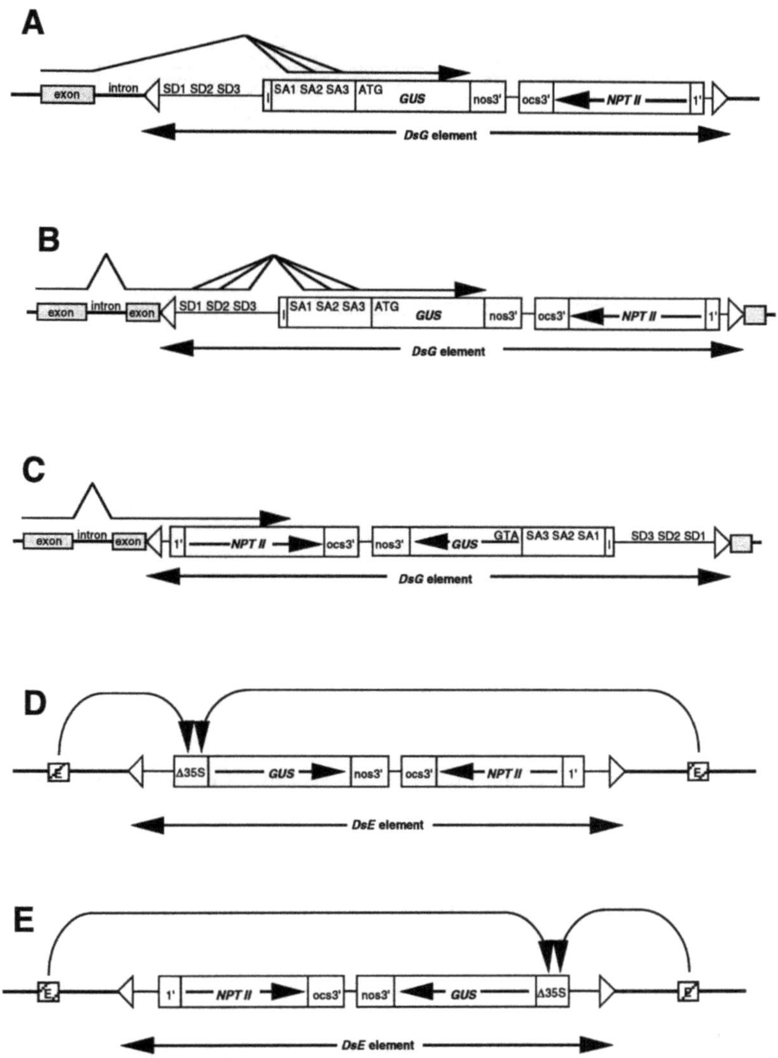

Fig. 2. Different possibilities for chromosomal insertion of *DsG* and *DsE* elements.

A: *GUS* expression from *DsG* element after insertion into an intron.

B: *GUS* expression from *DsG* element after insertion into an exon.

C: no *GUS* expression from *DsG* element due to insertion in opposite orientation.

D and E: *GUS* expression from *DsE* element after insertion near a chromosomal enhancer, independent of orientation.

(SD1, SD2, and SD3) splice donor sides at 3' end of *Ds* element, each in a different reading frame.

(SA1, SA2, and SA3) splice acceptor sites, each in a different reading frame.

(E) Chromosomal enhancer.

For further details see Figure 1. This figure has been adapted from Sundaresan *et al.* (1995).

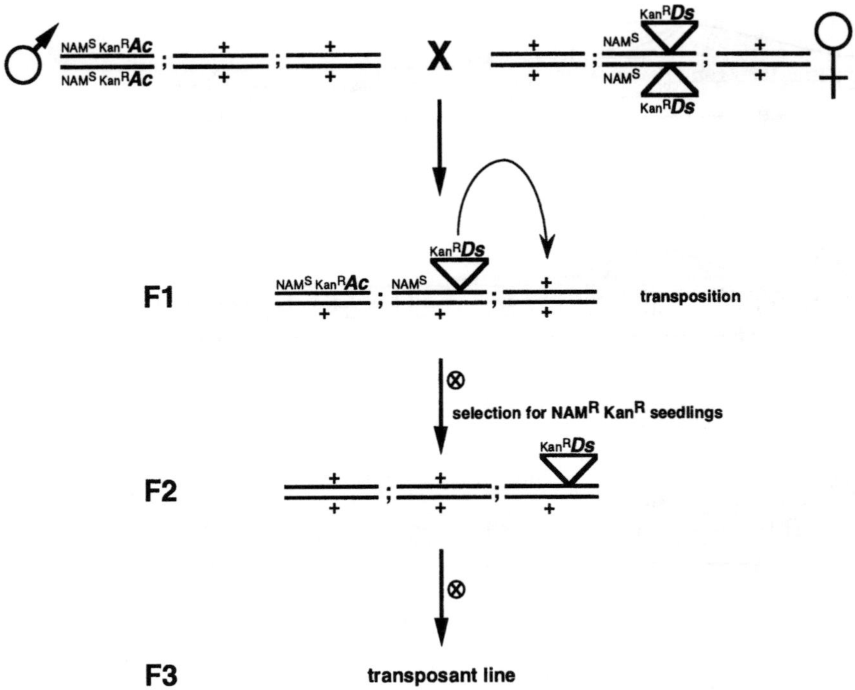

Fig. 3. Generation of transposants, i.e. lines carrying transposed *Ds* elements by selection for stable, unlinked transposition events.
(NAM^S) NAM sensitivity caused by *IAAH* gene.
(Kan^R) kanamycin resistance caused by *NPTII* gene.
Transposable element *Ds* is represented by a triangle.
This figure has been adapted from Sundaresan *et al.* (1995).

accomplished by selecting for the presence of the *Ds* element and, simultaneously, against both the *Ac* T-DNA and the *Ds* donor-T-DNA. Selection for the presence of the *Ds* element is possible due to the *NPTII* gene, conferring kanamycin resistance, that is located on both the *DsG* and *DsE* elements (Figure 1). Selection against the *Ac* and *Ds* donor-T-DNA is accomplished using the indole acetic acid hydrolase (*IAAH*) gene, present on the *Ac*-T-DNA, the *DsG* donor-T-DNA, and the *DsE* donor-T-DNA (Figure 1). The *IAAH* gene can be used as a counter selectable marker (Karlin-Neumann *et al.*, 1991), because it confers sensitivity to naphtalene acetamide (NAM), by converting it to the potent auxin naphtalene acetic acid (NAA), which causes severely stunted seedlings. Due to the selection against the presence of the *DsG* and *DsE* donor T-DNA, the recovery of transposed *Ds* elements depends on the recombination frequency between the *Ds* donor-T-DNA and the transposed *Ds* element. Only *Ds* elements that have transposed to locations unlinked or loosely linked to the donor-T-DNA are recovered.

Transposition of *Ds* elements is initiated by crossing *Ds* starter lines, homozygous for the *DsE* or *DsG* donor-T-DNA, to *Ac* starter lines, homozygous for the *Ac* T-

DNA (Figure 3). In the resulting F1 plants, *Ds* transposition can occur under the influence of *Ac* transposase. After allowing the F1 plants to self-fertilize, the F2 seed families are germinated on plates containing kanamycin and NAM, and the double resistant seedlings are recovered as transposants. Transposant lines are maintained as F3 seed batches, obtained by self-fertilization of selected F2 plants. By selecting against *Ac*, the *Ds* element in selected transposant plants will not be able to re-transpose. Thus, the transposant lines obtained this way represent a collection of stable *DsE* and *DsG* insertion lines that serves as source for further screening and characterization.

2 Materials and Methods

2.1 Mobilization of *Ds* elements

DsG elements were mobilized by crossing plants from *Ac* starter lines *Ac*1 and *Ac*2, both homozygous for the *Ac* T-DNA, to plants from *DsG* starter lines *DsG*1, *DsG*6, *DsG*7 and *DsG*8, all homozygous for the *DsG* T-DNA, in all possible pairwise combinations. In these crosses, *Ac* starter lines were used as the male and *Ds* starter lines as the female parental line. Likewise, *DsE* elements were mobilized by crossing plants from *Ac* starter lines *Ac*1 and *Ac*2 to plants from *DsE* starter lines *DsE*1, *DsE*2, *DsE*3 and *DsE*6, all homozygous for the *DsE* T-DNA, in all possible pairwise combinations. F1 seeds from the *Ac* x *DsG* and the *Ac* x *DsE* crosses were planted individually and the resulting F1 plants were allowed to self-fertilize. 1000-5000 seeds from each F1 plant were collected to establish independent F2-families.

2.2 NAM-Kan selection for transposants

750-1000 (15-20 mg) F2 seeds from each F2 family were surface-sterilized by successive washes with 70% ethanol for 10 min, diluted bleach solution (containing 0.9% sodium hypo chlorite, and 0.1% Tween 20) for 10 min, and twice with sterile water. The seeds were then suspended in 5 ml of liquid MS-agar (containing 0.46% (w/v) MS salts (Duchefa, (Murashige and Skoog, 1963) adjusted to pH 5.7 with KOH, 1% sucrose and 0.7% agar (Difco)), and plated onto square 12 x 12 cm selection plates containing MS-agar supplemented with 50 μg / ml kanamycin sulphate (Duchefa), and 3.5 μM NAM (α-naphtalene-acetamide, Sigma). After 1-4 days at 4 °C, the plates were incubated for 4 days in a growth chamber at 25 °C with 16h light / 8h dark photoperiod. Transposant seedlings resistant to both NAM and kanamycin, recognizable by their green cotyledons, normal size and normal root development, were transferred to 60 mm round selection plates and further incubated to verify the double resistance. After reaching the second-leaf stage, transposants were transplanted to soil and allowed to self-fertilize. Flowers and siliques, that contained immature seeds, from these F2 plants were screened for *GUS*-expression. Mature seeds (the F3 generation) were harvested and stored as a transposant line, i.e. a gene trap (referred to as Wageningen Gene Trap lines WGT1 through WGT373) or enhancer trap line (Wageningen Enhancer Trap lines WET1 through WET431).

2.3 Histochemical localization of *GUS* expression

For localization of *GUS* expression in seedlings, seeds from each gene trap and enhancer trap line were germinated in microtiter wells containing 400 µl of sterile water. After 5 days of incubation at 25 °C in the light, one volume of two times concentrated GUS staining solution was added, to make up final concentrations of 100 mM NaPi pH 7.2, 10 mM EDTA, 0.1% Triton X-100, 100 µg/ml chloramphenicol to inhibit bacterial growth, and 1 mg/ml X-Gluc (5-bromo-4-chloro-3-indolyl β-D-glucuronic acid). The seedlings were vacuum infiltrated with GUS staining solution for 1 hour, and the reaction was allowed to proceed for up to 48 hours at 37 °C in the dark. After the reaction, seedlings were cleared through several washes in 70% alcohol at 37 °C. GUS staining patterns were viewed using a Nikon binocular (Nikon Corp., Tokyo, Japan).

Localization of *GUS* expression in flowers, developing seeds and embryos was performed either directly in the F2 plant generation (i.e. mature transposant plants after transplantation from selection plates to soil) or in the F3 generation. Prior to planting in soil, F3 seeds were germinated on MS agar plates containing 50 µg/ml kanamycin to select for individuals carrying the gene or enhancer trap transposon. Flowers were sampled from the plants and incubated in GUS staining solution as described above. Siliques with immature seeds covering all stages of embryo development (typically 3 - 5 siliques per line) were sampled from the plants, opened longitudinally, and incubated in GUS staining solution as described above. After the reaction, flowers and siliques (containing immature seeds) were cleared for a minimum of 16 hours in Hoyers solution (100 g chloral-hydrate, 2.5 g Arabic gum, 15 ml glycerol, 30 ml water). Flowers and immature seeds were mounted in Hoyers solution on a microscope slide. GUS staining patterns were viewed with a binocular and with a Nikon Optiphot-2 equipped with Normarski optics. If GUS staining was observed in embryos, the staining reaction was repeated in GUS staining solution containing 1.25 mM, or even 5 mM each of potassium ferrocyanide and potassium ferricyanide (Jefferson *et al.*, 1987), to minimize diffusion of the reaction intermediates and thereby improve the specificity of the localization of *GUS* expression in embryos. Very weak GUS staining, however, was in most cases only visible in the primary staining reaction, i.e. in the absence of ferricyanide and ferrocyanide.

2.4 Histological sections

After the GUS staining reaction, immature seeds were transferred to FAA fixative (2% formaldehyde, 5% acetic acid, 65% ethanol). The fixative was vacuum infiltrated and the seeds were fixed for at least 3 days at 4 °C. After dehydration through an ethanol series, the seeds were infiltrated in Technovit 7100 resin (Heraeus Kulzer, Wehrheim, Germany) according to the manufacturer's instructions. In brief, subsequent changes of Technovit preparation solution (1 g hardener I, 2.5 ml PEG 400, 100 ml Technovit 7100) of increasing concentrations in 96% ethanol (1:3, 1:1, 3:1) were done for one hour, followed by a one hour and an overnight incubation in 100% Technovit preparation solution. Seeds were embedded in Technovit embedding was allowed to proceed for one hour at 37 °C. Serial sections (3 μm thick) were cut with a Reichert-Jung microtome, transferred to microscope slides, stained with 0.01% Ruthenium Red (Sigma) for 1-10 min, and mounted in Euparal (Agar Scientific, Stansted, UK). Sections were analyzed with a Nikon Optiphot-2 using bright-field and dark-field optics.

2.5 Southern blot analysis

Genomic DNA from individual transposant plants was isolated according to Bouchez *et al.* (1996). 1-2 μg of genomic DNA was digested with *Pst*I, separated on a 1% agarose gel and blotted onto a Nitran Plus membrane (Schleicher & Schuell, Keen, NH, USA). Blotting and hybridization were performed according to the manufacturer's recommendations. A 2.2 kb [α^{32}P-dATP] random prime labelled *GUS* fragment, covering the entire coding sequence, was used as probe. The blot was washed for 15 min with 2 x SSC, 0.1% SDS and for 15 min with 0.1 x SSC, 0.1% SDS at 65 °C (Sambrook *et al.*, 1989), before exposure to X-ray film (Amersham, 's Hertogenbosch, the Netherlands).

2.6 PCR-analysis

For PCR detection of the *Ds* element, either a set of *GUS* primers was used, with sequences GUS-1, 5'-AGA CTG TAA CCA CGC GTC TG-3' and GUS-2, 5'-CCG ACA GCA GTT TCA TCA ATC-3', or a combination of a *GUS* specific primer and a primer specific for the 3' end of the *Ds* element, with sequences GUS-4, 5'-GCT CTA GAT CGG CGA ACT GAT CGT TAA AAC-3' and Ds3, 5'-TAT TTA ACT TGC GGG ACG GAA ACG AAA AC-3'. For detection of both the *Ac* and the *Ds* donor T-DNA, *IAAH* specific primers were used, with the following sequences: NAM3, 5'-CAT TCC CCA CCT TGA CGA ACT G-3' and NAM4, 5'-GGT CTG AAT CCG CTA ATC CA-3'. PCR conditions for all primer pairs were 5 min at 94°C, followed by 35 cycles of 94°C (1 min), 55°C (1 min 30 sec) and 72°C (1 min 30 sec). PCR products were separated on a 1% agarose gel. Transposant plants are expected to be positive for the *GUS* gene, and negative for the *IAAH* gene, whereas *Ds* starter lines should be positive for both. *Ac* starter lines should be negative for *GUS* and positive for *IAAH.*

2.7 TAIL-PCR

Genomic DNA flanking *Ds* insertions was amplified by thermal asymmetric interlaced (TAIL) PCR, essentially as described by Liu *et al.* (1995). A set of three nested primers for the 5' end of *Ds*, Ds5-1, Ds5-2, and Ds5-3, was used in combination with one arbitrary primer, AD2 (Liu *et al.*, 1995), to amplify genomic DNA flanking the 5' end of either *DsG* or *DsE* insertions. The sequences of the primers are as follows: Ds5-1, 5'-CCG TTT ACC GTT TTG TAT ATC CCG-3'; Ds5-2 5'-CGT TCC GTT TTC GTT TTT TAC C-3'; Ds5-3, 5'-GGT CGG TAC GGA ATT CTC CC-3' and AD2, 5'-NGT CGA (G/C)(A/T)G ANA (A/T)GA A-3'.

After three subsequent rounds of TAIL-PCR, the primary, secondary and tertiary reaction products were separated on a 3% agarose gel. In successful reactions, the tertiary reaction product should be 71 bp smaller than the secondary product. Typically, the reaction products of successful amplifications ranged in size from 200 to 1500 bp. Secondary and tertiary reaction products were either sequenced directly using Ds5-2 and Ds5-3 as sequencing primer, respectively, or cloned into the pGEM-T vector (Promega, Madison, WI, USA) and subsequently sequenced using T7 and SP6 sequencing primers.

3 Results

3.1 Generation of independent gene and enhancer trap lines

Gene trap and enhancer trap *Ds* elements were induced to transpose by crossing plants from *DsG* or *DsE* starter lines to plants from *Ac* starter lines. Table 1 outlines the different combinations of *Ac* and *DsG* or *DsE* starter lines used, and the numbers of independent F2 families that resulted from crosses of each combination. Each of two *Ac* starter lines were crossed to each of four *DsG* and each of four *DsE* starter lines, resulting in 2610 F1 plants heterozygous for *DsE* and *Ac*, and 1975 F1 plants heterozygous for *DsG* and *Ac*. Self-fertilization of the F1 plants and collection of F2 seed from each individual F1 plant yielded *DsG* and *DsE* F2 families. In the F2 generation, selection was performed by plating approximately 750-1000 F2 seeds on NAM-Kan plates, as described in Materials and Methods. Out of 1975 *DsG* F2 families, 373 independent gene-trap lines were established. Out of 2610 *DsE* F2 families, 431 independent enhancer-trap lines were recovered. Thus, 19% of the *DsG* carrying F1 plants (373/1975) and 17% of the *DsE* carrying F1 plants (431/2610) generated $NAM^R Kan^R$ F2 progeny (transposants) in a frequency high enough to allow detection of at least one transposant per selection plate.

Over 80% of the F1 plants had no $NAM^R Kan^R$ progeny, suggesting that in these plants there were either no *Ds* transpositions, only transpositions without re-integration, or only transpositions to sites closely linked to the donor locus, which are known to occur frequently in the *Ac-Ds* transposon system (Bancroft and Dean, 1993; Belzile and Yoder, 1992; Dooner and Belachew, 1989; Greenblatt, 1984; Jones *et al.*, 1990; Osborne *et al.*, 1991). F1 plants that did have $NAM^R Kan^R$ progeny typically generated between 0.1% and 3% (i.e. 1-30 per 1000 F2 seedlings) $NAM^R Kan^R$ F2 seedlings. In accordance with the results obtained by Sundaresan and co-workers (Sundaresan *et al.*, 1995) , the majority of F1 plants that did have $NAM^R Kan^R$ progeny, yielded between 0.1% and 1% of $NAM^R Kan^R$ F2 seedlings. It should be noted that, since only one-sixteenth (6.25%) of the progeny of an F1 plant is expected to be NAM^R, and the frequency of forward transposition, defined as the proportion of F2 plants in which the *Ds* element has excised from the donor locus, with the 35S-*Ac* transposase fusion used ranges from 5 - 50% in the F2 generation (Long *et al.*, 1993), a single transposition event is unlikely to result in more than 3% of $NAM^R Kan^R$ progeny.

Table 1. Numbers of F2 families generated from different combinations of *Ac* and *DsE* or *DsG* starter lines, respectively, and numbers and frequencies of recovered transposant lines.

starter lines	No. of F2 families generated	No. of transposant lines	% of F2 families yielding a transposant line
DsE1 x Ac1	477	107	22
DsE1 x Ac2	584	65	11
DsE2 x Ac1	225	38	17
DsE2 x Ac2	455	55	12
DsE3 x Ac1	70	39	56
DsE3 x Ac2	657	93	14
DsE6 x Ac1	60	13	22
DsE6 x Ac2	82	21	26
Total	2610	431	17
DsG1 x Ac1	519	103	20
DsG1 x Ac2	409	82	20
DsG6 x Ac1	95	32	34
DsG6 x Ac2	349	20	6
DsG7 x Ac1	123	38	31
DsG7 x Ac2	419	89	21
DsG8 x Ac1	12	5	42
DsG8 x Ac2	49	4	8
Total	1975	373	19

3.2 Molecular analysis of transposants

To ascertain whether the NAM selection effectively selected against both the *Ac* T-DNA and the *Ds* T-DNA, and whether the kanamycin selection resulted in selection for inheritance of a *Ds* element, DNA from independent transposant seedlings was checked for the presence of the *IAAH* gene (which is located on both the *Ac* and *Ds* T-DNAs) and the *GUS* gene (which is located only on the *Ds* element). PCR using different sets of *IAAH* and *GUS* specific primers showed that all transposant lines contained the *GUS* gene, confirming inheritance of the *Ds* transposon, and not the *IAAH* gene, providing evidence for the absence of both *Ac* and *Ds* T-DNAs (data not shown). As expected, both the *GUS* gene and the *IAAH* gene were detected in *DsG* and *DsE* starter lines, whereas only the *IAAH* gene was detected in *Ac* starter lines. These results prove that the NAM-Kan selection procedure effectively selected for transposed *Ds* elements and against *Ac* and *Ds* T-DNAs. The selection against the *Ds* T-DNA results in selection against nearby re-insertion of the *Ds* element (Sundaresan *et al.*, 1995).

Figure 4 shows Southern blot analysis of DNA from 21 selected gene trap and enhancer trap transposant plants. Genomic DNA was digested with *Pst*I, which cuts once in the *NPTII* coding sequence within the *Ds* element. Probing with the entire *GUS* coding sequence showed that 20 of the transposants carried a single copy of the *Ds* element, whereas 1 transposant (WET115) contained 2 *Ds* elements inserted at different chromosomal locations. The different fragment lengths observed in the lanes in Figure 4 are indicative of the different chromosomal locations, and thus different flanking genomic sequences of *Ds* insertions in the selected transposants. A single transposed *DsG* or *DsE* element is essential for efficient screening of the transposant lines for specific *GUS* expression patterns, since more than one transposed *Ds* element, at different chromosomal locations, could result in overlapping or combined GUS expression patterns, under the control of regulatory regions at different sites in the genome. This would of course complicate interpretation of *GUS* expression patterns during the primary screen.

3.3 Screening transposants for *GUS* expression patterns

All 373 gene trap (WGT) transposants and 431 enhancer trap (WET) transposants were examined for *GUS* expression patterns at various stages of plant development: seedling, flower, silique, developing seed and embryo. The results of the GUS staining data are summarized in Table 2. 27% of the WGT lines (100 out of 373) displayed *GUS* expression at some stage of the plant life cycle. The frequency of *GUS*-expressing WET lines amounted to 81% (317 out of 431), much higher than the frequency of *GUS* expressing WGT lines. In 58% of the *GUS* expressing WET lines (184 out of 317), GUS staining was only found in pollen grains. Due to the high frequency of this staining pattern in this and other screens employing a minimal 35S-promoter-*GUS* fusion for detection of enhancer action (Klimyuk *et al.*, 1995), and taking into account other reports on possible artefactual GUS staining in pollen (Klimyuk *et al.*, 1995; Mascarenhas and Hamilton, 1992; Uknes *et al.*, 1993), lines displaying this "pollen only" staining pattern are put in a separate class. It is unclear

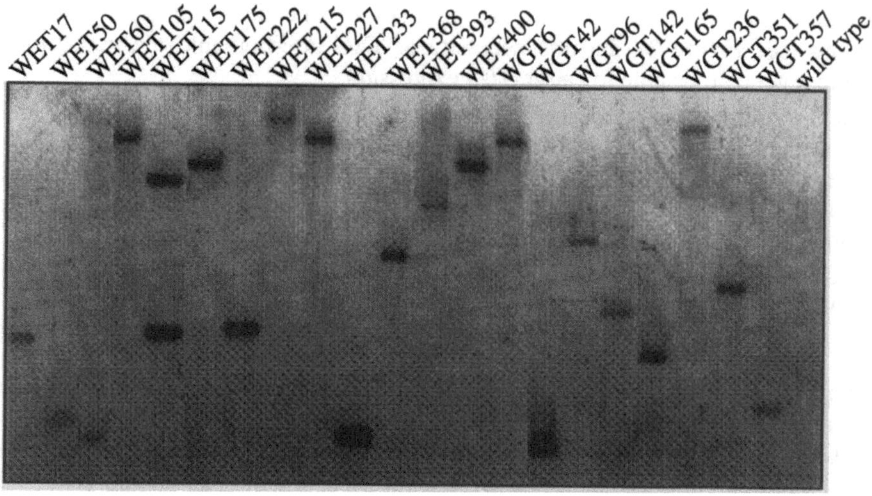

Fig. 4. Southern blot of DNA from 21 selected gene trap and enhancer trap transposants. The DNAs were digested with *Pst*I, and the probe used was the entire *GUS* coding sequence.

what the significance of the pollen GUS staining is. The fact that it occurs at much higher frequencies in enhancer trap lines than in gene trap lines (Table 2) suggests that it is not only the result of artefactual expression of the *GUS* gene independent of its genetic context, but that it can also be caused in some way by the minimal 35S promoter included in the enhancer trap *Ds* element. If the "pollen only" staining lines are excluded, the over-all GUS staining frequency of the WET lines (34%) is roughly similar to that of the WGT lines (25%). The lower staining frequency of the WGT

Table 2. Numbers and frequencies of *GUS* expressing WGT and WET lines.

	WGT-lines	WET-lines
No. screened	373	431
No. showing *GUS* expression (%)	100 (27%)	317 (81%)
No. showing "pollen only" expression	7	184
No. other than "pollen only" (%)	93 (25%)	133 (34%)

lines could be explained by both the orientation dependency of gene trap *Ds* insertions (see Figure 2), and the necessity of *DsG* insertion into the transcribed region of a gene, to be able to cause *GUS* expression.

Table 3 shows a summary of the frequencies of GUS staining patterns of WGT and WET lines found at different developmental stages and in different plant organs. Since our screen was primarily focused on detection of *GUS* expression in embryos and seedlings, GUS staining in other plant organs, such as the different flower organs, the silique and the seed coat, is put in a single category. The fact that "pollen only" staining occurs at high frequency among WET, but not among WGT lines, would complicate comparison of the staining frequencies in WGT and WET lines. Therefore, the class of "pollen only" staining lines is not included in Table 3. In a total of 39 lines, 12 WGT and 27 WET lines, *GUS* expression was detected in the embryo. The number of embryo-staining lines was rather low as compared to the total number of staining lines: only 13% of the GUS-positive WGT lines and 20% of the GUS-positive WET lines showed staining in the embryo. This corresponds to overall frequencies of 3-4% and 6-7% *GUS* expression in embryos, among gene and enhancer trap lines, respectively. Most of the lines that showed GUS staining in the embryo also had GUS staining in the seedling: 9 out of 12 for the embryo staining WGT lines, and 23 out of 27 for the WET lines. In some cases, the seedling staining

Table 3. Summary of *GUS* expression in WGT and WET lines at different developmental stages.

	WGT-lines		WET-lines	
Total no. of *GUS* staining lines[a]	93		133	
Stain observed in:				
embryo (% of total staining lines[a])	12	(13%)	27	(20%)
from which also in seedling		9		23
from which embryo-specific		0		1
seedling (% of total staining lines[a])	48	(52%)	80	(60%)
from which seedling specific		17		24
flower / silique / seed coat (% of total staining lines[a])	72	(77%)	102	(77%)

[a]Lines with "pollen only" expression are not included.

pattern corresponded precisely to that in the embryo, whereas in other cases it was either completely different or resembled the embryo pattern only partially (data not shown). None of the WGT lines, and only one WET line showed embryo-specific *GUS* expression, taken as *GUS* expression only detectable in the embryo and not at any other stage of development or in any other plant organ. In this line, WET393, *GUS* expression is restricted to the suspensor (see Figure 5), which is *senso stricto* not even part of the embryo proper. In all other embryo staining lines, *GUS* expression was not restricted to the embryo, but also seen at other developmental stages, or in more than one organ or tissue. Among this class are also lines with a specific staining pattern in the embryo, that is perpetuated in the seedling. Other lines show, for example, GUS staining in the embryo and also in one or more flower organs. *GUS* expression in different organs or at different developmental stages could point to genes that are expressed in different developmental programs, or towards genes which are expressed in similar cell types or tissues in different plant organs (Sundaresan *et al.*, 1995). The frequency of lines that show *GUS* expression in seedlings is higher than the frequency of embryo staining lines. Among both the WGT and WET lines, more than half of the staining lines show *GUS* expression at the seedling stage (52% and 60%, respectively), and in approximately one third of these (17/48 and 24/80, respectively), *GUS* expression is restricted to the seedling stage. The majority of the GUS positive WGT and WET lines (77% for both) shows GUS staining somewhere in the flower, silique and / or seed coat. It should be noted that these include lines with staining patterns in, for example, a single flower organ, as well as lines with expression in different organs, and lines that are also expressed at the embryo and / or seedling stage.

Figure 5 shows an example of an enhancer trap line with a restricted *GUS* expression pattern, that was found in our screen for GUS staining patterns in the embryo. In this line, WET393, *GUS* expression is restricted to the basal cells of the suspensor from the pre-globular to the late heart stage (Figure 5 A-C). In later stage embryos, from the torpedo stage up to the mature embryo stage, *GUS* expression was seen in the entire suspensor, but not in the embryo proper (Figure 5 D-F).

3.4 Screening transposants for mutant phenotypes

During our screen for *GUS* expression patterns in the embryo, we observed two WGT lines with morphological defects in the embryo (not shown). One of these lines segregated for embryos with strongly reduced cotyledons, and exhibited a good correlation between mutant phenotype and cotyledon specific *GUS* expression pattern. Our screening protocol for detection of *GUS* expression patterns in the developing embryo was optimised for detection of GUS staining in the embryo, and did not allow visualisation of subtle morphological aberrations in the embryo. This was caused by the fact that after the GUS staining reaction, clearing of the seed coat with Hoyers solution was much less efficient than without prior GUS staining. Therefore, it is likely that subtle phenotypic aberrations in the embryo were missed during our primary screen. A detailed screen of the GUS positive WGT and WET lines for phenotypically visible mutations in the developing embryo is currently being performed. In the case of gene trap lines, *DsG* insertion within the transcribed region of a gene is a prerequisite for *GUS* expression. Mutant embryo phenotypes caused by disruption of genes that are expressed in the embryo are, therefore, expected

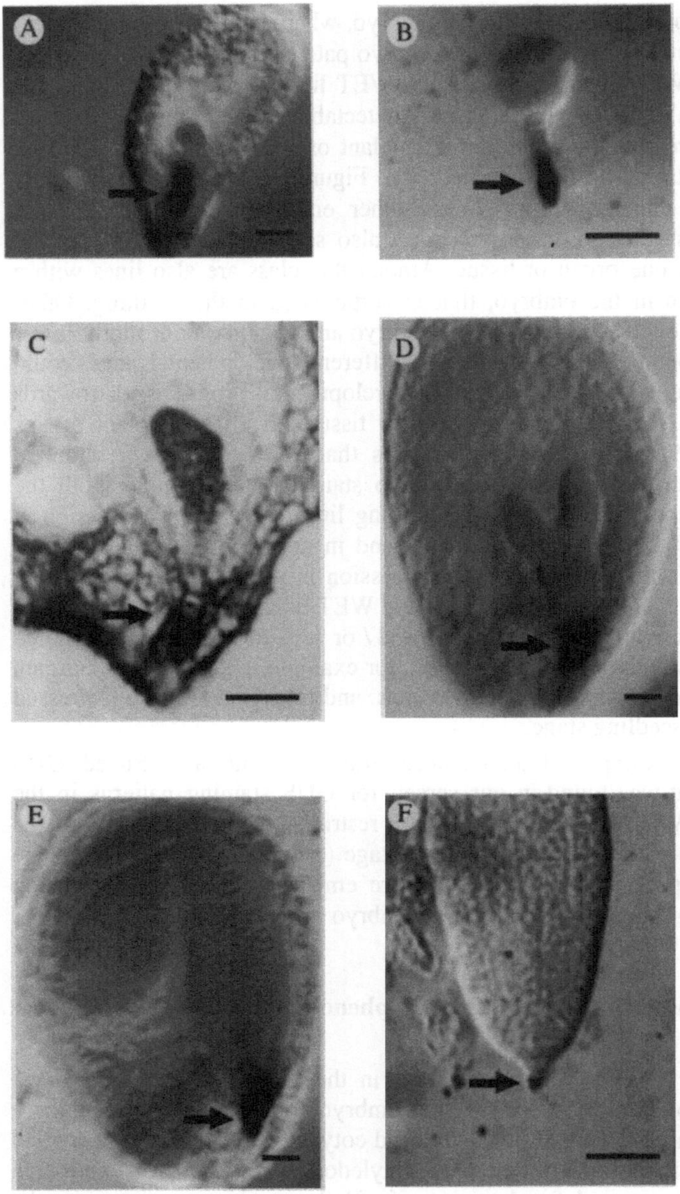

Fig. 5. *GUS* expression pattern of enhancer trap line WET393.

A: developing seed with globular stage embryo.

B: transition stage embryo.

C: histological section of a developing seed with transition stage embryo.

D: developing seed with early torpedo stage embryo.

E: developing seed with bent cotyledon stage embryo.

F: basal region of bent cotyledon stage embryo after dissection out of the seed. The upper suspensor cell is still attached to the embryo proper.

Arrow points at GUS stained cells in the basal part of the suspensor in A-C, and in the entire suspensor, but not in the embryo proper, in D-F. Bar = 50 μm.

to occur at higher frequencies among embryo-staining WGT lines than among embryo-staining WET lines.

Morphologically visible phenotypes were more easily observed at developmental stages later than the embryo stage. Although not studied in detail, putative mutant phenotypes were observed in 26 out of 366 WGT lines (7%), and in 16 out of 400 WET lines (4%). Assuming an equal transposition behaviour for the *DsG* and *DsE* transposons, the frequency of insertion into a gene, possibly causing a mutant phenotype, is expected to be equal among gene trap lines and enhancer trap lines. Putative mutants included pigmentation mutants (chlorotic leaves), dwarfs, plants with retarded development, male steriles, plants with aberrant floral morphology, bushy plants and plants with aborted seeds. Detailed analysis, including verification of cosegregation of the mutant phenotype and the *Ds* element is in progress for some of the putative morphological mutants.

3.5 Amplification of genomic DNA flanking *DsG* and *DsE* insertions

Genomic DNA flanking *DsG* and *DsE* insertions has been generated by TAIL-PCR (Liu *et al.*, 1995) for 27 of the 39 WGT and WET lines with *GUS* expression in the embryo. Typically, 50 bp to 1.5 kb of flanking DNA was obtained using this procedure. In our hands, the success rate of the TAIL-PCR procedure was on average 60%, meaning that in a successful experiment, flanking DNA was generated for approximately 60% of the lines. The sequence of flanking DNA from two gene trap lines showed that in both lines, the *DsG* element had inserted within the putative open reading frame of a gene. For three out of four enhancer trap lines from which flanking DNA was sequenced, no open reading frame could be detected in the sequence of the DNA flanking the *DsE* insertion. If an open reading frame is detected in the sequence of the flanking DNA, it may be possible to predict the function of the tagged gene from its sequence. This was the case for one of the two gene trap insertions of which flanking DNA was sequenced. Taken together, the TAIL-PCR procedure proved efficient in generating flanking DNA from WGT and WET insertions. In general, the generation of flanking DNA provides sequence information and molecular probes which can be used to determine the map position of the *Ds* insertions, and in library screening. For a selected number of lines, this work is currently in progress.

4 Discussion

In this study, we have used the *Ac* / *Ds* based gene / enhancer trap system developed by Sundaresan and co-workers (Sundaresan *et al.*, 1995) to detect genes that exhibit position-, cell-type- or tissue-specific expression in the *Arabidopsis thaliana* embryo. The modification of insertional mutagens to contain a reporter gene overcomes the limitation of conventional insertional mutagens that only allow detection of genes that give a phenotype when disrupted. The reporter gene carried by a gene or enhancer trap element permits the detection of genes by expression pattern rather than by mutant phenotype only. Previously, Fedoroff and Smith (1993), Smith and Fedoroff

(1995), and Klimyuk *et al.* (1995) have described the successful use of an enhancer trap *Ds* element for detection of plant enhancers. A serious limitation on the use of these elements as enhancer traps is imposed by the preferential transposition of *Ac* / *Ds* elements to closely linked sites. Enhancer sequences located on the *DsE* donor T-DNA could directly cause *GUS* expression in lines carrying a *DsE* insertion close to the donor T-DNA. Klimyuk *et al.* (1995) found that almost half of the lines that showed *GUS* expression revealed expression patterns similar or identical to those of the *DsE* starter lines they originated from. The parental GUS staining resulting from T-DNA located enhancers could partly or totally obscure novel GUS staining patterns conferred by endogenous enhancers. For this reason, and to obtain a more random distribution of *Ds* insertions across the genome, the mutagenesis scheme applied in this study includes a one-step plate selection against the *Ds* donor T-DNA, thus enriching for unlinked transpositions (Sundaresan *et al.*, 1995). At the same time, this procedure selects against lines carrying unexcised *Ds* elements, against the presence of the *Ac* T-DNA, and for re-insertion of the *Ds* element after excision from the donor T-DNA. This alleviates the need of large-scale PCR or Southern hybridization analyses, necessary in other systems (Klimyuk *et al.*, 1995) for selection of transposant lines. As an inevitable consequence of the relatively infrequent transposition of *Ds* elements to unlinked or loosely linked sites, only about one-fifth of the F1 plants yielded F2 progeny that survived our selection scheme: 19% for gene trap lines, and 17% in case of enhancer trap lines. These frequencies are somewhat lower than the frequency of plants with unique transposition events reported by Klimyuk *et al.* (1995), who obtained 87 of these plants starting from 314 F1 plants (28%). However, the mutagenesis scheme used in that study neither included selection against local transposition events, nor against the *Ac* T-DNA.

From the frequencies of recovered transposant lines resulting from crosses of each combination of starter lines (Table 1), it is apparent that particular combinations, such as for example *DsG*6 *Ac*2, and *DsG*8 *Ac*2, gave a very poor recovery of transposants. Such a poor recovery of $NAM^R Kan^R$ progeny could arise if the *Ac* and *Ds* T-DNA loci from two starter lines were linked in repulsion. In this case, F2 progeny would only survive NAM-Kan double selection if recombination would occur between the two loci. From recombinant inbred mapping (Lister and Dean, 1993), it is known that *DsG*1 , *DsE*1, *DsE*2, and *DsE*3 are not linked to either *Ac*1 or *Ac*2 (Sundaresan *et al.*, 1995). For the other starter lines used in this study, such mapping data are currently unavailable. The fact that, in the case of the *DsG*6 *Ac*2 and *DsG*8 *Ac*2 combinations, only 14 transposant lines were recovered from nearly 400 F2 families, clearly points out the importance of testing each combination of *Ds* and *Ac* starter lines before proceeding with large scale mutagenesis. Determination of the map position of T-DNA donor loci might also help to select for more optimal combinations of *Ac* and *Ds* starter lines. In future large scale mutagenesis programs, the over-all yield of transposant lines might be improved by selecting combinations of starter lines based on the frequencies found in this study.

Molecular analysis of transposants obtained through the described selection procedure showed that 95% of the insertions generated by *Ds* transposition were single copies. This almost eliminates the possibility of detecting *GUS* expression patterns arising from more than one independent or multiple tandem insertions. This is in contrast to T-DNA promoter trap insertion methods (Kertbundit *et al.*, 1991; Koncz *et al.*, 1989; Topping *et al.*, 1994; Topping and Lindsey, 1995; Topping *et al.*, 1991), in which only 50-60% of the lines carried T-DNA at a single locus (Lindsey *et al.*, 1993). Moreover, these loci often contained rearranged or multiple tandem insertions of the T-DNA, which could influence reporter gene expression and complicate amplification of flanking DNA sequences. From the results of our screen, it is clear that the selection against local transpositions, and the fact that the vast majority of our lines has a single insert, greatly facilitates screening of transposant lines for *GUS* expression patterns.

4.1 *GUS* expression in gene / enhancer trap transposants

The fraction of transposants that showed *GUS* expression at some stage of the plant life cycle was 27% for gene trap lines and 81% for enhancer trap lines. After correction for the very abundant class of "pollen only" expressing lines found among enhancer traps, these frequencies were 25% for gene traps, and 34% for enhancer traps. Sundaresan and co-workers obtained a similar frequency of *GUS* expressing gene trap lines (26%) (Sundaresan *et al.*, 1995). Our frequency of 34% for *GUS* expression among enhancer trap lines is somewhat lower than the 48% reported by Sundaresan *et al.* (1995). This could be due to differences in the developmental stages at which transposants were essayed for *GUS* expression between the two different screens. Although not explicitly mentioned in their paper, Sundaresan and co-workers also found very frequent pollen staining among their collection of enhancer trap lines (R. Martienssen, pers. comm.).

Klimyuk *et al.* (1995) reported an over-all GUS staining frequency of 60% among a collection of enhancer trap lines. However, almost half of the *GUS* expressing lines revealed expression patterns similar or identical to the parental staining patterns, and besides those, the majority of the GUS-stained lines showed common staining patterns, such as pollen, stigma or stipule specific staining. Only a limited number of lines (11%) showed unique expression patterns. It could be possible that, in some lines, the high level of parental GUS staining obscures weaker levels of *GUS* expression under the control of endogenous enhancers.

In our collection of WGT and WET lines, the frequency of GUS staining in embryos was relatively low as compared to the total frequency of GUS staining lines. The overall frequencies of *GUS* expression in embryos (3-4% among WGT lines and 6-7% among WET lines) compare well with the 3-4% frequency of *GUS* expression in embryos obtained with a T-DNA-based promoter trap approach, in which lines with T-DNA insertions at a single locus were studied (Topping *et al.*, 1994).

The overall frequencies found by us for *GUS* expression in seedlings were 13% and 19%, for WGT and WET lines, respectively. This is significantly higher than the frequencies of *GUS* expression in embryos. The low number of lines that show *GUS* expression in the embryo may reflect a relatively low number of genes expressed in the embryo, or may partly be caused by a low expression level of the *GUS* reporter gene in some lines, in combination with a low accessibility of the embryo for the GUS substrate. Surprisingly, only one line out of 39 lines in which GUS staining was observed in the embryo, showed *GUS* expression during embryogenesis only, and not during post-embryonic development. This enhancer trap line has restricted *GUS* expression in the basal cells of the suspensor, and during later stages of embryogenesis in all suspensor cells, indicating that in fact no real embryo-specific

GUS expressing line was recovered. All other lines displaying *GUS* expression in the embryo also show GUS staining at other developmental stages, such as the seedling, flower, silique, or seed coat. These include lines with similar expression patterns in the embryo and the seedling, such as lines displaying hypocotyl or cotyledon specific staining in both embryo and seedling, and lines staining in tissues that exist in the embryo, seedling and mature plant, such as the shoot meristem and vascular tissue. In these lines, *GUS* expression seems to remain associated with a specific tissue or position in the plant throughout development. Thus, in our screen, far more tissue- and position-specific than embryo-specific expression patterns were detected. Examples of *Arabidopsis* genes exhibiting such tissue- or position-specific expression pattern include *AtLTP1* (Thoma *et al.*, 1994; Vroemen *et al.*, 1996), *ATML1* (Lu *et al.*, 1996) , *SCARECROW* (Di Laurenzio *et al.*, 1996), and *STM* (Long *et al.*, 1996).

The detection of lines with expression patterns in multiple organs was not surprising: such expression patterns could be explained by the occurrence of common cell types or activities, such as cell division (Springer *et al.*, 1995) or photosynthesis, in different organs, or by the repeated use of the same gene products in different developmental programs. Examples of the latter are commonly found in animal systems. For example, the components of the ras signalling pathway are involved in specifying cell fates during *Drosophila* embryogenesis, wing vein formation, eye development and oogenesis (e.g. reviewed by Ruohola-Baker *et al.* (1994)). So far, none of the genes identified in embryo mutant screens in plants, such as *Bio-1* (Schneider *et al.*, 1989), *EMB30/GNOM* (Busch *et al.*, 1996; Shevell *et al.*, 1994), *FUSCA-1* (Castle and Meinke, 1994), *PROLIFERA* (Springer *et al.*, 1995), *KNOLLE* (Lukowitz *et al.*, 1996), *STM* (Long *et al.*, 1996), *SCARECROW* (Di Laurenzio *et al.*, 1996), and *CLAVATA1* (Clark *et al.*, 1997), exhibit an expression pattern that is restricted to the embryo. A rare example of a gene expressed exclusively during early plant embryogenesis is the carrot *SERK* gene (Schmidt *et al.*, 1997), whose expression ceases after the early globular embryo stage.

4.2 Mutant phenotypes in gene / enhancer trap transposants

The frequency of aberrant phenotypes observed among the transposant lines was 4 and 7% for WET and WGT lines, respectively. This is consistent with frequencies found in previous screens using *Ac / Ds* systems in *Arabidopsis* (2.1%, (Bancroft *et al.*, 1993), 8%, (Altmann *et al.*, 1995), 5%, (Bhatt *et al.*, 1996)). This frequency is significantly lower than the frequency of transposant lines that show *GUS* expression. This can partly be attributed to the fact that the screening conditions applied were predominantly aimed at the detection of GUS staining and, in most cases, did not allow visualisation of very subtle phenotypic aberrations. Nevertheless, from our data it is clear that the frequency of mutants is low as compared to the frequency of expression patterns. This difference is not surprising for a number of reasons. First, many insertions, especially enhancer trap insertions, may result in *GUS* expression without gene disruption, if the insertion is upstream, downstream, or in a non-essential region of the coding region of the gene. In these cases, the generation of mutant alleles could be possible by inducing secondary transposition using *Ac*. Second, even in case of gene disruption, the resulting phenotype might be subtle, or only visible under non-standard growth conditions, and would therefore be missed in our screen. Finally, the tagged gene might be functionally redundant, so that even in case of a severe disruption of gene function, no phenotype arises. Together, these data imply that gene / enhancer trapping is particularly useful for the identification of genes that would be missed in genetic screens for mutant phenotypes.

4.3 Gene isolation

Analysis of the genomic regions responsible for the observed *GUS* expression patterns has not been completed yet, but it is evident that TAIL-PCR amplification of genomic DNA flanking the *DsG* and *DsE* insertions is a straightforward method. In this context, an advantage of gene traps over enhancer traps is that in *GUS* expressing gene trap lines, the *DsG* element should be inserted within the coding region of the gene of which expression is visualised by the *GUS* reporter gene. This greatly facilitates isolation of the tagged gene, and sequence information may directly allow prediction of its function (Springer *et al.*, 1995). In fact, both gene trap insertions from which we sequenced flanking DNA were inserted within the putative open reading frame of a gene, and for one of these, gene function could be predicted based on sequence homology to known genes. Although enhancer traps might also insert into the coding region of a gene, they do not rely on it for *GUS* expression. Flanking DNA sequences from four of our enhancer trap insertions revealed only one putative open reading frame. Little is known about the distance over which enhancers can act in plants. A large physical distance between an enhancer trap insertion and the gene(s) activated by the enhancer would seriously complicate gene isolation. An enhancer trap insertion conferring *GUS* expression in the root endodermis of *Arabidopsis* had inserted approximately 1 kb upstream of the *SCARECROW* gene, whose expression is also restricted to the root endodermis (Di Laurenzio *et al.*, 1996). This suggests that isolation of genes detected by enhancer trapping would be feasible using flanking DNA probes for genomic library screening, followed by expression analysis of the genes found to be close to the site of insertion.

4.4 Conclusions

We have used gene trap and enhancer trap transposons to detect genes expressed during embryo development in *Arabidopsis*. Based on the data presented here, it appears that gene and enhancer trapping are particularly useful in the study of embryo development in different ways. First, the *GUS* expression patterns represent markers for specific cell-types, tissues, organs, and regions in the developing embryo. Such markers can be valuable for establishment of cell fate or position in embryo mutant backgrounds, and can supplement existing markers. Secondly, gene / enhancer trap insertions allow isolation of genes expressed during embryogenesis, without the requirement for a visible phenotype caused by gene disruption. This is particularly important if gene disruption does not cause a mutant phenotype, or if the mutant phenotype is embryo- or seedling-lethal. Thirdly, an advantage of gene / enhancer trapping over differential screening approaches is provided by the fact that gene / enhancer traps directly provide detailed information on the expression pattern of a gene, which can be an important criteria for selection of lines of interest. From our screen, enhancer traps appear to be more efficient in detecting expression patterns than gene traps. On the other hand, gene trap insertions only confer *GUS* expression if inserted into the transcribed region of a gene. This can greatly facilitate the cloning of the gene responsible for the observed *GUS* expression pattern, and may directly cause disruption of the tagged gene, possibly resulting in a mutant phenotype. By contrast, enhancer trap insertions that confer *GUS* expression might be outside of the coding region of the gene(s) activated by the enhancer. If so, gene disruption can only be achieved after remobilization of the *DsE* element, or by sequence based detection of a T-DNA or transposon insertion into the gene responsible for the observed *GUS* expression pattern. Taken together, the use of gene and enhancer trapping provides a powerful tool to dissect the molecular events involved in *Arabidopsis* embryogenesis.

5 Acknowledgements

We thank Rob Martienssen, Patricia Springer, Ueli Grossniklaus and Venkatesan Sundaresan for the *Ac, DsG,* and *DsE* starter lines, and for sharing their experience with the transposon system used in this study. We thank Marijke Hartog, Ellen Meijer and Henk Kuiper for assistance in the generation of transposant lines, Bertrand Dubreucq for assistance in screening and TAIL-PCR, Tony van Kampen for sequencing, and Harald Wolkenfelt for assistance in screening. The research reported here was supported by a grant awarded by the Wageningen Agricultural University to Casper Vroemen and by a grant awarded by the National Science Foundation to Venkatesan Sundaresan.

6 Citations

Altmann, T., Felix, G., Jessop, A., Kauschmann, A., Uwer, U., Peña-Cortés, H. and Willmitzer, L. (1995) *Ac/Ds* transposon mutagenesis in *Arabidopsis thaliana*: mutant spectrum and frequency of *Ds* insertion mutants. *Mol. Gen. Genet.* **247**, 646-652.

Bancroft, I. and Dean, C. (1993) Transposition pattern of the maize element *Ds* in *Arabidopsis thaliana. Genetics* **134,** 1221-1229.

Bancroft, I., Jones, J.D.G. and Dean, C. (1993) Heterologous transposon tagging of the *DRL1* locus in Arabidopsis. *Plant Cell* **5**, 631-638.

Bellen, H.J., O'Kane, C.J., Wilson, C., Grossniklaus, U., Pearson, R.K. and Gehring, W.J. (1989) P-element-mediated enhancer detection: a versatile method to study development in *Drosophila. Genes & Development* **3,** 1288-1300.

Belzile, F. and Yoder, J.I. (1992) Pattern of somatic transposition in a high copy *Ac* tomato line. *Plant J.* **2,** 173-179.

Benfey, P.N., Ren, L. and Chua, N.H. (1989) The CaMV 35S enhancer contains at least two domains which can confer different developmental and tissue-specific expression patterns. *EMBO J.* **8,** 2195-2202.

Bhatt, A.M., Page, T., Lawson, E.J.R., Lister, C. and Dean, C. (1996) Use of Ac as an insertional mutagen in Arabidopsis. *Plant J.* **9,** 935-945.

Bier, E., Vaessin, H., Shepard, S., Lee, K., McCall, K., Barbel, S., Ackerman, L., Carretto, R., Uemera, T., Grell, E., Jan, L.Y. and Jan, Y.N. (1989) Searching for pattern and mutation in the *Drosophila* genome with a P-*lacZ* vector. *Genes & Development* **3**, 1273-1287.

Bouchez, D., Vittorioso, P., Courtial, B., and Camilleri, C. (1996) Kanamycin rescue: A simple technique for the recovery of T-DNA flanking sequences. *Plant Molecular Biology Reporter* **14**, 115-123.

Busch, M., Mayer, U. and Jürgens, G. (1996) Molecular analysis of the *Arabidopsis* pattern formation gene *GNOM*: gene structure and intragenic complementation. *Mol. Gen. Genet.* **250**, 681-691.

Castle, L.A. and Meinke, D.W. (1994) A Fusca Gene of Arabidopsis Encodes a Novel Protein Essential for Plant Development. *Plant Cell* **6**, 25-41.

Clark, S.E., Williams, R.W. and Meyerowitz, E.M. (1997) The *CLAVATA1* gene encodes a putative receptor kinase that controls shoot and floral meristem size in Arabidopsis. *Cell* **89**, 575-585.

Devic, M., Albert, S. and Delseney, M. (1996) Induction and expression of seed-specific promoters in *Arabidopsis* embryo-defective mutants. *Plant J.* **9**, 205-215.

Di Laurenzio, L., Wysocka-Diller, J., Malamy, J.E., Pysh, L., Helariutta, Y., Freshour, G., Hahn, M.G., Feldmann, K.A. and Benfey, P.N. (1996) The *SCARECROW* gene regulates an asymmetric cell division that is essential for generating the radial organization of the Arabidopsis root. *Cell* **86**, 423-433.

Dooner, H.K. and Belachew, A. (1989) Transposition pattern of the maize element *Ac* from the *bz-m2(Ac)* allele. *Genetics* **122**, 447-457.

Fedoroff, N.V. and Smith, D.L. (1993) A versatile system for detecting transposition in *Arabidopsis. Plant J.* **3**, 273-289.

Friedrich, G. and Soriano, P. (1991) Promoter traps in embryonic stem cells: A genetic screen to identify and mutate developmental genes in mice. *Genes & Development* **5**, 1513-1523.

Goebl, M.G. and Petes, T.D. (1986) Most of the yeast genomic sequences are not essential for cell growth and division. *Cell* **46**, 983-992.

Gossler, A., Joyner, A.L., Rossant, J. and Skarnes, W.C. (1989) Mouse embryonic stem cells and reporter constructs to detect developmentally regulated genes. *Science* **244**, 463-465.

Greenblatt, I.M. (1984) A chromosome replication pattern deduced from pericarp phenotypes resulting from movements of the transposable element *modulator* in maize. *Genetics* **108**, 471-485.

Jefferson, R.A., Kavanagh, T.A. and Bevan, M.W. (1987) GUS fusions: ß-glucuronidase as a sensitive and versatile gene fusion marker in higher plants. *EMBO J.* **6**, 3901-3907.

Jones, J.D.G., Carland, F., Lim, E., Ralston, E. and Dooner, H.K. (1990) Preferential transposition of the maize element *Activator* to linked chromosomal locations in tobacco. *Plant Cell* **2**, 701-709.

Jürgens (1995) Axis formation in plant embryogenesis: cues and clues. *Cell* **81**, 467-470.

Karlin-Neumann, G.A., Brusslan, J. and Tobin, E. (1991) Phytochrome control of the *tms2* gene in transgenic *Arabidopsis*: A strategy for selecting mutants in the signal transduction pathway. *Plant Cell* **3**, 573-582.

Keller, J., Lim, E. and Dooner, H.K. (1993) Preferential transposition of *Ac* to linked sites in Arabidopsis. *Theor. Appl. Genet.* **86**, 585-588.

Kerr, W.G., Nolan, G.P., Serafini, A.T. and Herzenberg, L.A. (1989) Transcriptionally defective retroviruses containing *lacZ* for the in situ detection of endogenous genes and developmentally regulated chromatin. *Cold Spring Harbor Symp. Quant. Biol.* **54**, 767-776.

Kertbundit, S., DeGreve, H., Deboeck, F., van Montagu, M. and Hernalsteens, J.-P. (1991) *In vivo* random β-glucuronidase gene fusion in *Arabidopsis thaliana. Proc. Natl Acad. Sci. USA* **88**, 5212-5216.

Klimyuk, V.I., Nussaume, L., Harisson, K. and Jones, J.D.G. (1995) Novel GUS expression patterns following transposition of an enhancer trap *Ds* element in *Arabidopsis. Mol. Gen. Genet.* **249**, 357-365.

Koncz, C., Martini, N., Mayerhofer, R., Koncz-Kalman, Z., Korber, H., Redei, G.P. and Schell, J. (1989) High-frequency T-DNA-mediated gene tagging in plants. *Proc. Natl Acad. Sci. USA* **86**, 8467-8471.

Lindsey, K., Wei, W.B., Clarke, M.C., Mcardle, H.F., Rooke, L.M. and Topping, J.F. (1993) Tagging Genomic Sequences That Direct Transgene Expression by Activation of a Promoter Trap in Plants. *Transgenic Res.* **2**, 33-47.

Lister, C. and Dean, C. (1993) Recombinant inbred lines for mapping RFLP and phenotypic markers in *Arabidopsis. Plant J.* **4**, 745-750.

Liu, Y.-G., Mitsukawa, N., T., O. and Whittier, R.F. (1995) Efficient isolation and mapping of *Arabidopsis thaliana* T-DNA insert junctions by thermal asymmetric interlaced PCR. *Plant J.* **8**, 457-463.

Long, D., Martin, M., Sundberg, E., Swinburne, J., Puangsomlee, P. and Coupland, G. (1993) The Maize Transposable Element System Ac/Ds as a Mutagen in Arabidopsis - Identification of an Albino Mutation Induced by Ds Insertion. *Proc. Natl Acad. Sci. USA* **90**, 10370-10374.

Long, J.A., Moan, E.I., Medford, J.I. and Barton, M.K. (1996) A member of the KNOTTED class of homeodomain proteins encoded by the *STM* gene of *Arabidopsis. Nature* **379**, 66-69.

Lu, P., Porat, P., Nadeau, J.A. and O'Neill, S.D. (1996) Identification of a meristem L1 layer-specific gene in Arabidopsis that is expressed during embryonic pattern formation and defines a new class of homeobox genes. *Plant Cell* **8**, 2155-2168.

Lukowitz, W., Mayer, U. and Jürgens, G. (1996) Cytokinesis in the *Arabidopsis* embryo involves the syntaxin-related KNOLLE gene product. *Cell* **84**, 61-71.

Ma, H., Yanofsky, M.F. and Meyerowitz, E.M. (1990) Molecular cloning and characterization of *GPA1*, a G protein α subunit gene from *Arabidopsis thaliana. Proc. Natl. Acad. Sci.* **87**, 3821-3825.

Mascarenhas, J.P. and Hamilton, D.A. (1992) Artifacts in the localization of GUS activity in anthers of petunia transformed with a CaMV 35S-GUS construct. *Plant J.* **2**, 405-408.

Mayer, U., Torres Ruiz, R.A., Berleth, T., Miséra, S. and Jürgens, G. (1991) Mutations affecting body organization in the *Arabidopsis* embryo. *Nature* **353**, 402-407.

Mlodzik, M., Hiromi, Y., Weber, U., Goodman, C.S. and Rubin, G.M. (1990) The *Drosophila seven up* gene, a member of the steroid receptor gene superfamily, controls photoreceptor cell fates. *Cell* **60**, 211-224.

Mordhorst, A.P., Toonen, A.J. and de Vries, S.C. (1997) Plant Embryogenesis. *Critical Reviews In Plant Science* **in press,**

Murashige, T. and Skoog, F. (1963) A revised medium for rapid growth and bioassays with tobacco tissue culture. *Physiol. Plant.* **15**, 473-497.

Nussaume, L., Harrison, K., Klimyuk, V., Martienssen, R., Sundaresan, V. and Jones, J.D.G. (1995) Analysis of splice donor and acceptor site function in a transposable gene trap derived from the maize element Activator. *Mol. Gen. Genet.* **249**, 91-101.

O'Kane, C.J. and Gehring, W.J. (1987) Detection *in situ* of genomic regulatory elements in *Drosophila. Proc. Natl Acad. Sci. USA* **85**, 9123-9127.

Osborne, B.I., Corr, C.A., Prince, J.P., Hehl, R., Tanksley, S.D., McCormick, S. and Baker, B. (1991) *Ac* transposition from a T-DNA can generate linked and unlinked clusters of insertions in the tomato genome. *Genetics* **129**, 833-844.

Ruohola-Baker, H., Jan, L.Y. and Jan, Y.N. (1994) The role of gene cassettes in axis formation during *Drosophila* oogenesis. *Trends Genet.* **10**, 89-94.

Sambrook, J., Fritsch, E.F. and Maniatis, R. (1989) *Molecular cloning, a laboratory manual.* Cold Spring Harbor Laboratory Press, Cold Spring Harbor, NY, USA.

Schmidt, E.D.L., Guzzo, F., Toonen, M.A.J. and de Vries, S.C. (1997) A leucin-rich repeat containing receptor-like kinase marks somatic plant cells competent to form embryos. *Development* **124**, 2049-2062.

Schneider, T., Dinkins, R., Robinson, K., Shellhammer, J. and Meinke, D.W. (1989) An embryo-lethal mutant of *Arabidopsis thaliana* is a biotin auxotroph. *Dev. Biol.* **131**, 161-167.

Shevell, D.E., Leu, W.M., Gillmor, C.S., Xia, G.X., Feldmann, K.A. and Chua, N.H. (1994) EMB30 is essential for normal cell division, cell expansion, and cell adhesion in Arabidopsis and encodes a protein that has similarity to Sec7. *Cell* **77**, 1051-1062.

Smith, D.L. and Fedoroff, N.V. (1995) *LRP1*, a gene expressed in lateral and adventitious root primordia of *Arabidopsis. Plant Cell* **7**, 735-745.

Springer, P.S., Mccombie, W.R., Sundaresan, V. and Martienssen, R.A. (1995) Gene trap tagging of PROLIFERA, an essential MCM2-3-5-like gene in *Arabidopsis. Science* **268**, 877-880.

Sundaresan, V., Springer, P., Volpe, T., Haward, S., Jones, J.D.G., Dean, C., Ma, H. and Martienssen, R. (1995) Patterns of gene action in plant development revealed by enhancer trap and gene trap transposable elements. *Genes & Development* **9**, 1797-1810.

Swinburne, J., Balcells, L., Scofield, S.R., Jones, J.D.G. and Coupland, G. (1992) Elevated levels of *Activator* transposase mRNA are associated with high frequencies of *Dissociation* excision in Arabidopsis. *Plant Cell* **4**, 211-224.

Thoma, S., Hecht, U., Kippers, A., Botella, J., De Vries, S. and Somerville, C. (1994) Tissue-specific expression of a gene encoding a cell wall-localized lipid transfer protein from Arabidopsis. *Plant Physiol.* **105**, 35-45.

Topping, J.F., Agyeman, F., Henricot, B. and Lindsey, K. (1994) Identification of molecular markers of embryogenesis in Arabidopsis thaliana by promoter trapping. *Plant J.* **5**, 895-903.

Topping, J.F. and Lindsey, K. (1997) Promoter trap markers differentiate structural and positional components of polar development in Arabidopsis. *Plant J.* **in press,**

Topping, J.F. and Lindsey, K. (1995) Insertional mutagenesis and promoter trapping in plants for the isolation of genes and the study of development. *Transgenic Res.* **4**, 291-305.

Topping, J.F., Wei, W. and Lindsey, K. (1991) Functional tagging of regulatory elements in the plant genome. *Development* **112**, 1009-1019.

Uknes, S., Dincher, S., Friedrich, L., Negrotto, D., Williams, S., Thompson-Taylor, H., Potter, S., Ward, E. and Ryals, J. (1993) Regulation of pathogenesis-related protein-1a gene expression in tobacco. *Plant Cell* **5**, 159-169.

Vroemen, C.W., Langeveld, S., Mayer, U., Ripper, G., Jürgens, G., Van Kammen, A. and De Vries, S.C. (1996) Pattern formation in the Arabidopsis embryo revealed by position-specific lipid transfer protein gene expression. *Plant Cell* **8**, 783-791.

Wessler, S.R., Baran, G. and Varagona, M. (1987) The maize transposable element *Ds* is spliced from RNA. *Science* **237**, 916-918.

Wilson, C., Pearson, R.K., Bellen, H.J., O'Kane, C.J., Grossniklaus, U. and Gehring, W.J. (1989) P-element mediated enhancer detection: An efficient method for isolating and characterizing developmentally regulated genes in *Drosophila. Genes & Development* **3**, 1301-1313.

Yadegari, R., Depaiva, G.R., Laux, T., Koltunow, A.M., Apuya, N., Zimmerman, J.L., Fischer, R.L., Harada, J.J. and Goldberg, R.B. (1994) Cell differentiation and morphogenesis are uncoupled in Arabidopsis raspberry embryos. *Plant Cell* **6**, 1713-1729.

Clonal analysis in plant development

Francesco Salamini

Max-Planck-Institut für Züchtungsforschung, D-50829 Köln, Germany

Keywords: Clonal analysis, plant chimeras, plant sectoring, sector boundary analysis, pattern determination

Clonal analysis is of value in understanding the cellular dinamics of morphogenetic processes. In their review, Dawe and Freeling (1991) have grouped the methods useful in plants to this type of analysis, under the headings chimeras, spontaneous sectors, sectors induced at specific stages of development and sector boundary analysis. The presentation given at the course has been completed by adding principles and concepts of phenomena described as pattern determination (Spena and Salamini, 1995).

Plant chimeras are the visible manifestation of the existence of genetic differences within layered meristems. The chimeric state is lost during sexual reproduction, because plant gametes originate from the LII layer only. Plant chimeras have been widely used to understand development processes, establishing, for example, that 1) cell position has a relevant role in plant development; 2) the layered structure of plant apices; 3) the contribution of cell layers to the development of leaf primordia; 4) the ontogenetic variations of the contribution of cell layers to mature organs; 5) the influence of a cell layer on the neighbour cells.

Spontaneous sectors can be induced by several methods, but the excision of transposable elements is the phenomenon more useful to the scope. Spontaneous sectors have been analyzed in maize to follow the radially expanding development of cellular clones in the endosperm, using as marker genetic situations affecting amylose synthesis. In *Anthirrium majus* similar experiments have established that the determination of petal specific cells is cell autonomous. The excision of the *Ac* (*Activator*) element of maize from the *P* (*pericarp color*) locus has helped in understanding the contribution of LI and LII layers to the maize gynoecioum. The role of early cell division in maize leaf development has been established by using an epidermal marker and the transposition of the *Spm* (*Suppressor mutator*) transposable element (Cerioli et al., 1994).

In plants, sectors can be obtained at specific stages of organ development by using agents which induce mutations or chromasomal aberrations, like X-rays. Genes affecting authocyanin or chlorophyll pigmentation are the markers of use. The method has provided fate maps of cells of the embryo or of meristematic tissues and can estimate the

NATO ASI Series, Vol. H 104
Cellular Integration of Signalling Pathways in Plant Development
Edited by F. Lo Schiavo, R. L. Last, G. Morelli, and N. V. Raikhel

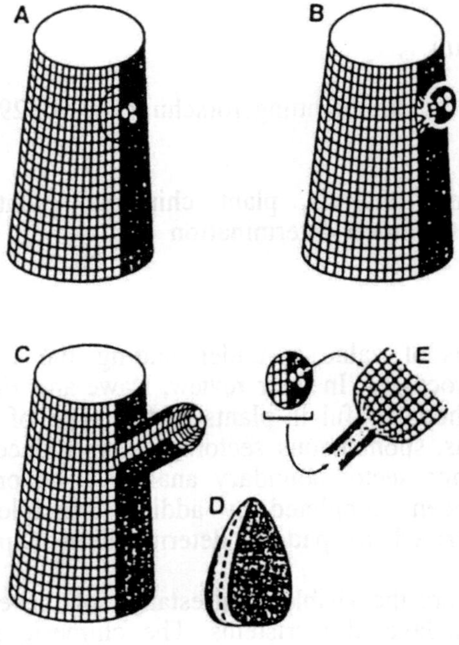

Fig. 1: Schematic representation summarizing the contribution of different cell lineages to the formation of the ear shoot of maize. Putative initials of the reproductive tissues of the ear are left uncoloured, as well as all tissues originating from them. In this figure the number of ear initials is 3. In the 24 cases studied, the cell lineages originating from the ear initials were either homogeneously green or yellow-green. The few cases of ear sectoring were due to the penetration of the ear by otherwise peduncle-specific cellular clones, or to late reversion events. A, B, C = successive stages of lateral meristem formation and of ear shoot development. Leaf, prophyll and husk primordia are omitted from the scheme. In A, founder cells of the lateral meristerm are all those included in the circle. In B, the cells included in the circle represens the founder cells of the prophyll and of the peduncle. They are formed both by green and yellow-green founder cells. This explains the origin of the sectored prophyll in D. E = Summary of the contribution of founder cells to the ear shoot. Founder cells contribute predominantly to the formation of the peduncle of the ear. The three white cells are the initials which, upon an apical type of growth, lead to the formation of the ear.

number of cells present in a primordium which will generate the incoming organ. Experiments based on the induction of sectors at specific stages have been useful to clarify the dinamics of formation of the apical meristem in the maize embryo and the contribution of embryogenic cells to adult structures (which is the base for the production of fate maps).

Sector boundary analysis was first proposed in *Drosophila*: when a mosaic border runs across areas important for morphogenesis, the structures originating at the border can be solid or variegated. When solid, they originate from one cell; when variegated, the number of cells at the base of the structure can be calculated as the reciprocal of the size of the smaller sector recorded. In plants, the method was first used by Christianson (1986) in the attempt to extend the concept of compartment of Garcia-Bellido et al. (1973) to the development of the cotton embryo. We have recently adopted the same procedure to understand the clonal formation of the lateral meristem generating in maize the ear shoot (Uhrig et al., 1996; Fig. 1).

In plant development, the absence of a sequestered germ-line, the continuous embryogenic state of meristems, and the late commitment of cells to a specific fate are typical phenomena. In these organisms the cellular feature with the most profound implications on plant cell growth and differentiation is the presence of cell walls. This condition prevents cell rotation and migration, and increases the morphogenetic relevance of early cell divisions that already define the adult cell arrangements. Morphogenetic events are thus necessary which create something functioning like cartesian coordinates in the orientation of the future symmetry of the plant. This "something" is called and treated as pattern determination. Studies on pattern determination during the development of the maize shoot and anthers and of the cotton embryo indicate that the plan of the first or the second cell division of founder cells can be adopted for the orientation of organ simmetry.

Citations

Cerioli S., A. Marocco, M. Maddaloni, M. Motto, F. Salamini, 1994. Early event in maize leaf epidermis formation as revealed by cell lineage studies. Development 120: 2113-2120.

Christiansen M., 1986. Fate maps of the organizing shoot apex in *Gossypium*. Am. J. Bot. 73: 947-958.

Dawe K.K., M. Freeling, 1991. Cell lineage and its consequences in higher plants. The Plant J. 1: 3-8.

Garcia-Bellido A., P. Ripoll, G. Morata, 1973. Developmental compartmentalization of the wing disc of *Drosophila*. Nature New Biol. 245: 251-253.

Spena A., F. Salamini, 1995. Genetic tagging of cells and cell layers for studies of plant development. Methods in Cell Biol. 49: 331-354.

Uhrig H., A. Marocco, H.P. Döring, F. Salamini, 1996. The clonal origin of the lateral meristem generating the erar shoot of maize. Plant 201: 9-17.

Laser Ablation In Arabidopsis Roots: A Tool To Study Cell-To-Cell Communication

Claudia van den Berg[1], Willem Hage[2], Peter Weisbeek[1] and Ben Scheres[1]

[1]Dpt. of Molecular Cell Biology, Utrecht University, Padualaan 8, 3584 CH Utrecht, The Netherlands
[2]Netherlands Institute for Developmental Biology, Uppsalalaan 8, 3584 CT Utrecht, The Netherlands

Keywords: laser ablation, root meristem, cell fate, *Arabidopsis*

1 Introduction

Developmental processes have been widely studied by manipulations of cells and tissues in intact organisms. These techniques are only in the last decades being complemented by the powerful approach of genetic manipulations. In animals, several experimental tools have been deployed that are directed towards unravelling the processes that cause differential cell fate. First, transplantation experiments of cells or tissues to heterotropic sites can reveal many inductive processes and a well known example was the 1924 Spemann and Mangold experiment using two differently pigmented species of newt. A transplanted dorsal blastophore lip into a host embryo induced a second embryo jointed to its host, showing that axis formation reveals multiple inductive processes (reviewed by Gilbert, 1988). Second, disconnection of blastomeres at early embryonic stages shows whether early fate determinants are spatially separated (Laufer et al., 1980; Priest and Thomson, 1987). Third, injection of vital dyes in single cells labels its descendants (clonal analysis), and reveals at what stage cells are committed or still can contribute to distinct cell types (Keller 1975, 1976).

A fourth method to unravel signalling processes involved in cell determination is the selective killing of cells in a developing organism. This powerful tool is widely used in *C. elegans*, which has a regular cell lineage, and for which a precise cell fate map has been established (Sulston and White, 1980; Sulston et al., 1983; Bowerman et al., 1992; Mello et al., 1992; Chamberlin and Sternberg, 1993; Hutter and Schnabel, 1995). Ablation of the anchor cell for instance, demonstrated that this cell induces the formation of the vulva (reviewed by Horvitz and Sternberg, 1991).

In vivo manipulations have also been applied in the analysis of plant development. For example, isolated single differentiated phloem cells of a mature carrot plant can give rise to an entire new plant, showing that at least some plant cells have totipoteny (Stewart et al., 1964). Bisecting the shoot meristem of *Vicia faba* by a vertical incision resulted in the regeneration of two new meristems (reviewed by Steeves and Sussex, 1989). Furthermore, grafting of organs or parts of whole plants from one plant species onto another is a technique traditionally used by plant breeders to combine favourable properties. For instance, tasty grape species can be grafted onto species having strong

NATO ASI Series, Vol. H 104
Cellular Integration of Signalling Pathways in Plant Development
Edited by F. Lo Schiavo, R. L. Last, G. Morelli, and N. V. Raikhel

stems. Shoot meristems grafted to undifferentiated callus clumps induce xylem formation and auxin addition mimics this response (Wetmore and Rier, 1963). Removing the quiescent centre in maize roots and replacing the excised tissue with an auxin soaked bead showed that hormones can play a role in maintaining the integrity of an organ (Feldman, 1979). Taken together, a wealth of information has been gained, but these manipulations provided no information about developmental processes between single cells.

We developed a laser ablation system in plants to study fate determination at the single cell level. We used the *Arabidopsis thaliana* root as a model system. Due to its cellular regularity, the *Arabidopsis* root has proven to be a suitable organ to study mechanisms regulating pattern formation (Dolan et al., 1993; Galway et al., 1994; Masucci et al., 1996; Di Laurenzio et al., 1996; reviewed by Scheres et al., 1996 and Schiefelbein et al., 1997).

Here we discuss the laser ablation system that we have developed, to kill single cells in the seedling root. We describe plant growth conditions and the use of a fluorescent dye to sensitise cells for ablation and to monitor the dead cell and it's surrounding cells. We discuss this method next to a recently developed tool with similar effects: genetic ablation.

2 Materials and Methods

2.1 Plant growth conditions

Seeds of *Arabidopsis thaliana* ecotype 'Columbia' were surface-sterilised with 20% commercial bleach (5% sodium hypochloride) for 10 minutes and imbibed for 5 days at 4 °C in the dark in sterile water. Seedlings were grown on Petri plates containing 0.8% Duchefa plant agar. Plates were incubated in a near-vertical position in a plant chamber at 22 C, 80% humidity with a 16 hours light/8 hours dark cycle.

2.2 Laser ablations

For laser ablation experiments, we used a confocal laserscanning microscope (Bio-Rad, mrc-600, 25mW argon-ion laser) mounted on an inverted microscope (Zeiss Axiovert). Plants were placed in small chambers containing 10 µg/ml propidium iodide (Sigma) as follows: Two small coverslips (18*18 mm) were sealed on 3 sides with paraffin on a larger coverslip (24*50 mm) (Fig. 1). Plants were placed into these chambers, taking care that the cotyledons were not submerged. One chamber could contain 6-12 plants. Propidium iodide outlines all individual cells (see Fig. 3b) and presensitises the cells for ablation (see Results).

A thus mounted and labelled root was imaged on the confocal microscope. For the ablation procedure, an MPL macro (macro programming language, Bio-Rad, for the mrc-600) was called in. With a mouse cursor, a spot was chosen in the image where the ablation had to take place. Test ablations were performed using fluorescent labelled paper fibres, in which the position of the mirrors was corrected to obtain an exact match of the bleached spot with the screen co-ordinates. Pressing the mouse button, the macro parked the scanning mirrors at the defined spot. The laser shutter was opened and the laser beam was focused for one second on the defined location, using the 'park' option (ND filters were manually removed to obtain 100% laser power). Repeatedly calling in this macro made it possible to ablate more cells if wanted. When all roots in the chamber were treated, the whole chamber was placed in a small plastic humid box to limit evaporation. The roots were re-examined within an hour, to check whether the ablations were successful. The plants were transferred to Petri plates and grown for the desired time under the conditions described above. The plants could be re-examined repeatedly using the same procedure.

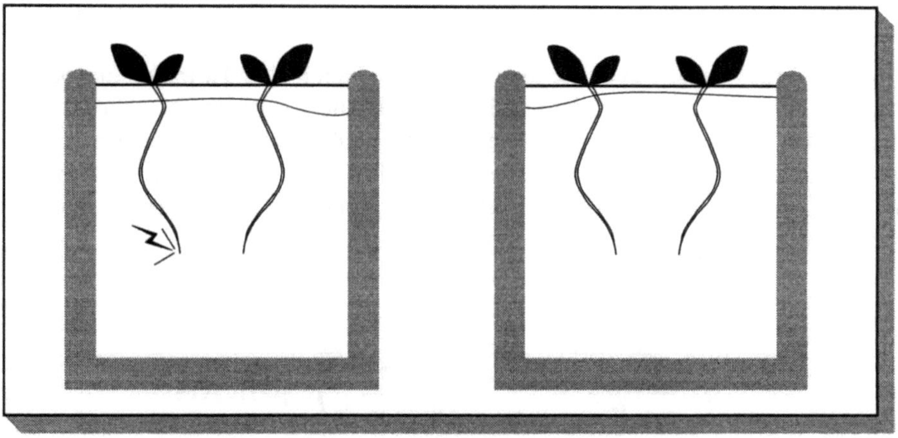

Fig. 1. Schematic drawing of the ablation chamber. Seedlings are placed in a chamber made of small coverslips sealed on a larger one, containing propidium iodide. The plants can thus be examined using an inverted microscope and ablated.

2.3 Light microscopy

Roots were stained in 0.5 mg/ml X-gluc (Biosynth AG) for 2 hours and fixed overnight in 70% ethanol. Plants were dehydrated and embedded in Technovit 7100 (Kulzer, Hereaus). Seedlings were embedded using the 'platelet' method as described (Scheres et al., 1994). 4 µm sections were made on a Reichert-Jung 1140 rotary microtome carrying a disposable Adamas steel knife. Sections were mounted in water and photographed using a Zeiss Axioscop fluorescence microscope (excitation: 450-490 nm; emission: above 520 nm), using Kodak Ektapress 400 film. Sections were then stained with 0.5% Astra Blue (Merck) or 0.05% Ruthenium Red (Sigma) in water and mounted in DePeX. Sections were photographed using a Zeiss Axioscop using Kodak Technical Pan film.

3 Results

3.1 Laser ablation system

In the *Arabidopsis* seedling root, the fate of every cell is predictable (Dolan et al., 1993; Scheres et al., 1994; Fig. 2). Furthermore, due to their transparency, *Arabidopsis* roots can be visualised with the confocal laserscanning microscope (CLSM) using transmission optics (Fig. 3a). This enabled us to develop a system to ablate individual, defined cells in *Arabidopsis* roots. The laser beam can be focused on a single cell (see Methods) and without sensitising dyes, a cell can be killed in about 10 minutes (not shown). This approach is very unpractical, for two reasons: i) movement of the root is very likely to occur during such a long time period, ii) the cellular pattern, although visible, is not very clear using transmission optics only. We tested different dyes to better visualise the cellular architecture of the root, and noticed that incubation in 10 µg/ml propidium iodide results in a very clear picture in which all the cells are outlined (Fig. 3b).

A possible disadvantage can be the toxicity of propidium iodide. However, incubation of plants in propidium iodide during the laser ablation experiment (ablation and checking, total time one hour or less) had no measurable effects on the plants. The growth rate and survival were similar to that of plants incubated in water or grown continuously on plant medium (Fig. 4). Therefore, propidium iodide has no apparent toxic effects on seedlings when used temporarily.

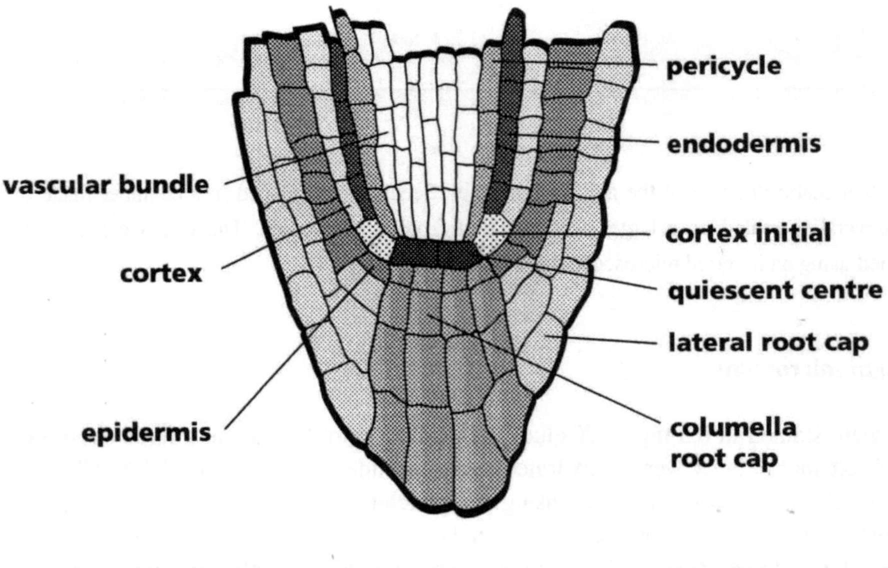

Fig. 2. Schematic representation of the *Arabidopsis* root. The cellular pattern is laid down during embryogenesis and both the fate and the embryonic progenitors of all the cells in the root are predictable. The initial cells are the most basal cells of all the cell types and contact the quiescent centre cells.

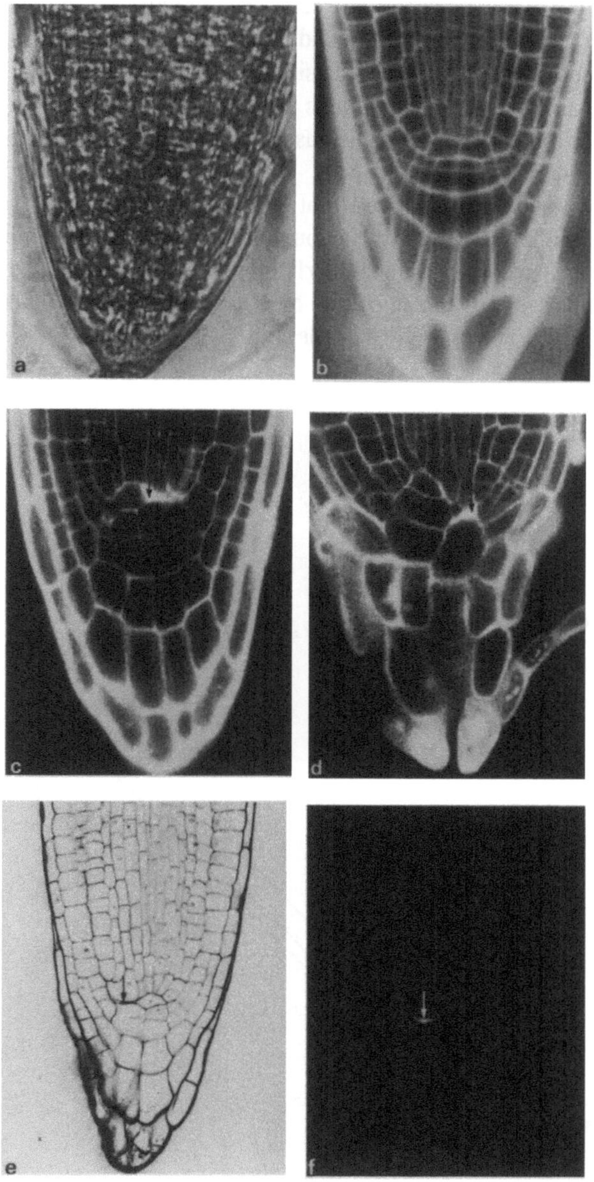

Fig. 3. a-d Scans of seedling root made with the CLSM. Roots are stained with propidium iodide. a, Transmission. b, Fluorescence. c, Ablation of a quiescent centre cell (arrow). Propidium iodide fills the ablated cell. d, Three days after ablation of a quiescent centre cell. The dead cell (arrow) is compressed to the outside of the root. Neighbouring cells invade the former position of the dead cell. e, f Section of root two days after ablation of a cell using light (e) and fluorescence (f) microscopy. The dead cell (arrows) is still visible after fixation and embedding (f).

We observed that the effectivity of laser pulses to kill cells was variable. Seedlings grown from the same seed batch and under the same conditions, show a variation in effectivity concerning cell ablation: plants were either sensitive or resistant to the cell ablation treatment. We have varied growth conditions, ablation times and the concentration of propidium iodine but obtained no increase in the ratio of sensitive versus resistant plants (data not shown).

Within the group of sensitive plants, we observed that propidium iodide greatly presensitises the cell for ablation: a one-second pulse is enough to kill a cell. Longer or shorter pulses (ranging from 0.5 till 3 seconds) were similarly effective as a one-second pulse. We noted an additional important advantage of this dye, it visualises intact and dead cells differently: propidium iodide penetrates dead cells (Fig. 3c).

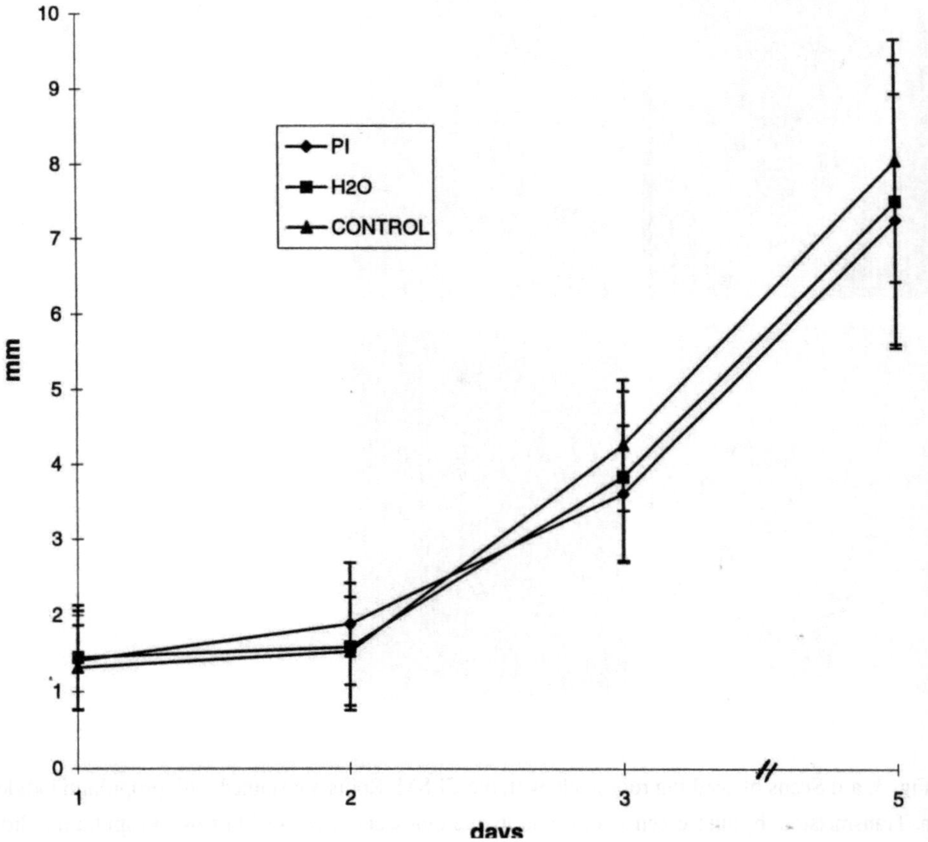

Fig. 4. Incubation of plants in propidium iodide has no effects on root growth. Plants were placed in containers containing 10 μg/ml propidium iodide, sterile water or remained on plant medium. After incubation for one hour, all plants were transferred to a fresh agar plate. Root growth after incubation was measured at different time points.

All these observations were taken into account to design optimal laser ablation experiments as follows: Plants are grown for two or three days on plant medium (see Methods). For the ablation experiments, plants are placed in the chamber made of coverslips (Fig. 1) containing propidium iodide. In this position the plants are examined with the confocal microscope. Ablation of cells is performed by focusing the laserbeam for 1 sec. on an individual cell. Following successful ablation, propidium iodide immediately enters the dead cell (Fig. 3c). The plants are then grown on normal agar medium for a desired amount of days and the roots are again examined similarly as described above. The ablated cell re-stains (Fig 3d) showing that this cell is still permeable to the dye. Moreover, the ablated cell becomes progressively more compressed and is ultimately pressed towards the outside of the root. After fixation, embedding and sectioning, the dead cell can still be visualised with a fluorescence microscope (Fig. 3f) and the cellular architecture of the root can be viewed using light microscopy (Fig. 3e). In conclusion, the laser pulse can selectively kill cells within the root, and the fate of these cells can be followed both by CLSM and histological methods.

3.2 Applications

The cellular organisation of the *Arabidopsis* seedling is set up during embryogenesis and is further perpetuated after germination. For this perpetuation, both a group of proliferating cells and cells that specifically differentiate into cell types are needed. The mechanisms that regulate these processes are thus far not understood. We wanted to investigate these processes by addressing the following questions.

First, how is cell fate in the root regulated? We know that a precisely defined invariant group of embryonic cells is responsible for generation of the root meristem (Scheres et al., 1994), thus cells could be restricted in their fate at embryogenesis. On the other hand, surgical experiments showed that plant cells are highly flexible (reviewed by Steeves and Sussex, 1989).

Second, how is the indeterminate nature of the root meristem maintained? The balance between cell proliferation and cell differentiation should be tightly regulated to guarantee continuation of root growth. We wanted to address these questions by applying specific laser ablations.

The activity of initial cells perpetuates the rigid cellular architecture of the *Arabidopsis* root meristem (Fig. 1). By characteristic divisions, they generate daughter cells that terminally differentiate. Ablation of an initial cell results in compression of the dead cell towards the outside of the root. Neighbouring cells of more internal layers invade the position of the dead cell and their fate can be studied using anatomical criteria and cell type specific markers. Ablation of a cortical initial cell results in the replacement of the dead cell by abutting pericycle initials (Fig. 5a, c). Pericycle initials are smaller than cortical initials so two pericycle cells invade into the former cortical initial position. These pericycle cells start making asymmetric periclinal divisions characteristic of cortical cells (compare Fig. 5b with 5c). Furthermore, staining for an endodermal specific property (the casparian strip) reveals that the progeny of invading pericycle cells have endodermal characteristics (van den Berg et al., 1995). Therefore pericycle initial cells switch fate when their relative position in the meristem is altered. Cortical and pericycle cells are clonally separated at early stages of embryogenesis (Scheres et al., 1994). This separation does apparently not lead to irreversible determination. On the contrary, meristem cells in the root appear flexible and their fate is spatially controlled (van den Berg et al., 1995).

The first zygotic division separates future vascular from future columella cells (Scheres et al., 1994). To determine whether vascular cells are able to switch to columella cell fate, we ablated the quiescent centre. The dead quiescent centre cells are pressed towards the root tip (Fig 6b, c, e, f) and vascular cells move in the position of the former quiescent centre and of the columella. In these vascular-derived cells, a vascular specific ß-glucuronidase enhancer trap construct (Fig. 6a) is not expressed (Fig. 6b). However, in these cells a 35S CAMV promoter B2 subdomain fused to ß-glucuronidase, which is normally expressed in columella cells (Fig. 6d), is switched on (Fig. 6e). This reveals that vascular cells switch fate according to positional cues.

In conclusion, laser ablations in the root show that positional signals determine the fate of cells.

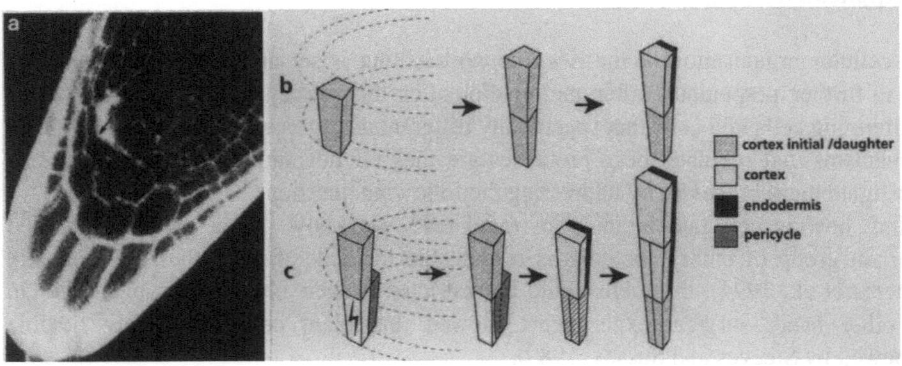

Fig. 5. a, CLSM scan of a root upon ablation of one cortical initial (arrow). The dead cell is pressed to the outside of the root and pericycle cells invade the position of the dead cell. b, Schematic drawing of the characteristic divisions of cortical initials and their daughters in wild type. Cortical initials divide anticlinally and generate daughters, which divide periclinally forming cortical and endodermal cells. c, Schematic drawing of ablation of a cortical initial. The dead cell is pressed aside and its position is occupied by pericycle cells. They switch fate and behave as cortical initials.

In the *Arabidopsis* root meristem, initial cells and their more mature daughters have a different differentiation status. An illustration of this is the starch granule localisation in the columella root cap. Starch granules are present in all columella cells except the columella initials (Fig. 7a). To analyse the basis of this differential differentiation status, we ablated again the quiescent centre. We performed starch staining when the dead cells are compressed but before they are pushed away. The columella initials now contain starch granules (Fig. 7b). Ablation of the quiescent centre apparently results in a change of the differentiation of columella initials. In conclusion, the quiescent centre controls the differentiation of the columella initials (van den Berg et al., 1997).

Taken together, our ablation experiments demonstrate the presence of several distinct cell-to-cell communication processes in the *Arabidopsis* root meristem.

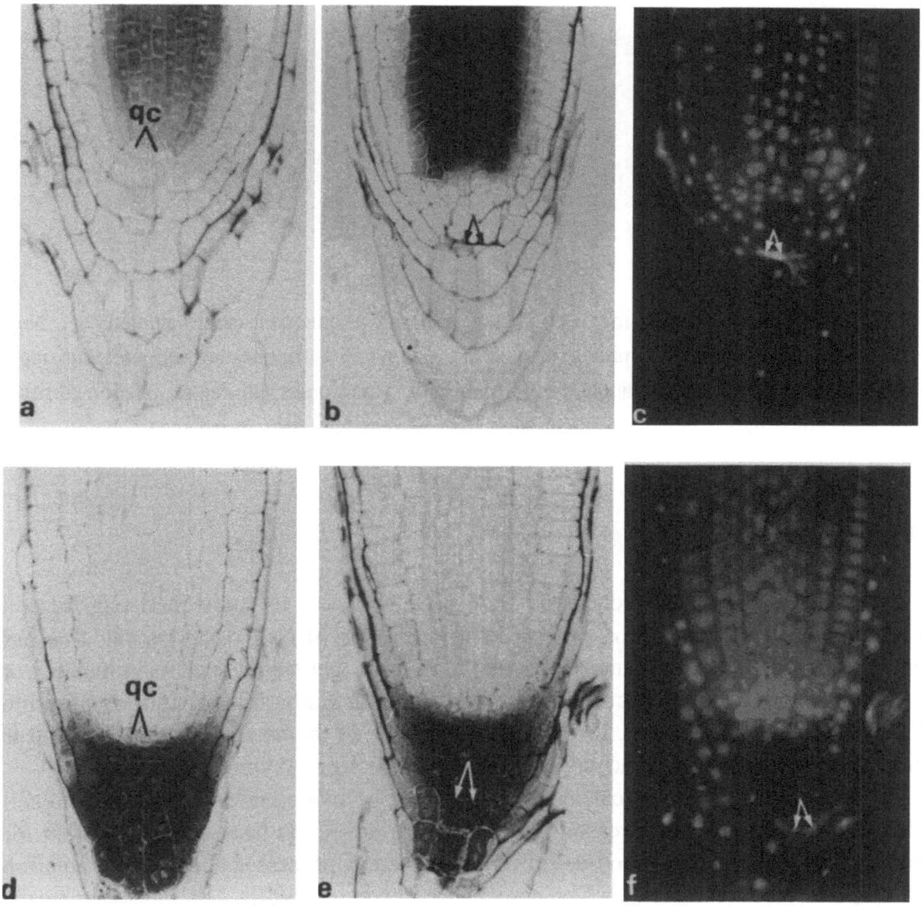

Fig. 6. β -glucuronidase reporter gene expression in wild type plants in a, vascular tissue and d, root cap. Following quiescent centre ablation, vascular cells move into the columella position and switch off the vascular marker (b) and turn on the columella marker (e) (arrows: dead cells). c, f, Similar section as b, e respectively, examined under the fluorescence microscope to visualise the dead cells (arrows).

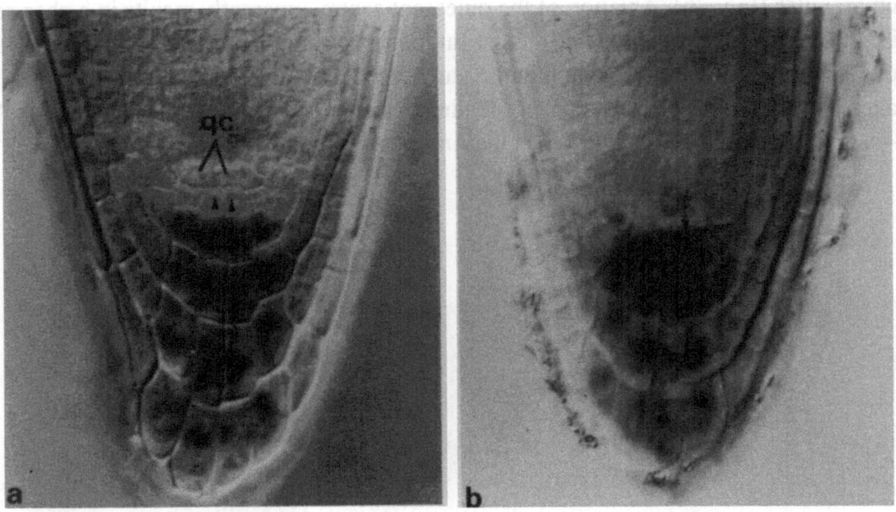

Fig. 7. Starch granule staining in a, non-ablated root and b, quiescent centre ablated root. Note that in wild type, columella initials do not express starch (arrowheads) and are thus more undifferentiated compared to daughter columella cells. Upon quiescent centre ablation (arrow), columella cells progress in their differentiation stage as shown by their starch accumulation.

4 Discussion

In the past, microsurgical techniques were used in plants to study meristematic cell programming and regeneration processes. For analysis of root development, this has been performed mainly in plant species that have large meristems, such as in maize. For example, various surgical excisions were performed to study the size and organisation of the root meristem (Clowes, 1953). This method bears the inherent difficulty that it is hard to verify whether the exact amount of cells of the correct types is excised.

The disadvantages of surgical manipulations can be overcome by more specific cell ablation techniques. Several reasons have led us to develop a laser ablation system for *Arabidopsis* roots to study cell-to-cell communication processes. First, laser ablation can be performed very rapidly. The complete procedure to kill a cell and examine the consequences is accomplished in only a few days. Second, laser ablations are very specific: single cells can be killed with a laser beam. Third, *Arabidopsis* roots are both transparent and small and therefore all target cells are accessible to the beam. Fourth, both the ancestors and the fate of all the cells in the root meristem are known. Therefore, by laser ablation the control of cell fate can be studied at single cell resolution.

We do not know the cause of cell death upon laser ablation of plants incubated in propidium iodide. Cell death could be a physical consequence of the heat caused by the laser beam due to propidium iodide quenching. On the other hand, entry of propidium iodide into the cell, induced by transient membrane leaking due to the laser pulse, could be toxic to the cell resulting in its death.

The argon-ion laser beam that we used has a relative low energy level and was sometimes unable to kill a cell. The use of a higher energy laser beam (e.g., a UV laser) appears to be more efficient, but a possible disadvantage is that too many cells may be targeted (F. Berger pers. comm.).

We have not observed any a-specific aberrant cell divisions on surrounding cells following ablation (van den Berg et al., 1995, 1997). However, surrounding cells could still elicit stress response reactions that cannot be detected by this criterion. In the future, a stress response could be monitored by transgenic plants containing suitable marker gene fusions, such as heat-shock or other stress-induced promoters.

Another way to kill cells is by genetic ablation (reviewed by Day and Irish, 1997). Laser ablation can be applied to remove one or a small group of cells, whereas genetic ablation is more suitable to kill a whole tissue or all cells of a given type. For genetic cell ablation two different tools are needed. First a cell-autonomous cytotoxic protein is required which causes cell death (Mariani et al., 1990; Koning et al., 1992; Day et al., 1995). These proteins mostly work by arresting protein synthesis in several ways (Pappenheimer, 1977; Yamaizumi et al., 1987; Siegall et al., 1989). Second, a cell type or tissue type specific promoter is needed to kill specific cells (Thorsness et al., 1991; Czakó et al., 1992; Kandasamy et al., 1993; Goldman et al., 1994; Day et al., 1995. Lately, an increasing number of specific promoters has become available through gene cloning and enhancer/promoter trapping (Topping et al., 1994; Sundaresan et al., 1995; Malamy and Benfey, 1997).

Possible side effects can however occur when the transgenic plant repeatedly activates the promoter that is hooked up with a cytotoxic gene during regeneration of missing tissue. More problems using genetic ablation can be the high basal level of the promoter (other cells are killed also), the temporal control of promoters (cells are killed too early or too late) or low expression in other cells.

The choice of laser or genetic ablation is largely dependent on the type of experiment. Genetic ablation is by large the most suitable method to kill many cells, to kill cells that are inaccessible to a laser beam and to kill a tissue that should be excluded permanently.

On the other hand, laser ablation is suitable if tissues are transparent, and does not require specific genetic tools, such as well-defined promoter sequences.

5 Acknowledgements

We thank Dick Smit and Frouke Kuyer for art work and Frits Kindt, Ronald Leitho, Wil Veenendaal and Piet Brouwer for photography.

6 Citations

Bowerman, B., Eaton, B.A. and Priess, J.R. (1992) *skn-1*, a maternally expressed gene required to specify the fate of ventral blastomeres in the early *C. elegans* embryo. Cell 68, 1061-1075.

Chamberlin, H.M. and Sternberg, P.W. (1993) Multiple cell interactions are required for fate specification during male spicule development in *Caenorhabditis elegans*. Development 118, 297-324.

Clowes, F.A.L. (1953) The cytogenerative centre in roots with broad columellas. New Phytol. 52: 48-57.

Czakó, M., Jang, J.-C., Harr Jr., J.M. and Márton, C. (1992) Differential manifestation of seed mortality induced by seed specific expression of the gene for diphtheria toxin A chain in *Arabidopsis* and tobacco. Mol. Gen. Genet. 235: 33-40.

Day, C.D., Galgoci, B.F.C. and Irish, V.F. (1995) Genetic ablation of petal and stamen primordia to elucidate cell interactions during floral development. Development 121: 2887-2895.

Day, C.D. and Irish, V.F. (1997) Cell ablation and the analysis of plant development. Trends in Plant Science 2, 106-111.

Di Laurenzio, L., Wysocka-Diller, J., Malamy, J.E., Pysh, L., Helariutta, Y., Freshour, G., Hahn, M.G., Feldmann, K.A. and Benfey, P.N. (1996) The *SCARECROW* gene regulates an asymmetric cell division that is essential for generating the radial organization of the *Arabidopsis* root. Cell 86: 423-433.

Dolan, L., Janmaat, K., Willemsen, V., Linstead, P., Poethig, S., Roberts, K. and Scheres, B. (1993) Cellular organisation of the *Arabidopsis thaliana* root. Development 119: 71-84.

Feldman, L.J. (1979) Cytokinin biosynthesis in roots of corn. Planta 145: 315-321.

Galway, M.E., Masucci, J.D., Lloyd, A.M., Walbot, V., Davis, R.W. and Schiefelbein, J.W. (1994) The *TTG* gene is required to specify epidermal cell fate and cell patterning in the *Arabidopsis* root. Dev. Biol. 166: 740-754.

Gilbert, S.F. (ed.) (1988) Developmental biology. 2nd. edn. Sinauer Associates 297-299.

Goldman, M.H.S., Goldberg, R.B. and Mariani, C. (1994) Female sterile tobacco plants are produced by stigma-specific cell ablation. EMBO J. 13: 2976-2984.

Horvitz, H.R. and Sternberg, P.W. (1991) Multiple intercellular signalling systems control the development of the *Caenorhabditis elegans* vulva. Nature 351: 535-541.

Hutter, H. and Schnabel, R. (1995) Establishment of left-right asymmetry in the *Caenorhabditis elegans* embryo: a multistep process involving a series of inductive events. Development 121, 3417-3424.

Kandasamy, M.K., Thorsness, M.K., Rundle, S.J., Goldberg, M.L., Nasrallah, J.B. and Nasrallah, M.E. (1993) Ablation of papillar cell function in *Brassica* flowers results in the loss of stigma receptivity to pollination. Plant Cell 5: 263-275.

Keller, R.E. (1975) Vital dye mapping of the gastrula and neurula of *Xenopus laevis*. I Prospective areas and morphogenetic movements of the superficial layer. Dev. Biol. 42: 222-241.

Keller, R.E. (1976) Vital dye mapping of the gastrula and neurula of *Xenopus laevis*. II Prospective areas and morphogenetic movements in the deep layer. Dev. Biol. 51: 118-137.

Koning, A., Jones, A., Fillatti, J.J., Comai, L. and Cassner, M.W. (1992) Arrest of embryo development in *Brassica napus* mediated by modified *Pseudomonas aeruginosa* exotoxin A. Plant Mol. Biol. 18: 247-258.

Laufer, J.S., Bazzicalupo, P. and Wood, W.B. (1980) Segregation of developmental potential in early embryos of *Caenorhabditis elegans*. Cell 19: 569-577.

Malamy, J.E. and Benfey, P.N. (1997) Organization and cell differentiation in lateral roots of *Arabidopsis thaliana*. Development 124: 33-44.

Mariani, C., de Beuckeleer, M., Truettner, J., Leemans, J, and Goldberg, R.B. (1990) Induction of male sterility in plants by a chimaeric ribonuclease gene. Nature 347: 737-741.

Masucci, J.D., Rerie, W.G., Foreman, D.R., Zhang., M., Galway, M.E., Marks, M.D. and Schiefelbein, J.W. (1996) The homeobox gene *GLABRA 2* is required for position-dependent cell differentiation in the root epidermis of *Arabidopsis thaliana*. Development 122: 1253-1260.

Mello, C.C., Draper, B.W., Krause, M., Weintraub, H. and Priess, J.R. (1992) The *pie-1* and *mex-1* genes and maternal control of blastomere identity in early *C. elegans* embryos. Cell 70, 163-176.

Pappenheimer, A.M., Jr (1977) Diphtheria toxin. Ann. Rev. Biochem. 46: 69-94

Priest, J.R. and Thomson, J.N. (1987) Cellular interactions in early *C. elegans* embryos. Cell 48: 241-250.

Scheres, B., Wolkenfelt, H., Willemsen, V., Terlouw, M., Lawson, E., Dean, C. and Weisbeek, P. (1994) Embryonic origin of the *Arabidopsis* primary root and root meristem initials. Development 120: 2475-2487.

Scheres, B., McKhann, H., van den Berg, C., Willemsen, V., Wolkenfelt, H., de Vrieze, G. and Weisbeek, P. (1996) Experimental and genetic analysis of root development in *Arabidopsis thaliana*. Plant and Soil 187: 97-105.

Schiefelbein, J.W., Masucci, J.D. and Wang, H. (1997) Building a root: The control of patterning and morphogenesis during root development. Plant Cell 9: 1089-1098.

Siegall, C.B., Chaudhary, V.K., FitzGerald, D.J. and Pastan, I. (1989) Functional analysis of domains II, Ib and III of *Pseudomonas* exotoxin. J. Biol. Chem. 264: 14256-14261.

Steeves, T.A. and Sussex, I.M. (1989) in: Patterns in plant development 2nd edn 91-97 (Cambridge Univ. Press, New York).

Steward, F.C. Mapes, M.O., Kent, A.E. and Holsten, R.D. (1964) Growth and development of cultured plant cells. Science 143: 20-27.

Sundaresan, V., Springer, P., Volpe, T., Haward, S., Jones, J., Dean, C., Ma, H. and Martienssen, R. (1995) Patterns of gene action in plant development revealed by enhancer trap and gene trap transposable elements. Genes Dev. 9: 1797-1810.

Sulston, J.E. and White, J.G. (1980) Regulation and cell autonomy during postembryonic development of *Caenorhabditis elegans*. Dev. Biol. 78, 577-597.

Sulston, J.E., Schierenberg, E., White, J.G. and Thomson, J.N. (1983) The embryonic cell lineage of the nematode *Caenorhabditis elegans*. Dev. Biol. 100: 64-119.

Thorsness, M.K., Kandasamy, M.K., Nasrallah, M.E. and Nasrallah, J.B. (1991) A *Brassica* S-locus gene promoter targets toxic gene expression and cell death to the pistil and pollen of transgenic *Nicotiana*. Dev. Biol. 143: 173-184

Topping, J.F., Agyeman, F., Henricot, B. and Lindsey, K. (1994) Identification of molecular markers of embryogenesis in *Arabidopsis thaliana* by promoter trapping. Plant J. 5: 895-903.

van den Berg, C., Willemsen, V., Hage, W., Weisbeek, P. and Scheres, B. (1995) Cell fate in the *Arabidopsis* root meristem determined by directional signalling. Nature 378: 62-65.

van den Berg, C., Willemsen, V., Hendriks, G., Weisbeek, P. and Scheres, B. (1997) Short-range control of cell differentiation in the *Arabidopsis* root. Nature (in press).

Wetmore, R.H. and Rier, J.P. (1963) Experimental induction of vascular tissues in callus of angiosperms. Am. J. Bot. 509: 418-430.

Yamaizumi, M. et al. (1978) One molecule of diphtheria toxin fragment A introduced into a cell can kill the cell. Cell 15: 245-250

HOMEODOMAIN-LEUCINE ZIPPER PROTEINS IN THE CONTROL OF PLANT GROWTH AND DEVELOPMENT

Giorgio Morelli[1], Simona Baima[1], Monica Carabelli[2], Manlio Di Cristina[1], Sabrina Lucchetti[1], Giovanna Sessa[2], Corinna Steindler[2], and Ida Ruberti[2].

[1] Unità di Nutrizione Sperimentale, Istituto Nazionale della Nutrizione, Via Ardeatina 546 00178 Rome Italy
[2] Centro di studio per gli Acidi Nucleici c/o Dip. di Genetica e Biologia Molecolare, Università La Sapienza , P.le Aldo Moro 5 00185 Rome Italy

Introduction

The homeobox (HB) was originally identified as a region of sequence similarity shared by several genes that when mutated caused homeotic transformations in Drosophila development (Gehring, 1987). Homeobox genes have subsequently been identified in many animal species and found to play key roles in diverse developmental processes, including the control of pattern formation in insect and vertebrate embryos, and the specification of cell fates in many tissues (Affolter *et al.*, 1990; McGinnis and Krumlauf, 1992). The HB sequence encodes a 60 amino acid sequence referred to as the homeodomain (HD), which is responsible for sequence-specific recognition of DNA (Gehring *et al.*, 1994).

In contrast with animal systems, the functions of homeobox genes in plants are just beginning to be elucidated. A large number of HB genes have been identified in several species including maize (Vollbrecht *et al.*, 1991; Bellmann and Werr, 1992, Kerstetter *et al.*, 1994), *Arabidopsis* (Ruberti *et al.*, 1991; Lincoln *et al.*, 1994; Rerie *et al.*, 1994; Quaedvlieg *et al.*, 1995; Reiser *et al.*, 1995), rice (Matsuoka *et al.*, 1993; Meijer *et al.*, 1997), parsley (Korfhage *et al.*, 1994), sunflower (Chan and Gonzales, 1994), barley (Muller *et al.*, 1995), carrot (Kawahara *et al.*, 1995), and tomato (Meissner and Theres, 1995; Tornero *et al.*, 1996). These HB genes can be distinct in two large groups encoding homeodomain and homeodomain-leucine zipper (HD-Zip) proteins, respectively. Some insight in the function of HD proteins came from the analysis of *KNOTTED-1* and *KNOTTED*-like genes in maize (Smith *et al.*, 1992; Schneeberger *et al.*, 1995). Their pattern of RNA accumulation suggested that they may define patterning events in the shoot apical meristem (Jackson *et al.*, 1994). A

NATO ASI Series, Vol. H 104
Cellular Integration of Signalling Pathways in Plant Development
Edited by F. Lo Schiavo, R. L. Last, G. Morelli, and N. V. Raikhel

role for HD proteins in meristem activity is further suggested by the recent finding that the *Arabidopsis SHOOT-MERISTEMLESS* gene, required to maintain a functional meristem, encodes a KNOTTED-like protein (Long *et al.*, 1996).

HD-Zip proteins seem to be unique to higher plants. DNA binding studies indicated that HD-Zip proteins interact with DNA recognition elements in a fundamentally different fashion from the classic HD proteins, thereby constituting a distinct class of regulatory proteins (Sessa *et al.*, 1993; Sessa *et al.*, 1997). A number of HD-Zip proteins have been identified in *Arabidopsis* and they are thought to constitute a large class of proteins. The HD-Zip proteins were grouped into four different families, named HD-ZIP I, II, III and IV (Sessa *et al.*, 1994; Figure 1). Here we review recent results obtained with selected members of the four HD-ZIP families.

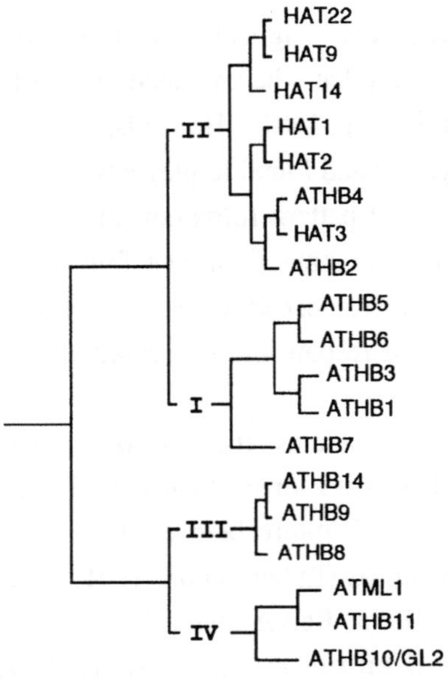

Figure 1. Dendrogram of the *Arabidopsis* HD-Zip sequences. The dendrogram shows the order of the pairwise alignment of the *Arabidopsis* HD-Zip proteins characterised so far. The distance along the horizontal axis is proportional to the differences between sequences; the distance along the vertical axis has no significance.
The alignment was made with the HD sequences. ATHB-1, -2 (Ruberti *et al.*, 1991; Carabelli *et al.*, 1993); ATHB-3 (Mattsson *et al.*, 1992); ATHB-4 (Carabelli *et al.*, 1993; Steindler *et al.*, 1997); ATHB-5, -6, -7 (Söderman *et al.*, 1994); ATHB-8 (Sessa *et al.*, 1994; Baima *et al.*, 1995); ATHB-9 (Sessa *et al.*, 1994; EMBL Data Library accession number Y10922); ATHB-10/GLABRA2 (Sessa *et al.*, 1994; Di Cristina *et al.*, 1996); ATHB-11 (Sessa *et al.*, 1994); ATHB-14 (EMBL Data Library accession number Y11122); HAT1, 2, 3, 7, 9, 14 (Schena and Davis, 1994); HAT4, 5, 22 (Schena and Davis, 1992, 1994); ATML1 (Lu *et al.*, 1996).

HD-Zip I protein family

ATHB-1 encodes for a protein of 272 amino acids in length, with a calculated molecular mass of 31 kDa (Ruberti *et al.*, 1991). *In vivo* and *in vitro* studies showed that ATHB-1 binds DNA as a dimer and it recognises a 9 bp pseudopalindromic DNA sequence (Sessa *et al.*, 1993; Aoyama *et al.*, 1995). Moreover, ATHB-1 was found to positively regulate reporter gene activity in particle bombardment experiments with tobacco and *Arabidopsis* leaves, confirming that the protein represents a transcription factor (Aoyama *et al.*, 1995; G. Sessa, unpublished results).

Expression studies revealed that *ATHB-1* is developmentally regulated during *Arabidopsis* growth. Moreover, the *ATHB-1* gene was found to be rapidly induced by wounding, flooding and ethylene treatments. RNA analysis showed that the *ATHB-1* gene expression increases in the leaf with plant age, reaching the highest steady state level at bolting. This increase is determined, in part, by ethylene as shown by expression analysis in *ein2-1*, *etr1-1* and *ctr1-1* mutants (S. Lucchetti, unpublished results).

In transgenic tobacco plants, overexpression of ATHB-1 or its chimeric derivatives with heterologous transactivating domains of the yeast transcription factor GAL4 or herpex simplex virus transcription factor VP16 conferred deetiolated phenotypes in the dark. Expression of ATHB-1 or the two chimeric derivatives also affected the development of palisade parenchima under normal growth conditions, resulting in light green sectors in leaves and cotyledons, whereas other organs in the transgenic plants remained normal. Both developmental phenotypes were induced by glucocorticoid in transgenic plants expressing a chimeric transcription factor comprising the ATHB-1 DNA binding domain (corresponding to the HD-Zip domain), the VP16 transactivating domain, and the glucocorticoid receptor domain. Moreover, the cotyledons of transgenic seedlings with severe inducible phenotypes seemed to have no palisade parenchima at all, and they showed abnormal expansion (Aoyama *et al.*, 1995).

All three phenotypes induced by overexpression of ATHB-1 or its chimeric derivatives are related to leaf development. It is likely that these phenotypes were caused by an ectopic activation of target genes containing specific cis elements recognized by the HD-Zip domain of ATHB-1, suggesting that the target genes of ATHB-1 are closely linked to leaf development. However, it is unclear whether the phenotypes are caused by ectopic expression of the same set of genes or of different genes which could be target of distinct HD-Zip proteins belonging to the HD-ZIP I family. To investigate the function of ATHB-1 further, we are currently characterizing *Arabidopsis* transgenic plants with elevated and reduced levels of ATHB-1. These studies should help to clarify the role of ATHB-1 in *Arabidopsis* leaf development.

Several genes encoding ATHB-1—related proteins have been identified in *Arabidopsis* (see Figure 1). Interestingly, one of them (*ATHB-7*) is strongly and rapidly induced by warer stress, by a mechanism which is dependent on abscisic acid as well as *ABI1*. It has been proposed that ATHB-7 might have a role in the control of plant growth in relation to water availability (Soderman *et al.*, 1996).

HD-Zip II protein family

ATHB-2 (also know as *HAT4*, Ruberti *et al.*, 1991; Schena and Davis, 1992) has been shown to be regulated by light quality; the light regulation of the *ATHB-2* gene is unique in that it is not induced by red-rich day light, nor by the light-dark transition, but is instead induced by changes in the ratio of red (R) to far-red (FR) light, both in seedlings and in adult plants (Carabelli *et al.*, 1993 & 1996; Steindler *et al.*, 1997). These changes which normally occur at dawn and dusk (end-of-day far-red) also occur during the day-time in the proximity of other vegetation and in canopy shade. The latter two situations induce the so-called shade-avoidance responses that produce dramatic chenges in the development of a plant. In fact, these situations result in a redirection of resources and growth potential from leaf and storage organs into increased extension growth of internode and petioles in a plant effort to optimize light capture (Smith, 1995).

By performing experiments in mutant plants lacking functional phytochrome A and/or phytochrome B, we have demonstrated that the *ATHB-2* gene is reversibly regulated by changes in the R:FR ratio largely through the action of a phytochrome other than A or B and secondarily by phytochrome B whose action is more obvious during the night than during the day (Carabelli *et al.*, 1996; Steindler *et al.*, 1997).

The phytochrome-mediated responses of *ATHB-2* in green plants strongly suggested a direct involvement of ATHB-2 in light-mediated growth phenomena during neighbour detection and shade avoidance (Carabelli *et al.*, 1993 & 1996; Steindler *et al.*, 1997). Consistently, plants overproducing ATHB-2 show developmental phenotypes characteristic of plants grown in low R:FR ratio: elongated petioles, reduced leaf area, early flowering, and reduced number of rosette leaves (Schena *et al.*, 1993).

Several genes encoding ATHB-2—related proteins have been identified in *Arabidopsis* (see Figure 1). Alignment of full-length protein sequences confirmed that the HD-ZIP II proteins can be distinct into two subfamilies: the first one comprising HAT22 and HAT9, the second one HAT1, ATHB-4, HAT3 and ATHB-2 (see also Figure 1). ATHB-2 exhibits greater sequence identity to members of the second subfamily (45.0-48.2%) than to members of first one (33.7-35.5%). Interestingly, we found that *ATHB-4*, as *ATHB-2*, is reversibly regulated by changes in the R:FR ratio (Carabelli *et al.*, 1993; Figure 2). In contrast, the *HAT22* mRNA levels are similar in high and low R:FR ratio (Figure 2). It will be interesting to determine what sort of functional interactions might exist between ATHB-2 and its relatives.

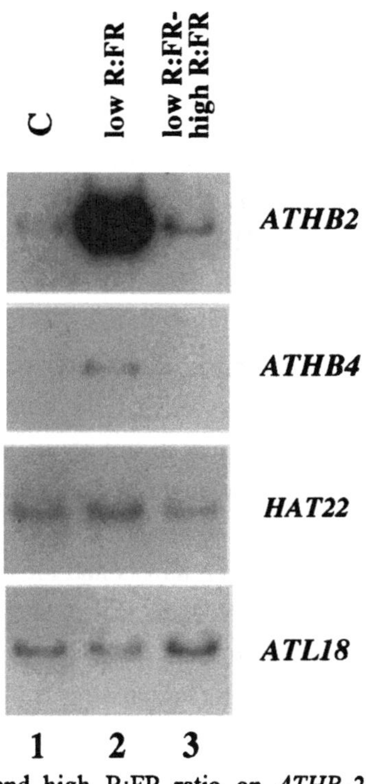

Figure 2. Effect of low and high R:FR ratio on *ATHB-2*, *-4*, and *HAT22* mRNA levels. *Arabidopsis* plants were grown in 16 h light/8 h dark cycle for 15 d under high R:FR ratio. Total RNA was prepared either from plants collected in the middle of the day (control plants, C, lane 1), or from plants exposed to a low R:FR ratio for 60 min (low R:FR, lane 2), or from plants exposed to a low R:FR ratio for 60 min and then returned to a high R:FR ratio for 60 min (low R:FR - high R:FR, lane 3). A cDNA probe encoding the *Arabidopsis* L18 ribosomal protein was used as a control (Baima *et al.*, 1995a).

HD-ZIP III protein family

The *Arabidopsis ATHB-8* gene encodes a 95 kDa protein characterised by the presence of a HD-Zip domain to its N-terminal end (Sessa *et al.*, 1994; Baima *et al.*, 1995b). *ATHB-8* is a member of a small gene family, designated *HD-ZIP III*. This family consists of four highly related HD-Zip protein coding genes (see Figure 1). Two of these genes, *ATHB-9* and *-14*, have been cloned and sequenced. The predicted ATHB-9 and -14 proteins contain 841 amino acids with a calculated molecular mass of 93.7 kDa and 852 amino acids with a calculated molecular mass of 94.9 kDa, respectively. Alignment of amino acid sequences of ATHB-9 and -14 revealed a strong homology throughout their entire length (90% similarity with 85% identity). In addition, pairwise amino acid comparisons of ATHB-8, -9 and -14 showed 60% homology between ATHB-8 and -9 and 62% between ATHB-8 and -14. A search of homology between ATHB-8 and the proteins encoded by the open reading frames contained in the EST databases revealed a high degree of similarity with the peptide encoded by the T88239 cDNA of *Arabidopsis thaliana.* (*Newman et al*, 1994). T88239 is likely to be a partial cDNA, and encodes for a peptide of 131 amino acids. Pairwise amino acid comparisons showed that T88239 shares 69% identity with ATHB-8, 54% with ATHB-9, and 53% with ATHB-14. The cDNA T88239 has been named *ATHB-15*, and its characterization is in progress (Sessa *et al.*, 1994; C. Steindler, unpublished data).

In situ mRNA analysis of *Arabidopsis* plants showed that the expression of *ATHB-8* is mainly restricted to provascular cells of the embryo and plant organs. Moreover, the analysis of *ATHB-8-GUS* in tobacco transgenic plants revealed that the gene is expressed in a polar fashion in the region of revascularization of wounded stems (Baima *et al.*, 1995b).

Ectopic expression of *ATHB-8* in *Arabidopsis* and tobacco plants resulted in a peculiar phenotype consisting of curled cotyledons, leaves and petals; short internode length in juvenile plants (tobacco); short flowering stem (*Arabidopsis*); delayed senescence (*Arabidopsis*). A trumpet-like structure arising from the fusion of the firts pair of true leaves was observed in transgenic tobacco lines with a more severe phenotype. Finally, an alteration of the polar auxin transport was detected in the *ATHB-8 Arabidopsis* transgenic plants compared to the wild-type (Baima *et al.*, manuscript in preparation).

The expression studies in combination with reverse genetic experiments in *Arabidopsis* and tobacco strongly suggest that ATHB-8 may be involved in the regulation of auxin distribution.

HD-ZIP IV protein family

Gene mapping experiments and sequence comparison analysis revealed that ATHB-10 (Sessa *et al.*, 1994) corresponds to GLABRA2 (GL2), described as a homeodomain protein involved in trichome development (Rerie *et al.*, 1994). In a preliminary partial sequence analysis we suggested that the *ATHB-10* gene also encodes a putative leucine zipper motif (Sessa *et al.*, 1994). This leucine zipper motif is formed by two subdomains of five and three heptad repeats separated by a loop of 10 amino acid residues. Consistent with the hypothesis of a-helix formation, no helix-disrupting proline or glycine residues are present in these heptad repeats. However, the first repeat is not canonical because it contains a glutamic acid residue at position d1. We recently demonstrated that the *ATHB-10/GL2* gene does indeed encode a functional leucine zipper domain (Di Cristina *et al.*, 1996).

Expression studies revealed that *ATHB-10/GL2* is expressed not only in trichome-bearing organs, but also in the root. The analysis of wild-type and mutant plants showed that the *ATHB-10/GL2* gene expression in the aerial part of the plant and in the root is affected by mutations at the *TTG* locus (Di Cristina *et al.*, 1996).

Morphological analysis of the *gl2-1* and *-2* mutants revealed that the gene is necessary not only for local outgrowth of the trichome, but also for the regulation of the root hair development in a subset of epidermal cells. Mutations in the *ATHB-10/GL2* gene specifically alter the differentiation of the hairless epidermal cells, causing them to produce root hairs (Di Cristina *et al.*, 1996; Masucci *et al.*, 1996).

In situ mRNA analysis and GUS reporter gene fusion studies showed that the *ATHB-10/GL2* gene is preferentially expressed in the differentiating hairless cells of the wild-type, during a period in which epidermal cell identity is believed to be establihed (Masucci *et al.*, 1996).

In conclusion, ATHB-10/GL2 is an HD-Zip protein involved in two epidermal cell differentiation processes which are under the control of the *TTG* locus: trichome development and root hair formation. The finding that the expression of *ATHB-10/GL2* is positively regulated in both shoot and root by TTG, implies that ATHB-10/GL2 acts as a positive regulator of trichome differentiation and as a negative regulator of root hair formation in a cell position-dependent manner to ensure that a subset of differentiating epidermal cells do not form root hairs (Di Cristina *et al.*, 1996; Masucci *et al.*, 1996)

Two genes encoding ATHB-10/GL2—related proteins have been identified in *Arabidopsis* (see Figure 1). Interestingly, one of them (*ATML1*) is expressed specifically in the protodermal cells of the 16-cell embryo, and in the L1 layer of the meristem from the very earliest stages of meristem patterning throughout the complet diploid life cycle of the plant (Lu *et al.*, 1996).

It is tempting to speculate that members of the HD-ZIP IV family are involved in the regulation of the epidermal cell fate.

Conclusions

Molecular and genetic studies of homeobox-leucine zipper genes presented here have begun to provide some insights into the possible roles of particular HD-Zip proteins in plant growth and development.

Proteins of the HD-ZIP I and II families are likely to be growth regulators whose expression is controlled by distinct environmental stimuli. *ATHB-1*, encoding for a HD-ZIP I protein, is rapidly induced by wounding, flooding and ethylene treatments, and seems to be involved into several aspects of leaf development (Aoyama *et al.*, 1995; S. Lucchetti, unpublished data); *ATHB-7*, encoding for another HD-ZIP I protein, is strongly induced by water stress, by a mechanism which is dependent on abscisic acid (Soderman *et al.*, 1996). Moreover, *ATHB-2* and *-4*, encoding HD-ZIP II proteins, are rapidly and strongly induced by changes in the R:FR ratio, and seem to have a direct role in light-regulated growth phenomena, such as neighbour detection and shade avoidance (Carabelli *et al.*, 1993 & 1996; Schena *et al.*, 1993; Steindler *et al.*, 1997).

Proteins of the HD-ZIP IV family seem to be involved in the regulation of epidermal cell differentiation. *ATHB-10/GL2* is expressed in specific epidermal cells in the shoot and in the root, and is required for regulation of trichome development and root hair formation (Rerie *et al.*, 1994; Di Cristina *et al.*, 1996; Masucci *et al.*, 1996). *ATML1*, encoding for another HD-ZIP IV protein, is specifically expressed in the L1 layer of the shoot apical meristem (Lu *et al.*, 1996). It will be interesting to investigate whether a gene encoding an ATHB-10/GL2-related protein is specifically expressed in the epidermal initials of the root meristem, and to determine what sort of functional interactions might exist between various HD-ZIP IV proteins.

Finally, several observations suggested a role for ATHB-8, an HD-ZIP III protein, in the regulation of auxin distribution (Baima *et al.*, 1995; Baima *et al.*, manuscript in preparation). Future studies on the other HD-ZIP III protein encoding genes (*ATHB-9*, *-14*, and *-15*) should enhance our understanding of the roles of the HD-ZIP III proteins in the auxin-regulated developmental processes

Acknowledgments

This work was supported, in part, by the Fondazione Istituto Pasteur-Fondazione Cenci Bolognetti, Università di Roma La Sapienza and by Piano Nazionale 'Biotecnologie Vegetali', Ministero per il Coordinamento delle Politiche Agricole, Alimentari e Forestali.

References

Affolter, M., Schier, A. and Gehring, W.J. (1990). Homeodomain proteins and the regulation of gene expression. *Current Biol.* **2**, 485-495.

Aoyama, T., Dong, C.-H., Wu, Y., Carabelli, M., Sessa, G., Ruberti, I., Morelli, G.and Chua, N.-H. (1995). Ectopic expression of the *Arabidopsis* transcriptional activator Athb-1 alters leaf cell fate in tobacco. *Plant Cell* **7**, 1773-1785.

Baima, S., Sessa, G., Ruberti, I. and Morelli, G. (1995a). A cDNA encoding *Arabidopsis thaliana* cytoplasmic ribosomal protein L18. *Gene* **153**, 171-174.

Baima, S., Nobili, F., Sessa, G., Lucchetti, S., Ruberti, I. and Morelli G. (1995b). The expression of the *Athb-8* homeobox gene is restricted to provascular cells in *Arabidopsis thaliana*. *Development* **121**, 4171-4182.

Bellman, R. and Werr, W. (1992). Zmhox1a, the product of a novel maize homeobox gene, interacts with the Shrunken 26 bp feedback control element. *EMBO J.* **11**, 3367-3374.

Carabelli, M., Sessa, G., Baima, S., Morelli, G. and Ruberti, I. (1993). The *Arabidopsis Athb-2* and *-4* genes are strongly induced by far-red-rich light. *Plant J.* **4**, 469-479.

Carabelli, M., Morelli, G., Whitelam, G. and Ruberti, I. (1996). Twilight-zone and canopy shade induction of the *Athb-2* homeobox gene in green plants. *Proc. Natl. Acad. Sci. USA* **93**, 3530-3535.

Chan, R.L. and Gonzales, D.H. (1994). A cDNA encoding an HD-Zip protein from sunflower. *Plant Physiol.* **106**, 1687-1688.

Di Cristina, M., Sessa, G., Dolan, L., Linstead, P., Baima, S., Ruberti, I. and Morelli, G. (1996). The *Arabidopsis* Athb-10 (GLABRA2) is an HD-Zip protein required for regulation of root hair development. *Plant J.* **10**, 393-402.

Gehring, W.J. (1987) . Homeoboxes in the study of development. *Science* **236**, 1245-1252.

Gehring, W.J., Qian, Y.Q., Billeter, M., Furukubo-Tokunaga, K., Schier, A.F., Resendez-Perez, D., Affolter, M., Otting, G., and Wuthrich, K. (1994). Homeodomain-DNA recognition. *Cell* **78**, 211-223.

Jackson, D., Veit, B. and Hake, S. (1994). Expression of maize *KNOTTED-1* related homeobox genes in the shoot apical meristem predicts patterns of morphogenesis in the vegetative shoot. *Development* **120**, 405-413.

Kawahara, R., Komamine, A. and Fukuda, H. (1995) Isolation and characterization of homeobox-containing genes of carrot. *Plant Mol. Biol.* **27**, 155-164.

Kerstetter, R., Vollbrecht, E., Lowe, B., Veit, B., Yamaguchi, J. and Hake, S. (1994). Sequence analysis and expression patterns divide the maize *knotted1*-like homeobox genes into two classes. *Plant Cell* **6**, 1877-1887.

Korfhage, U., Trezzini, G.F., Meier, I., Hahlbrock, K. and Somssich, I.E. (1994). Plant homeodomain protein involved in transcriptional regulation of a pathogen defense-related gene. *Plant Cell* **6**, 695-708.

Lincoln, C., Long, J., Yamaguchi, J., Serikava, K. and Hake, S. (1994). A *knotted1*-like homeobox gene in *Arabidopsis* is expressed in the vegetative meristem and dramatically alters leaf morphology when overexpressed in transgenic plants. *Plant Cell* **6**, 1859-1876.

Long, J. A., Moan, E.I., Medford, J.I. and Barton, M.K. (1996). A member of the KNOTTED class of homeodomain proteins encoded by the *STM* gene of *Arabidopsis*. *Nature* **379**, 66-69.

Lu, P., Porat, R., Nadeau, J.A. and O'Neil, S.D. (1996). Identification of a meristem L1 layer-specific gene in *Arabidopsis* that is expressed during embryonic pattern formation and define a new class of homeobox genes. *Plant Cell* **8**, 2155-2168.

Masucci, J.D., Rerie, W.G., Foreman, D.R., Zhang, M., Galway, M.E., Marks, M.D. and Schiefelbein, J.W. (1996). The homeobox gene *GLABRA2* is required for position-dependent cell differentiation in the root epidermis of *Arabidopsis thaliana*. *Development* **122**, 1253-1260.

Matsuoka, M., Ichikawa, H., Saito, A., Tada, Y., Fujimura, T. and Kano-Murakami, Y. (1993) Expression of a rice homeobox gene causes altered morphology of transgenic plants. *Plant Cell* **5**, 1039-1048.

Mattson, J., Soderman, E., Svensson, M., Borkird, C. and Engstrom, P. (1992). A new homeobox-leucine zipper gene from *Arabidopsis thaliana*. *Plant Mol. Biol.* **18**, 1019-1022.

McGinnis, W. and Krumlauf, R. (1992). Homeobox genes and axial patterning. *Cell* **68**, 283-302.

Meijer, A.H., Scarpella, E., van Dijk, E.L., Qin, L., Taal, A.J.C., Rueb, S., Harrington, S.E., McCouch, S.R., Schilperoort, R.A. and Hoge, J.H.C. (1997).

Transcriptional repression by Oshox1, a novel homeodomain leucine zipper protein from rice. *Plant J.* **11**, 263-276.

Meissner, R. and Theres, K. (1995) Isolation and characterization of the tomato homeobox gene *THOM1*. *Planta* **195**, 541-547.

Muller, K.J., Romano, N., Gerstner, O., Garcia-Maroto, F., Pozzi, C., Salamini, F. and Rohde, W. (1995). The barley *Hooded* mutation caused by a duplication in a homeobox gene intron. *Nature* **374**, 727-730.

Newman T., de Bruijn F.J., Green P., Keegstra K., Kende H., Mclntosh L., Ohlrogge J., Raikhel N., Somerville S., Thomashow M., Retzel E., Somerville C. (1994). Genes galore: a summary of methods for accessing results from large-scale partial sequencing of anonymous Arabidopsis cDNA clones. *Plant Physiol.* **106**, 1241-1255.

Quaedvlieg, N., Dockx, J., Rook, F., Weisbeek, P., J., and Smeekens, S.C.M.(1995). The homeobox gene *ATH1* of *Arabidopsis thaliana* is derepressed in the photomorphogenic mutants *cop1* and *det1*. *Plant Cell* **7**, 117-129.

Reiser, L., Modrusan, Z., Margossian, L., Samach, A., Ohad, N., Haughn, G.W. and Fischer, R.L. (1995). The *BELL1* gene encodes a homeodomain protein involved in pattern formation in the *Arabidopsis* ovule primordium. *Cell* **83**, 735-742.

Rerie, W.G., Feldmann, K.A. and Marks, M.D. (1994). The GLABRA2 gene encodes a homeodomain protein required for normal trichome development in Arabidopsis. *Genes Dev.* **8**, 1388-1399.

Ruberti, I., Sessa, G., Lucchetti, S. and Morelli, G. (1991). A novel class of plant proteins containing a homeodomain with a closely linked leucine zipper motif. *EMBO J.* **10**, 1787-91.

Schena, M. and Davis, R. W. (1992). HD-Zip proteins: members of an *Arabidopsis* homeodomain protein superfamily. *Proc. Natl. Acad. Sci. USA* **89**, 3894-3898.

Schena, M., Lloyd, A.M. and Davis, R.W. (1993). The *HAT4* gene of *Arabidopsis* encodes a developmental regulator. *Genes Dev.* **7**, 367-379.

Schena, M. and Davis, R.W. (1994). Structure of homeobox-leucine zipper genes suggests a model for the evolution of gene families. *Proc. Natl. Acad. Sci. USA* **91**, 8393-8397.

Schneeberger, R.G., Becraft, P.W., Hake, S. and Freeling, M. (1995). Ectopic expression of the *knox* homeo box gene *rough sheath1* alters cell fate in the maize leaf. *Genes Dev.* **9**, 2292-2304.

Sessa, G., Morelli, G. and Ruberti, I. (1993). The Athb-1 and -2 HD-Zip domains homodimerize forming complexes of different DNA binding specificities. *EMBO J.* **12**, 3507-3517.

Sessa, G., Carabelli, M., Ruberti, I., Lucchetti, S., Baima S. and Morelli, G. (1994) Identification of distinct families of HD-Zip proteins in *Arabidopsis thaliana*. In Molecular-Genetic Analysis of Plant Development and Metabolism (eds. P. Puigdomenech and G. Coruzzi), pp. 411-426. Berlin: Springer.

Sessa, G., Morelli, G. and Ruberti, I. (1997). DNA binding specificity of the homeodomain-leucine zipper domain. *J. Mol. Biol.*, in press.

Smith, H. (1995). Physiological and ecological functions within the phytochrome family. Annu. Rev. of Plant Physiol. *Plant Mol. Biol.* **46**, 289-315.

Smith, L.G., Greene, B., Veit, B. and Hake, S. (1992). A dominant mutation in the maize homeobox gene, *Knotted-1*, causes its ectopic expression in leaf cells with altered fates. *Development* **116**, 21-30.

Soderman, E., Mattson, J., Svensson, M., Borkird, C. and Engstrom, P. (1994). Expression patterns of novel genes encoding homeodomain-leucine zipper proteins in *Arabidopsis thaliana*. *Plant Mol. Biol.* **26**, 145-154.

Soderman, E., Mattsson, J. and Engstrom, P. (1996). The *Arabidopsis* homeobox gene *Athb-7* is induced by water-deficit and by abscisic acid. *Plant J.* **10**, 275-281.

Steindler, C., Carabelli, M., Borello, U., Morelli, G. and Ruberti, I. (1997). Phytochrome A, phytochrome B and other phytochrome(s) regulate *ATHB-2* gene expression in etiolated and green *Arabidopsis* plants. *Plant, Cell and Env.* **20**, 759-763.

Steindler, C., Morelli, G. and Ruberti, I. (1997). Nucleotide sequence of the *Arabidopsis ATHB-4* gene encoding an HD-Zip proteins related to ATHB-2 (Accession No. Y09582) (PGR97-021). *Plant Physiol.* **113**, 664.

Tornero, P., Conejero, V. and Vera, P. (1996). Phloem-specific expression of a plant homeobox gene during secondary phases of vascular development. *Plant J.* **9**, 639-648.

Vollbrecht, E., Veit, B., Sinha, N. and Hake S. (1991). The developmental gene *Knotted-1* is a member of a maize homeobox gene family. *Nature* **350**, 241-243.

Self-Incompatibility: Self/Nonself Discrimination Between Pollen and Pistil

Joseph A. Verica and Teh-hui Kao

Department of Biochemistry and Molecular Biology, The Pennsylvania State University, University Park 16802, USA

Keywords. Pollen-pistil interactions, self-incompatibility, self/nonself recognition, RNases

1. Introduction to Self-Incompatibility

Pollination begins when pollen produced in anthers adheres to the surface of the stigma. The acceptance of the pollen leads to its germination on the stigma and subsequent tube growth through the style to the ovary. Each pollen tube then penetrates an ovule and the two sperm nuclei carried by the pollen tube are released, one fusing with the egg nuclei and the other with polar nuclei of the embryo sac, to complete double-fertilization. The process of sexual reproduction thus provides an ideal system for studying cell-cell communication and signal transduction - it requires continuing interactions between pollen or pollen tubes and various parts of the pistil before the fusion of the male gamete and the female gamete can take place (Heslop-Harrison, 1987). The importance of these interactions to fertilization is best illustrated by the existence of various types of prezygotic reproductive barriers. For example, in many cases of interspecific incompatibility, the pistil of one species fails to allow the pollen from a different species to germinate or to grow in the style (Frankel and Galun, 1977). Further, in the case of intraspecific incompatibility, the pistil of an individual has the ability to reject the "wrong" kind of pollen to prevent inbreeding. Thus, maintenance of “correct” cell-cell communication between pollen and pistil is essential for pollen tube growth and fertilization.

One type of intraspecific reproductive barrier that has been extensively studied at the physiological, cell biological, genetic, and more recently, molecular levels is self-incompatibility (SI). SI is a cell-cell recognition mechanism that allows the pistil to discriminate between self (genetically related) pollen and nonself (genetically unrelated) pollen, and to reject self pollen to prevent inbreeding.

Although flowering plants have adopted a wide variety of reproductive strategies through which outbreeding is promoted, SI is among the efficient ones, as evidenced by the its occurrence in a great

NATO ASI Series, Vol. H 104
Cellular Integration of Signalling Pathways in Plant Development
Edited by F. Lo Schiavo, R. L. Last, G. Morelli, and N. V. Raikhel

number of plants from a variety of distantly related plant families. In most families, SI is controlled by a single locus called the *S* locus; however, this is not always the case. For example, four locus systems (*Beta vulgaris* and *Ranunculus acris*) have also been observed (Lundqvist *et al.*, 1973; Lundqvist, 1990).

There are two major SI systems, gametophytic self-incompatibility (GSI) and sporophytic self-incompatibility (SSI). For GSI, the pollen *S* phenotype is determined by its own *S* genotype, and for SSI, the pollen *S* phenotype is determined by the *S* genotype of the pollen parent (for review see Franklin *et al.*, 1995). Figure 1 illustrates the genetic basis of these two systems and highlights the differences between them. This article focuses on recent developments concerning the molecules involved in both types of SI.

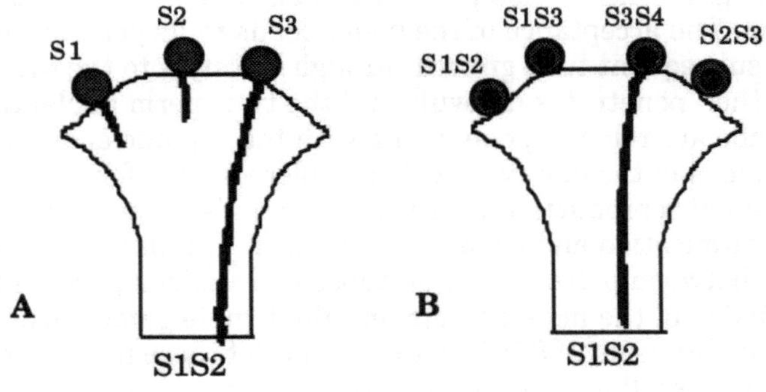

Figure 1. A, Gametophytic System. The pollen phenotype is determined by its own haploid genotype. If the pollen *S* allele matches either of the two alleles carried by the pistil, pollen is rejected. If the pollen *S* allele is different than those carried by the pistil, the pollen is accepted. B, Sporophytic System. The pollen phenotype is determined by the genotype of the pollen parent. If either of the *S* alleles of the pollen parent match either of the *S* alleles carried by the pistil, pollen is rejected. If both of the *S* alleles of the pollen parent are different from both alleles carried by the pistil, pollen is accepted.

2. Single-locus Gametophytic Incompatibility

GSI is displayed by a number of plant families, including the Onagraceae, Papaveraceae, Poaceae, Rosaceae, Scrophulariacea and Solanaceae. However, the mechanism by which GSI functions in these families varies.

2.1. Ribonuclease Systems

In the Rosaceae, Scrophulariacea and Solanaceae, molecular genetic studies revealed that stylar glycoproteins segregating with *S* alleles are ribonucleases (RNases) (McClure *et al.*, 1989; Broothaerts *et al.*, 1991; Singh *et al.*, 1991; Sassa *et al.*, 1992; Xue *et al.*, 1996). These proteins are referred to as S proteins or S RNases. This finding suggests a possible mechanism by which SI in these systems may operate. The RNase activity of S proteins has recently been demonstrated in transgenic *plants (Petunia inflata)* to be both necessary and sufficient for styles to reject self pollen. In a loss-of-function study, the RNase activity of S3 protein was abolished by replacing one of the catalytic histidine residues with asparagine. The *S3(H93N)* gene, when expressed in *S1S2* plants, failed to confer the ability to reject *S3* pollen (Huang *et al.*, 1994). In a gain-of-function approach, the wildtype *S3* gene, when expressed in *S1S2* plants, conferred the ability to reject *S3* pollen (Lee *et al.*, 1994).

Having demonstrated the role of S RNases in SI, it remained to be determined how the allele specificity was achieved. All S RNases identified to date are N-glycosylated glycoproteins. As such, it had been speculated that the specificity of S RNases could lie in either the glycan side chain or the amino acid sequence itself. This has been addressed by expressing a mutant *S3* gene, in which the lone glycosylation site was mutated, in *S1S2* plants. Plants expressing high levels of nonglycosylated S3 protein were subsequently shown to reject *S3* pollen, suggesting that specificity of S RNases lies in the amino acid sequence itself (Karunanandaa *et al.*, 1994).

Comparison of the amino acid sequences of S RNases shows that there are five conserved regions, two hypervariable regions and several scattered hypervariable residues (Ioerger *et al.*, 1991). It has been speculated that these hypervariable regions/residues may be involved in determining specificity; however, this remains to be shown. In addition to specificity determinants of stylar S RNases, there also appears to be a specificity determinant by which pollen is discriminated. This is supported by recent biochemical evidence. A ribonuclease deficient *S3 RNase, S3(H93R),* was expressed in *S2S3* plants and subsequently shown to have an allele specific dominant negative effect on the ability of engogenous S3 RNase to reject self-pollen. This finding suggests that S3(H93R) competes with endogenous S3 RNase for binding to a common molecule, perhaps the pollen *S* gene product (McCubbin *et al.*, 1997). The identity of the pollen *S* gene and the nature of its interaction with S RNases is currently unknown and is the focus of a great deal of research. Although all the players have not been identified, the available data allow for the formulation of a model for how the system functions (for review, see Newbigin *et al.*, 1993) (Figure 2).

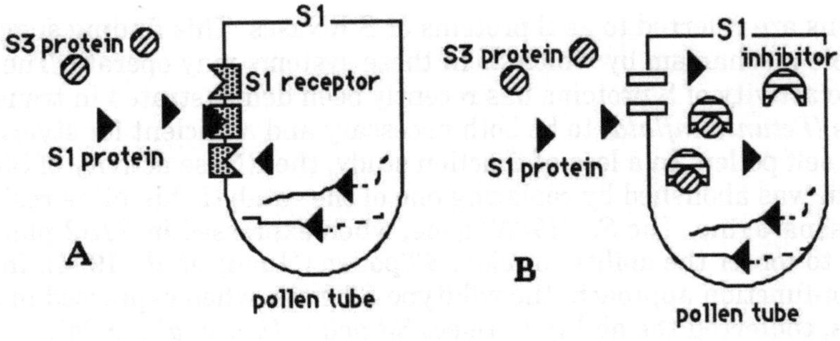

Figure 2. Pollen lands on the stigma, germinates, and begins to penetrate into the style. An active ribonuclease secreted by the style, the S RNase, is apparently responsible for the cytotoxic effects on incompatible pollen. A, Pollen S could function as a receptor on the pollen tube membrane that acts as a gate keeper, permitting entry only to S RNases of the same *S* genotype. B, Alternatively, all S RNases, regardless of genotype, may enter pollen tubes. Once inside, they may encounter the pollen S product which function as a ribonuclease inhibitor - inhibiting all nonself RNases. The answer to this question awaits the discovery of the pollen *S* gene.

2.2. Non-ribonuclease Systems

In the Papaveraceae, the genetic behavior of GSI appears identical to that of the Solanaceae. However, the biochemical mechanism employed is entirely different. In *Papaver rhoeas*, stigmatic glycoproteins displaying an inhibitory effect on *in vitro* pollen tube growth were identified and isolated from stigmatic extracts (Franklin-Tong *et al.*, 1988; Franklin-Tong *et al.*, 1989). The inhibitory effects of the stigmatic extracts were partially alleviated by treatment of pollen tubes with the metabolic inhibitors Actinomycin D (transcription inhibitor) and tunicamycin (glycosylation inhibitor) (Franklin-Tong *et al.*, 1990). These results suggest that both transcription and glycosylation of pollen proteins are required for the full inhibitory affects of the stigmatic extracts. Employing pollen tubes microinjected with the calcium sensitive dye Calcium Green-1, the stigmatic extracts were also shown to cause a transient increase in cytosolic free calcium in incompatible, but not compatible pollen tubes. Further, the rise in calcium levels was subsequently followed by the inhibition of pollen tube growth (Franklin-Tong *et al.*, 1993; Franklin-Tong *et al.*, 1995). Transient phosphorylation of pollen proteins has also been demonstrated during the SI interaction, implicating the involvement of protein kinases and phosphatases.

Degenerate oligonucleotides corresponding to the N-terminal amino acid sequences of proteins segregating with the *S1* allele were used to isolate the *S1* cDNA (Foote *et al.*, 1994). Just recently, the sequence of the *S3* allele was also reported (Walker *et al.*, 1996). Analysis of the *S* sequences shows that they do not share significant homology to other sequences in the database. To determine whether or not the cloned sequences displayed characteristics expected of the *S* gene, the *S1* gene was expressed in *E. coli* and the its protein product (S1e) was isolated and used in an *in vitro* bioassay. As expected, the bioassay showed that S1e had the same allele specific inhibitory effect on the *in vitro* growth of *S1* pollen as partially purified stigmatic extracts (Franklin-Tong *et al.*, 1995). It was also shown that S1e alone, despite being nonglycosylated, is sufficient to induce the transient calcium increase and subsequent inhibition of pollen tube growth in the absence of other stigmatic components.

The current data suggest that the SI response in *Papaver* involves a signaling cascade involving transcription, changes in cytosolic free calcium levels and protein phosphorylation. Presently, the identity of the pollen *S* gene in *Papaver* is unknown. In addition, it is not known how allele specificity in *Papaver* is controlled. To date, the sequences of only two *S* allele have been determined (Foote *et al.*, 1994; Walker *et al.*, 1996). Although overall sequence homology is only 55%, there do appear to be several variable regions which may function in specificity. With the isolation of more *S* allele sequences, a specificity domain(s) may become more apparent. In addition, the observation that recombinant S1e is functional in *in vitro* bioassays should provide a workable system to test the function of any identified domains. The current data allow for the formulation of a model of how GSI in poppy functions (Franklin-Tong and Franklin, 1995) (Figure 3).

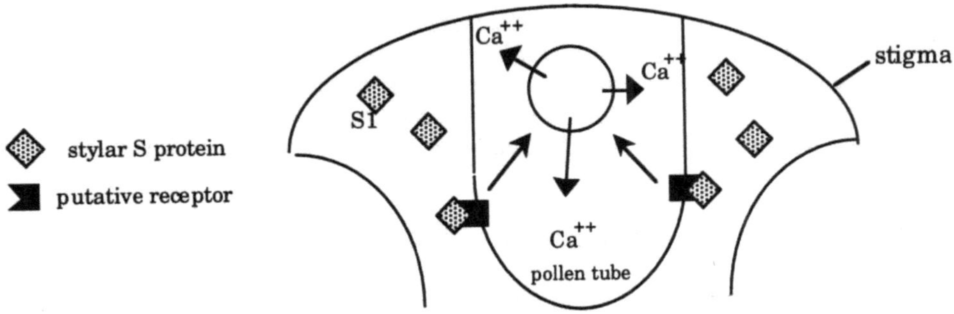

Figure 3. Self-incompatibility in *Papaver rhoeas*. After landing on the stigma, incompatible pollen is rejected on or just beneath the stigmatic surface. Stigmatic S protein is recognized by the pollen *S* gene product. Rather than being internalized, binding of S protein to the pollen S gene product results in the activation of a calcium-mediated signal transduction pathway which leads to the rejection of incompatible pollen.

3. Multiallelic Gametophytic Incompatibility

The grasses also display gametophytic self-incompatibility; however, in contrast to the Solanaceae and Papaveraceae, GSI in the Poaceae is controlled by two unlinked loci, *S* and *Z*. If both the *S* and *Z* alleles carried by the pollen match both *S* and *Z* in the stigma, the pollen is rejected (Lundqvist, 1956). The mechanism controlling GSI in this system is not well understood. Presently, there is no evidence indicating an associated RNase activity for either the *S* or *Z* alleles. Similar to the situation in *Papaver*, there is some evidence in rye (*Secale cereale*) suggesting that both Ca^{2+} and kinase activity may be involved in the SI response. Changes in phosphorylation were observed in an *in vitro* assay using either compatible and incompatible stigmatic extract; however, the response appeared to be much greater using incompatible extracts (Wehling *et al.*, 1994).

The identity of the pistil *S* or *Z* genes has not been determined; however, a 0.28 kb DNA fragment with homology to the *Brassica SLG* gene (see below) has been found to display complete *S* linked cosegregation in rye (Wehling *et al.*, 1994b). Whether or not this sequence represent all or part of the *S* gene is yet to be determined. In *Phalaris coerulescens*, several cDNAs corresponding to putative pollen *S* alleles have recently been identified (Li *et al.*, 1994). Comparison of the predicted amino acid sequences for these cDNA shows variability at their N-termini, suggesting a possible role in allele specificity, and a conserved region at their C-termini. The conserved region was found to share approximately 40% homology (including the active site) to thioredoxin H from tobacco. These genes have thus been termed S thioredoxins. The C-terminus of S thioredoxin was expressed in *E. coli* and shown to possess thioredoxin activity (Li *et al.*, 1995). Although S thioredoxins have not been shown to function in the SI response in *Phalaris*, the observation that a self-compatible mutant has several amino acids changes in the thioredoxin region would seem to be consistent with such a role (Li *et al.*, 1994). At present, it is not entirely clear how S thioredoxins may function in the SI response. A possible role may become clearer once the *Z* gene has been identified.

3. Sporophytic Self-Incompatibility

Sporophytic self-incompatibility (SSI) is displayed by several plant families including the Brassicaceae, Compositae and Convolvulaceae. Among these, the Brassicaceae is the most extensively studied. Molecular genetic analyses shows that at least two genes, *SLG* (*S* locus glycoprotein) and *SRK* (*S* locus receptor kinase) cosegregate with the *S* locus (Boyes and Nasrallah, 1993). *SLG* was the first of these genes to be isolated. *SLG* encodes an abundant glycoprotein which is localized to the walls of the papillae in stigmas and to the tapetum and developing microspores in anthers (Kandasamy *et al.*, 1989). *SRK* encodes a protein with a extracellular domain with extensive homology to *SLG* and a cytoplasmic domain with homology to serine/threonine kinases (Stein *et al.*, 1991). Bacterial expression of the kinase domains of *SRK* shows that it has ability to autophosphorylate on serine and threonine residues. Like SLG, SRK localizes to stigmatic papillae and pollen. Recently, it has been shown that SRK is targeted to the plasma membrane in transgenic tobacco cells, consistent with its hypothesized role in mediating SI through phosphorylation events at the stigma surface (Stein *et al.*, 1996). No activating ligand for SRK has yet been reported. However, it has recently been shown that the protein kinase domain of SRK-910 interacts with two thioredoxin-H-like proteins, THL1 and THL2, in the yeast two hybrid system (Bower *et al.*, 1996). Further, a recombinant THL1 protein produced in *E. coli* was shown to coprecipitate with the kinase domain of SRK-910. THL1 was also shown to possess thioredoxin activity and it is weakly phosphorylated by the kinase domain of SRK-910 (Bower *et al.*, 1996).

A great deal of data are available that strongly suggest that *SLG* and *SRK* mediate SSI. In addition to cosegregation with the *S* locus and the expression studies mentioned above, several plant lines altered in either *SLG* or *SRK* have been shown to be self-compatible (Goring *et al.*, 1993; Nasrallah *et al.*, 1994). Further, *SCF1*, a locus unlinked to *S*, has been shown to regulate *S* function via down regulation of *SLG* , resulting in self-compatibility (Nasrallah *et al.*, 1992). In addition, expression of an antisense *SLG* gene in a self-incompatible *Brassica campestris* plants resulted in the plants becoming self-compatible (Shiba *et al.*, 1995). Some reports concerning the role of *SLG* in SSI are not so clear. For example, the levels of *SLG* expression do not always correlate with the ability of plants to reject self pollen. Self-incompatible plants of the *S2* haplotype have been identified that express low levels of *SLG*. Conversely, self-compatible plants of the *S15* and *Sc* haplotypes were shown to express high levels of *SLG* (Gaude *et al.*, 1995).

A third *S* linked gene, *S locus Anther* (*SLA*), has been identified in the *S2* haplotype of *Brassica oleracea* (Boyes and Nasrallah, 1995). *SLA2* is transcribed from two promoters producing two complementary transcripts. Sequence analysis of *SLA2* shows that it could potentially encode two proteins of 7.5 kDa and 10 kDa respectively. RNA gel blot analysis shows that these two transcripts are specifically expressed in anthers and developing microspores, properties expected for the pollen *S* gene. A second suggestion that *SLA2* may represent the pollen *S* gene comes from the observation that the *SLA2* transcript was undetectable in self-compatible *B. napus* plants of an *S*2-like haplotype. However, sequences homologous to *SLA2* have not been found in other *S* haplotypes, as would be expected for the pollen *S* gene. Whether or not *SLA2* is the pollen *S* gene remains to be proven. The available evidence suggests the following model for SSI in *Brassica* (for review, see Nasrallah and Nasrallah, 1993) (Figure 4).

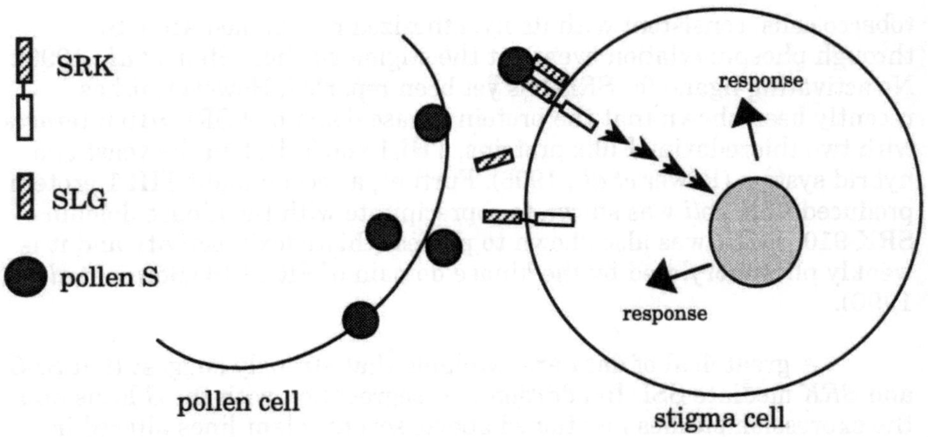

Figure 4. Sporophytic Self-incompatibility. The pollen S product, perhaps SLA or some other molecule(s), interacts with both SLG and SRK to produce a receptor complex. The receptor complex triggers a signal transduction cascade involving phosphorylation of target proteins, two of which may be THL1 and THL2. The cascade culminates in a localized response in the stigmatic papillae which prevents germination or growth of self-pollen.

References

Bower, M.S., Matias, D.D., Fernandes-Carvalho, E., Mazzurco, M., Gu, T. Rothstein, S.J. and Goring, D. (1996). Two members of the thioredoxin-h family interact with the kinase domain of a *Brassica S* locus receptor kinase. Plant Cell 8, 1641-50.

Boyes, D.C. and Nasrallah, J.B. (1993). Physical linkage of the *SLG* and *SRK* genes at the self-incompatibility locus of *Brassica oleracea*. Mol. Gen. Genet. 236, 269-73.

Boyes, D.C. and Nasrallah, J.B. (1995). An anther specific gene encoded by an *S* locus haplotype of *Brassica* is produces complementary and differentially regulated transcripts. Plant Cell 7, 1283-94.

Broothaerts, W., Janssens, G., Proost, P. and Broekaert W. (1995). cDNA cloning and molecular analysis of two self-incompatibility alleles from apple. Plant Mol. Biol. 27, 449-511.

Foote, H., Ride, J., Franklin-Tong, V.E., Walker, E., Lawrence, M.J. and Franklin, F.C.H. (1994). Cloning and expression of a distinctive class of self-incompatibility *S* gene from *Papaver rhoeas* L. Proc. Natl. Acad. Sci. USA 91, 2265-9.

Frankel, R. and Galun, E. (1977). Pollination Mechanisms, Reproduction and Plant Breeding. (Springer-Verlag, New York).

Franklin-Tong, V.H. and Franklin, F.C.H. (1992). Gametophytic self-incompatibility in *Papaver rhoeas* L. Sex. Plant Reprod. 5,1-7.

Franklin-Tong, V.E., Lawrence, M.J. and Franklin, F.C.H. (1988). An *in vitro* bioassay for the stigmatic product of the incompatibility gene in *Papaver rhoeas* L. New Phytol. 110:109-18.

Franklin-Tong, V.E., Ruuth, E., Marmey, P., Lawrence, M.J. and Franklin F.C.H. (1989). Characterization of a stigmatic component of *Papaver rhoeas* L. which exhibits the specific activity of a self-incompatibility *S* gene product. New Phytol. 112, 307-15.

Franklin-Tong, V.E., Lawrence, M.J. and Franklin F.C.H. (1990). Self-incompatibility in *Papaver rhoeas* L.: inhibition of incompatible pollen is dependent on pollen gene expression. New Phytol. 116, 319-24.

Franklin-Tong, V.E., Ride, J.P., Read, N., Trewavas, A.J., and Franklin F.C.H. (1993). The self-incompatibility response in *Papaver rhoeas* is mediated by cytosolic free calcium. Plant J. 4, 163-77.

Franklin, F.C.H., Lawrence, M.J., Franklin-Tong, V.E. (1995a). Cell and molecular biology of self-incompatibility in flowering plants. Int. Rev. Cytol. 158, 1-64.

Franklin-Tong, V.E., Ride, J.P. and Franklin F.C.H. (1995b). Recombinant stigmatic self-incompatibility (S-) protein elicits a Ca^{2+} transient in pollen of *Papaver rhoeas*. Plant J. 8, 299-307.

Gaude, T., Rougier, M., Heizmann, P., Ockendon, D.J. and Dumas, C. (1995) Expression level of the *SLG* gene is not correlated with the self-incompatibility phenotype in the class II *S* haplotypes of *Brassica oleracea*. Plant Mol. Biol. 27, 1003-14.

Goring, D.R., Glavin, T.L., Schafer, U. and Rothstein, S. (1993). An *S* receptor kinase in a self-compatible *Brassica napus* line has a 1-bp deletion. Plant Cell 5, 531-9.

Heslop-Harrison, J. (1987). Pollen germination and pollen tube growth. Int. Rev. Cytol. 107, 1-78.

Huang, S., Lee, H.-S., Karunanandaa, B. and Kao, T.-H. (1994). Ribonuclease activity of *Petunia inflata* S proteins is essential for rejection of self-pollen. Plant Cell 6, 1021-1028.

Ioerger, T.R., Gohlke, J.R., Xu, B. and Kao, T.-H. (1991). Primary structural features of the self-incompatibility protein in Solanaceae. Sex. Plant Reprod. 4, 81-87.

Kandasamy, M.K., Paolillo, C.D., Faraday, J.B., Nasrallah, J.B. and Nasrallah, M.E. (1989). The *S* locus specific glycoproteins of *Brassica* accumulate in the cell wall of developing stigma papillae. Devel. Biol. 134, 462-72.

Karunanandaa B, Huang S, Kao T-H. (1994). Carbohydrate moiety of the *Petunia inflata* S_3 protein is not required for self-incompatibility interactions between pollen and pistil. Plant Cell 6, 1933-40.

Lee, H.-S., Huang, S. and Kao, T.-H. (1994). S proteins control rejection of incompatible pollen in *Petunia inflata*. Nature 367, 560-563.

Li, X., Nield, J., Hayman, D. and Langridge, P. (1994). Cloning a putative self-incompatibility gene from the pollen of the grass *Phalaris coerulescens*. Plant Cell 6, 1923-32.

Li, X., Nield, J., Hayman, D. and Langridge, P. (1995). Thioredoxin activity in the C terminus of *Phalaris* S protein. Plant J. 8, 133-8.

Lundqvist, A. (1956). Self-incompatibility in rye I. Genetic control of the diploid. Hereditas 42, 239-348.

Lundqvist, A. (1990). Variability within and among population in a 4-gene system for sontrol of self-incompatibility in *Rannunculus polyanthemos*. Hereditas 113, 47-61.

Lundqvist, A., Osterbye, U., Larsen, K. and Linde-Laursen, I. (1973). Complex self-incompatibility systems in *Rannunculus acris* L. and *Beta vulgaris* L. Hereditas 74, 161-8.

McClure, B.A., Haring, V., Ebert, P.R., Anderson, M.A., Simpson, R.J., Sakiyama, F. and Clarke, A.E. (1989). Style self-incompatibility gene products of *Nicotiana alata* are ribonucleases. Nature 342, 955-7.

McCubbin, A., Chung, Y. and Kao, T.-H. (1997). A mutant S3 RNase of *Petunia inflata* lacking Rnase activity has an allele-specific dominant negative effect on self-incompatibility interactions. Plant Cell 8, 85-95.

Nasrallah, M.E., Kandasamy, M.K. and Nasrallah, J.B. (1992). A genetically defined trans-acting locus regulates *S*-locus function in *Brassica*. Plant J. 2, 497-506.

Nasrallah, J.B. and Nasrallah, M.E. (1993). Pollen-stigma signaling in the sporophytic self-incompatibility response. Plant Cell 5, 1325-35.

Nasrallah, J.B., Rundle, S.J. and Nasrallah, M.E. (1994). Genetic evidence for the requirement of the *Brassica S* locus receptor kinase in the self-incompatibility response. Plant J. 5, 373-84.

Newbigin E., Anderson M.A., Clarke A.E. (1993). Gametophytic incompatibility systems. Plant Cell 5, 1315-24.

Sassa, H., Hirano, H. and Ikehashi, H. (1992). Self-incompatibility-related RNases in styles of Japanese pear (*Pyrus serotina* Rehd.). Plant Cell Physiol. 33, 811-4.

Shiba, H., Hinata, K., Suzuki, A. and Isogai, A. (1995). Breakdown of self-incompatibility in *Brassica* by the antisense RNA of the *SLG* gene. Proc. Japan Acad. 71 (Ser. B), 223-5.

Singh, A., Ai, Y. and Kao, T.-H. (1991). Characterization of ribonuclease activity of three S-allele-associated proteins of *Petunia inflata*. Plant Physiol. 96, 61-8.

Stein, J.C., Howlett, B., Boyes, D.C., Nasrallah, M.E. and Nasrallah, J.B. (1991). Molecular cloning of a putative receptor kinase encoded at the *S*-locus *of Brassica oleracea*. Proc. Natl. Acad. Sci. USA 88, 8816-20.

Stein, J.C., Dixit, R., Nasrallah, M.E. and Nasrallah, J.B. (1996). SRK, the stigma-specific *S* locus receptor kinase of *Brassica*, is targeted to the plasma membrane in transgenic tobacco. Plant Cell 8, 429-45.

Walker, E.A., Ride, J.P., Kurup, S., Franklin-Tong, V.E., Franklin, F.C. (1996). Molecular analysis of two functional homologues of the *S3* allele of the *Papaver rhoeas* self-incompatibility gene isolated from differenct populations. Plant Molec. Biol. 30, 983-94.

Wehling, P., Hackauf, B. and Wricke, G. (1994a). Phosphorylation of pollen proteins in relation to self-incompatibility in rye (*Secale cereale* L.). Sex. Plant Reprod.7, 67-75.

Wehling, P., Hackauf, B. and Wricke, G. (1994b). Identification of *S*-locus linked PCR fragments in rye (*Secale cereale* L.) by denaturing gradient gel electrophoresis. Plant J. 5, 891-3.

Xue, Y., Carpenter, R., Dickinson, H.G. and Coen, E.S. (1996). Origin of allelic diversity in *Antirrhinum S* locus RNases. Plant Cell 8, 805-14.

Intercellular and Intracellular Trafficking: What We Can Learn from Geminivirus Movement

Sondra G. Lazarowitz[1], Brian M. Ward[1], Anton A. Sanderfoot[2] and Christina M. Laukaitis[1]

[1] Department of Microbiology, University of Illinois, Urbana, IL 61801 USA
[2] DOE Plant Research Lab, Michigan State University, E. Lansing, MI 48824 USA

Abstract. Bipartite geminiviruses such as squash leaf curl virus encode two movement proteins, BL1 and BR1, essential for virus movement. This reflects the nuclear replication of the viral single-stranded (ss)DNA genome, requiring that one movement protein (BR1) enters the nucleus and binds the viral genome, following which it interacts with the second movement protein (BL1) to facilitate transport of the viral genome to adjacent uninfected cells. Thus, geminiviruses offer the opportunity to investigate nuclear import and export in plant cells, as well as mechanisms for transporting macromolecules across the plant cell wall. The phloem-limitation of squash leaf curl virus also allows us to investigate virus movement within phloem and how a phloem-limited virus invades susceptible host plants.

Keywords. movement proteins, cell-cell communication, intracellular trafficking, nuclear export, nuclear shuttling, endoplasmic reticulum, geminiviruses, squash leaf curl virus

1 Introduction

To successfully infect a susceptible host, plant viruses must move locally cell to cell and enter the phloem to move long distance and establish a systemic infection This requires that plant viruses overcome the barrier posed by the plant cell wall, a feat they accomplish by encoding "movement proteins". The existence of such movement proteins (MPs) was originally inferred from genetic studies: in the absence of MP function (null mutant), plant viruses replicate and encapsidate progeny virus particles; however, they will not move out of the original inoculated cell, in effect being imprisoned within the cell (1, 2, 23). Thus, active expression of MPs eliminates the cell wall barrier to plant virus movement, suggesting that MPs must somehow function to fundamentally alter cell structure and thereby remove this hurdle.

Viewed from this perspective, the study of plant virus movement holds the promise of providing insights at the molecular level about the trafficking of macromolecules across the plant cell wall . In addition, MPs must transit through the cytoplasm in a directed manner in order to deliver the viral genome to and across the cell wall. Investigation of this orchestrated movement through the cytoplasm is beginning to define associations of MPs with elements of the cytoskeleton and endomembrane system of the cell, thus adding to our knowledge of the regulation of intracellular trafficking in plant cells (19, 31, 53). An added dimension to this issue of intracellular trafficking is provided by the DNA-containing geminiviruses, which

NATO ASI Series, Vol. H 104
Cellular Integration of Signalling Pathways in Plant Development
Edited by F. Lo Schiavo, R. L. Last, G. Morelli, and N. V. Raikhel

replicate in the nucleus (27, 43). This nuclear replication necessitates that a viral-encoded MP function as a nuclear shuttle protein to move the viral genome to and from the nucleus. Recent studies of the bipartite geminivirus squash leaf curl virus (SqLCV) have defined such an MP and provide a unique approach for investigating the regulation of nuclear export, as well as nuclear import, in plant cells (43, 45).

How do MPs interact with cellular components to direct the viral genome to the cell wall and then act to move the genome across the wall? Much of our knowledge comes from molecular and biochemical studies originally done for tobacco mosaic virus (TMV) and red clover necrotic mosaic virus. These are both RNA viruses that replicate in the cytoplasm of mesophyll cells. Using proteins overexpressed in *E. coli*, the single MP encoded by each of these viruses has the properties of a single strand nucleic acid binding protein *in vitro* (6, 16), and acts to increase the size exclusion limits of plasmodesmata when microinjected into tobacco mesophyll cells (16, 51). The latter effect on plasmodesmal gating is also observed when the MP is expressed in transgenic plants (11, 34, 57). The microinjection studies further suggest that these MPs themselves can rapidly move cell to cell (16, 51). Based on these studies it has been proposed that the single MP encoded by these two RNA viruses is a molecular chaperone that binds the viral RNA genome in a cooperative manner and targets it to plasmodesmata, where the MP then acts to increase the size exclusion limits and thereby facilitate movement of the viral genome to adjacent uninfected cells (7, 29, 30). Subsequent studies on the 3a MP encoded by cucumber mosaic virus, which is also a mesophyll replicating RNA virus, have used these same approaches and are in general agreement with this model (13). More recent studies using *in vitro* tubulin polymerization assays, immunofluorescence labeling, and the green fluorescent protein (GFP) from jellyfish (*Aequorea victoria*) fused to the MP of TMV as a sensitive *in vivo* cell-autonomous marker have shown the TMV MP (also known as the 30 kDa protein) to be associated with microtubules in transfected tobacco protoplasts and in infected cells, suggesting that this cytoskeletal association may be important for directing the TMV genome to the cell wall (19, 31). GFP-MP fusion studies are also consistent with earlier immunogold labeling studies that localized the TMV MP to plasmodesmata (32, 34), and suggest plasmodesmal localization of the MP encoded by cucumber mosaic virus as well (37, 38).

In contrast to the studies on TMV and similar viruses where plasmodesmal gating appears to be increased without any gross physical changes in the appearance of these intercellular channels (3, 12), a number of mesophyll infecting viruses have been reported to induce large tubular structures that contain virus-like particles and appear to extend from the cell wall at or near plasmodesmata. Immunogold-labeling studies primarily for the RNA viruses tomato spotted wilt virus, cowpea mosaic virus and tomato ringspot virus; and the pararetrovirus cauliflower mosaic virus (DNA genome is replicated via an RNA-templated mechanism) have localized the MP encoded by each of these viruses to these tubular structures (26, 28, 50, 54), and transient expression assays in protoplasts demonstrate that the MP is sufficient to induce these tubules (42, 48, 49). In addition, mutational studies of cowpea mosaic virus show that the viral coat protein is required to form the virus-like particles within the tubules, but not the tubules per se (55). From these studies it is proposed that in these systems virions or subviral particles move from cell to cell through tubules that are at least in part composed of the MP, possibly through altered plasmodesmata.

Recent molecular and cellular studies on the two movement proteins, BR1 and BL1, encoded by the phloem-limited bipartite geminivirus SqLCV have begun to define the functions and regulated interactions of these MPs, and their interactions with components within plant cells. As described below, these studies suggest that BR1 is a nuclear shuttle protein that functions to move the viral single strand DNA (ssDNA) genome into and out of the nucleus (41, 43). BL1 interacts with BR1 and is associated with unique tubular structures that appear to function as conduits for moving the viral genome across the cell wall and into adjacent uninfected cells (44, 53). Current studies with SqLCV engineered to contain GFP driven by the viral coat protein promoter are providing new insights into the association of virus movement with phloem development and the mechanism by which this phloem-associated virus invades the plant.

2 SqLCV BR1 and BL1

2.1 BR1 and BL1: Biochemical Properties and Localization in Infected Plants

As a typical bipartite geminivirus, SqLCV has a genome of two ~2.5 kb components of covalently-closed circular ssDNA and encodes two nonstructural MPs, BR1 and BL1, which are essential for systemic infection of all host plants (27, 47). Essential information on the tissue- and cell-specific locations of BR1 and BL1 in infected plants was obtained from immune localization studies and subcellular fractionation studies done on systemic leaves from infected cucurbit hosts, specifically pumpkin and squash (41, 53).

Immunogold-labeling of thin-sections localized BR1 to the nuclei of immature phloem cells (procambial cells) and phloem parenchyma cells (Table 2.1) (41). Neither BR1 nor BL1 (see below) was detected in non-phloem cells, consistent with the phloem-limitation of SqLCV infection. Consistent with the nuclear localization of BR1, *in vitro* DNA-cellulose binding assays, Southwestern gel blot assays and gel shift assays showed BR1 to be a typical ssDNA binding protein, having a higher affinity for ssDNA than for dsDNA or RNA (Table 2.1) (41). Thus, BR1 has the properties of a MP that would enter the nucleus to bind the viral ssDNA genome and form a complex for movement.

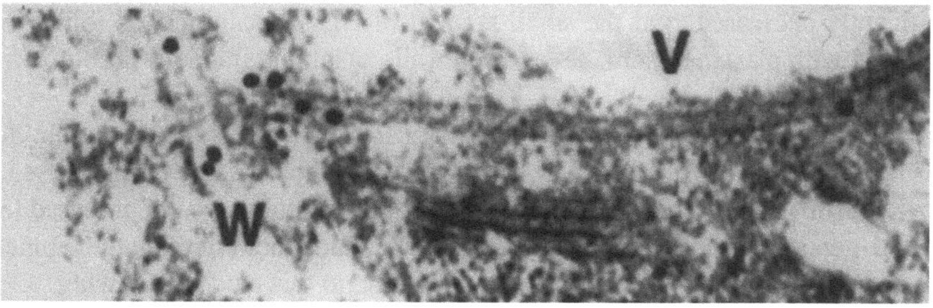

Fig. 2.1 Tubules in a procambial cell in a pumpkin leaf systemically infected with SqLCV. The section is labeled with anti-BL1 antisera and protein A gold. W, wall; V, vacuole.

Table 2.1. Properties and Localization of SqLCV BR1 and BL1

Movement Protein	DNA binding[a]	Infected Plants[b]	Tobacco Protoplasts[c]
BR1	+	nuclear	nuclear
BL1	–	tubules extending from and crossing cell wall	cortical cytoplasm

[a]Ability to bind DNA (ss or ds) and RNA *in vitro*.
[b]Immunogold labeling detected only in immature phloem cells (BL1 and BR1) or in phloem parenchyma cells (BR1).
[c]Detection by indirect immunofluorescent staining and confocal microscopy in protoplasts transiently expressing either MP individually.

Immunogold-labeling specifically localized BL1 to unique tubular structures that extended from and crossed the walls of procambial cells (Table 2.1 and Fig. 2.1). These tubules were not seen in uninfected plants, nor in any cells other than immature phloem cells. While the tubules did not resemble in appearance microtubules or actin filaments, they were similar to the cortical endoplasmic reticulum (ER) where it is compressed to form the desmotubule that runs through the center of a plasmodesma. Thus, to ask whether the BL1-associated tubules might be derived from the ER, thin-sections from systemically infected pumpkin leaves were labeled with anti-BiP antisera. Indeed, these tubules were also specifically labeled with anti-BiP antisera, suggesting that they are derived from the ER (53). Neither preimmune sera, nor anti-BR1 antisera stained the BL1-associated tubules.

The ER association of these BL1-containing tubules was independently confirmed in subcellular fractionation studies (53). When young systemically infected leaves from pumpkin or squash , prior to the appearance of severe symptoms, were subfractionated by differential centrifugation and aqueous two-phase partitioning, the majority of BL1 was found to co-fractionate with ER membrane- and BiP-containing fractions, based on enzyme marker assays and Western blot assays (Table 2.2) (53). A smaller fraction of BL1 was found to co-fractionate with plasma membrane-containing fractions from these young systemically infected leaves. It appeared that the tubules might be transitory in nature. When highly symptomatic leaves late in infection were fractionated, the levels of BL1 detected on Western blots was found to be significantly decreased, and the small amount of BL1 remaining was found to co-fractionate with plasma membrane- and cell wall-containing fractions (Table 2.2) (40, 53).

Based on these findings it was proposed that BR1 is a nuclear shuttle protein that binds newly replicated SqLCV ssDNA genomes in the nucleus and moves these to the cytoplasm. BL1, associated with unique ER-derived tubules, was proposed to trap BR1-genome complexes in the cytoplasm and guide these through the tubules to and across the walls of procambial cells. According to this model, following transport to adjacent uninfected phloem cells, the BR1-genome complexes would be released and target the viral genome back to the nucleus to begin a new round of infection. The predictions of this model, namely that BL1 and BR1 would interact in a regulated manner (bind and disengage), and that BR1 could shuttle into and out of

Table 2.2. Fractionation of BL1 from Systemically Infected Leaves[a]

Leaf Stage	Endoplasmic Reticulum	Plasma Membrane + Cell Wall	Golgi
Immature, prior to appearance of severe symptoms	65 -70%	30-35%	0%
Mature, highly symptomatic	0%	100%	0%

[a]Based on differential centrifugation and phase partitioning. Marker enzymes were antimycin A-insensitive NADH-dependent cytochrome *c* oxidase (ER), Triton-simulated ITPase (Golgi), and vanadate-sensitive K^+-Mg^{2+}-ATPase (plasma membrane).

the nucleus, were tested using transient expression assays in transfected tobacco protoplasts (43, 44).

2.2 BR1 and BL1 Cooperatively Interact in Protoplasts

Expression vectors in which *BL1* or *BR1* were cloned as transcriptional fusions to the cauliflower mosaic virus 35S promoter were electroporated into tobacco protoplasts derived from a suspension cell line (*Nicotiana tabacum* L cv. Xanthi), and each MP was detected at different times following transfection by indirect immunofluorescence staining and confocal microscopy (43, 44). To establish the validity of this model cultured cell system, the subcellular localizations of BR1 and BL1 when each was expressed individually were first examined. BR1 was found to localize to the nuclei of the transfected protoplasts, consistent with its localization in infected plants. BL1 localized to the cortical cytoplasm located at the periphery of expressing protoplasts, also consistent with its location in infected plants (Table 2.1 and Fig. 2.2a,b) (40, 44, 53). Hence, this transient expression assay is a valid model for studying the subcellular localization of BR1 and BL1.

Strikingly, when BR1 and BL1 were co-expressed in the same protoplasts, BL1 remained in the cortical cytoplasm, but all of BR1 was redirected to the cell periphery where it co-localized with BL1 (Fig. 2.2c-e) (43, 44). Thus, BR1 and BL1 do interact. Several BL1 mutants that mislocalized throughout the cytoplasm could still redirect BR1 from the nucleus, with BR1 now co-localizing with the particular BL1 mutant in the cytoplasm (44). This suggests that BL1 and BR1 directly interact.

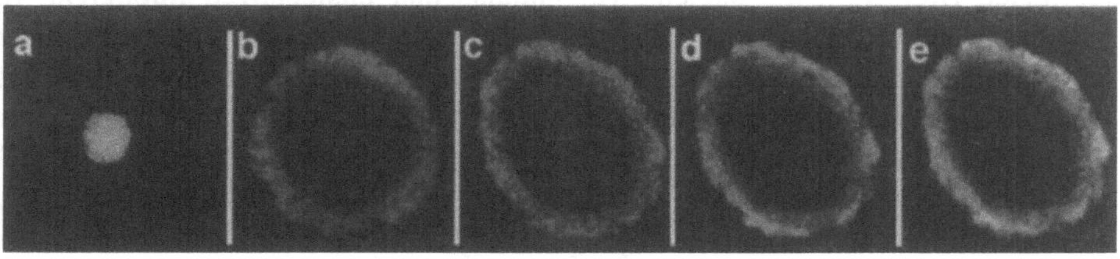

Fig. 2.2 Immunofluorescence staining of tobacco protoplasts expressing (a) BR1; (b) BL1; or (c, d) both BL1 and BR1. BR1 detected with secondary antibody coupled to fluorescein (green), and BL1 with secondary antibody coupled to TexasRed™ (red). (e) Superimposed images from "c" and "d" showing co-localization of BL1 and BR1 in this cell.

The relevance of these findings to the functions of BR1 and BL1 in infected plants was demonstrated through the analysis of alanine-scanning mutants of each MP, each of which had been characterized for their defects in infectivity, pathogenesis and host range properties (25). BL1 mutants that mislocalized throughout the cytoplasm of tobacco protoplasts were null in infectivity, indicating that correct targeting of BL1 to the cortical cytoplasm was essential for its function *in planta*. As a number of these BL1 mutants could still interact with BR1, these findings further demonstrated that the ability of BL1 to interact with BR1 and its correct targeting to the cell periphery were independently mutable functions (44). Similarly, BR1 mutants that mislocalized to the cytoplasm were null in their infectivity, indicating that correct targeting of BR1 to the nucleus was essential for its function in infectivity. Again, several of these cytoplasmically mislocalized BR1 mutants would still be redirected to the cortical cytoplasm by wild type BL1, indicating that nuclear targeting of BR1 and its ability to interact with BL1 were also independently mutable functions (43).

These results show that BL1 and BR1 interact cooperatively to facilitate movement of the viral genome (45). BR1 is a nuclear shuttle protein that can enter and exit from the nucleus, and as a ssDNA binding protein can thus shuttle the viral genome into and out of the nucleus. BL1 provides directionality to viral movement by trapping BR1 (presumably BR1-ssDNA complexes) in the cytoplasm and redirecting these to the cortical cytoplasm, where it acts to facilitate movement of the BR1-genome complexes into adjacent uninfected cells. The interaction of BL1 and BR1 is predicted to be regulated since the BR1-genome complexes should be released in the recipient cells to target the SqLCV genome back to the nucleus. Given that the interaction of the two MPs and the subcellular targeting of each could be mutated independently (see above), further analysis of our alanine scanning mutants allowed us to define these functional domains within BR1 and BL1.

2.2.1 Functional Domains within BL1

To identify functional domains within BL1, alanine-substituted missense mutants ("alanine-scanning") (9) of BL1 were expressed in protoplasts, either individually or in the presence of wild type BR1, and examined for their subcellular localization and ability to redirect BR1 from the nucleus, respectively (44). The defects observed for each BL1 mutant in these assays correlated well with their infectivity defects in cucurbit hosts. Three general types of BL1 mutants were identified in these transient expression assays: (1) mutants that mislocalized throughout the cytoplasm, but still redirected BR1 from the nucleus; (2) mutants that mislocalized throughout the cytoplasm and did not redirect BR1 from the nucleus; and (3) mutants that were correctly localized to the cell periphery, but were defective in their interaction with BR1. Mutants in the first two categories were null in infectivity; those in the third category retained partial infectivity or were null, correlating with their being partially or fully defective in their ability to interact with BR1 in the protoplast assay. BL1 mutants in the second category were judged likely to be misfolded and thus not useful for defining functional domains within BL1.

The BL1 mutants that mislocalized throughout the cytoplasm, but still redirected BR1 from the nucleus defined the subcellular targeting domain of BL1 (Fig. 2.3). Those mutants that were correctly localized to the cell periphery, but were defective in their interaction with BR1 were of two types: (1) mutants that were partially defective

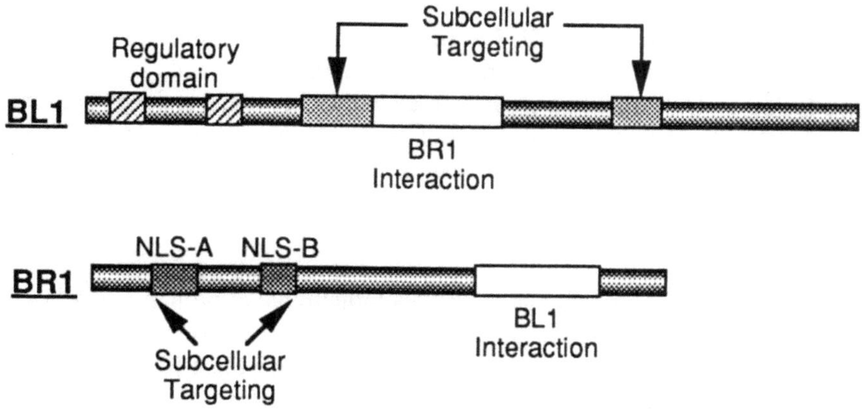

Fig. 2.3 Proposed functional domains of BL1 (top) and BR1 (bottom).

in their ability to redirect BR1 from the nucleus mapped to the central region of BL1 (Fig. 2.3); and (2) a separate clustering of mutants that mapped to the N-terminal region of BL1 which retained the ability to redirect BR1 to the cell periphery but were null in infectivity. It was suggested that these BL1 mutants might stably interact with BR1, failing to release it, as hypothesized by the model (44, 45) and time course studies show this to be the case (A.A. Sanderfoot and S.G. Lazarowitz, unpublished). Thus it appears that BL1 has a central domain required for BL1 to interact with BR1, which is flanked by residues essential for targeting BL1 to the cortical cytoplasm. In addition, we suggest that there is a regulatory domain within the N-terminal region of BL1 that regulates the activity of the central interactive domain (Fig. 2.3) (44, 45).

2.2.2 Functional Domains within BR1

In a similar manner, functional domains within BR1 were defined based on transient expression studies with alanine-scanning mutants in BR1 that either mislocalized to the cytoplasm but still interacted with BL1, or could target to the nucleus but were defective in their ability to interact with BL1. In addition, translational fusions of segments of *BR1* to β-glucuronidase (*GUS*) were tested in the transient expression assay for their ability to target GUS activity to the nucleus, thus defining the nuclear localization signals (NLSs) in BR1 (Fig. 2.3) (43).

Two NLSs were identified within the N-terminal half of BR1, based on the GUS fusion studies combined with site-directed mutational studies. NLS-A is a bipartite signal located within the 22-amino acid peptide encompassing residues 21-42 of BR1. NLS-B is an SV40-type NLS within the 16-amino acid peptide encompassing residues 81-96 of BR1 (43).

BR1 mutants which were partially defective in their ability to interact with BL1 mapped to the C-terminal ~80 amino acids of BR1. However, a complicating factor was that these mutants were also partially defective in their subcellular targeting, localizing to the perinuclear region prior to their final targeting to the nucleus or cytoplasm despite the fact that there are no NLSs within this region of BR1 (Fig. 2.3) (43). Thus, GUS fusion studies, in which the fusion protein was detected by indirect immunofluorescence staining and tested for its ability to interact with BL1, were done to clearly demonstrate the presence of the interactive domain within the C-terminal 80 amino acids of BR1. A fusion of GUS to the N-terminal 113 amino acids of BR1 localized to protoplast nuclei and remained nuclear in the presence of BL1. In contrast, a fusion protein consisting of GUS fused to the C-terminal residues 110-254 of BR1 was relocalized to the cortical cytoplasm in the presence of BL1. These findings, combined with our mutational studies, defined the interactive domain of BR1 as being located within the last 80 amino acids of the protein (Fig. 2.3) (43).

3 Future Directions

Genetic studies have demonstrated that movement proteins are of primary importance in defining the host range and pathogenic properties of plant viruses (8, 10, 24, 25, 33, 35, 36). Hence, there has been an interest in understanding the mechanism of MP function as a means to develop approaches for engineering disease-resistant plants (1, 4, 8). Molecular and cellular studies now make it clear that elucidating MP function is of more general interest as a potential means for defining pathways of intracellular and intercellular communication in plant cells. A number of specific associations of MPs with subcellular components have recently been identified. Such associations provide molecular probes for identifying the host cell components that constitute the signal transduction pathways for trafficking macromolecules into and out of the nucleus, through the cytoplasm and across cell walls.

The MPs encoded by TMV, CMV and red clover mosaic virus have been shown to affect the gating properties of plasmodesmata, and at least the first two MPs have been shown in immunogold labeling and/or GFP fusion studies to localize to plasmodesmata, which otherwise do not appear to be grossly altered in morphology (3, 12, 37, 38). Mutational studies further suggest that the C-terminal 73 amino acids of the TMV 30 kDa MP are required for targeting to plasmodesmata, and that the pertinent sequences act in *cis*, since the 30 kDa MP will potentiate movement of GUS through plasmodesmata between trichome cells when fused to GUS but not if expressed in *trans* in the same cell (5, 52). Such studies suggest the existence of a "plasmodesmal targeting sequence" within the 30 kDa protein. Precise identification of such a putative targeting signal and investigation of the cellular components in microtubules, the plasma membrane and/or plasmodesmata with which it interacts provides a novel approach for investigating the targeting of macromolecule to, and their regulated transport through plasmodesmata. The involvement of microtubules in this targeting process is suggested by studies demonstrating co-localization of the TMV 30 kDa MP to microtubules in tobacco protoplasts and infected plants, detected either by fluorescence of a fused GFP tag or by immunostaining for the MP itself (19, 31). Whether microtubule association and plasmodesmal targeting involve the same or distinct segments of the 30 kDa MP, and how interaction of this MP with plasmodesmata affects the gating properties are fascinating cell biological problems waiting to be solved.

Our studies of the two MPs encoded by the geminivirus SqLCV have shown that BR1 is a nuclear shuttle protein, and that BL1 is associated with unique ER-derived tubules that extend from and cross the walls of developing phloem cells (41, 43, 53). In light of the viral genome being DNA and the phloem-limitation of viral infection, these functions make sense. Two classic NLSs have been identified within the N-terminal 100 amino acids of BR1 (43), thus further study of this protein can provide information on the mechanism of nuclear import in plant cells, an area that is only now being addressed using biochemical approaches (21, 22). However, what is unique and exciting is the potential of BR1 as a model for investigating regulated nuclear export, an area about which little is known for any eukaryotic organism. Our understanding of the mechanism of nuclear export has lagged behind that of nuclear import largely because of the difficulty in demonstrating export. Nuclear import appears to be more rapid than nuclear export; hence, at equilibrium proteins that shuttle into and out of the nucleus appear to be nuclear localized based on standard immune assays (17, 18, 46). BR1 is no exception here. Thus, to observe shuttling one must perturb this equilibrium, and the assays used to accomplish this are very difficult ones based on heterokaryon formation or microinjections into nuclei of HeLa cells or *Xenopus laevis* oocytes. In this regard, our finding that BL1 is a cytoplasmic trap for BR1 provides us with a convenient assay to investigate the export of BR1 from the nucleus as a means for defining what nuclear export signals within BR1 may be involved and with what proteins within the nucleoplasm and at the nuclear pore BR1 will interact during its export. Current studies of BR1 are directed at identifying a nuclear export signal (NES) in BR1 of the type recently found in HIV Rev protein, protein kinase inhibitor protein, and the transcription factor TFIIIA from *Xenopus* (14, 15, 56). The yeast two-hybrid system is also being used to identify cellular proteins that interact with BR1 (J.E. Hill and S.G. Lazarowitz, unpublished).

The association of BL1 with ER-derived tubules in immature phloem cells suggests that this MP may create channels through the cell wall in much the same way as plasmodesmata appear to be formed (20, 39). Viewed in this manner, BL1 provides a unique approach to study the mechanism by which the cortical ER is recruited to form the desmotubule and coordinated with cell plate formation to lead to the development of intercellular channels through the wall. It also provides a unique view of the mechanism by which a phloem-limited virus may move cell-to-cell and invade the leaf. While it has been assumed that invasion by phloem-limited viruses would follow the sink-to-source transition, as shown for mesophyll-infecting viruses, our results for SqLCV BL1 suggest that movement of this phloem-limited virus may be coordinated with phloem development (53) and thus that the virus assures invasion of the leaf as maturation of the vasculature is completed.

This potential coordination of virus movement with phloem development is currently being investigated through the use of GFP as a probe for visualizing SqLCV movement (C.M. Laukaitis and S.G. Lazarowitz, unpublished). Using an SqLCV mutant in which the coat protein coding sequence has been replaced with that encoding GFP (the coat protein is not required for SqLCV movement) we can use long wave UV light to visualize SqLCV infection, which we find not unexpectedly to be limited to the veins of pumpkin and squash (Fig. 3.1). In addition, we are using confocal microscopy to follow infection at the cellular level. These studies are allowing us to track virus movement through the infected host plant . Our results support the hypothesis that movement of this phloem-limited virus is coordinated with maturation of the phloem. We are also using direct interactive cloning and the yeast two-hybrid system to identify cellular proteins that interact with BL1, which should begin to shed light on the associations BL1 makes with the ER to form tubules and what associations such tubules make with the cell plasma membrane and wall.

Fig. 3.1 Systemic leaf from squash infected with SqLCV in which the coat protein gene has been replaced with that encoding GFP. Fluorescence shows restriction of SqLCV infection to the veins. GFP is visualized using a long wave UV light (excitation 389 nm).

In viewing the models proposed for virus movement, it may appear at first glance that there are three discrete models in which that for SqLCV falls between that of TMV and related viruses, and the model proposed for viruses like tomato spotted wilt and cauliflower mosaic virus which form tubules containing virus-like particles in mesophyll cells. We suggest that the situation may not be so complicated, and that there is a continuum here in which each virus has adapted the cytoskeletal and endomembrane systems of the cell according to the cellular environment it finds itself in to direct its genome through the cytoplasm to the cortical cytoplasm and across the wall. An association of MPs with microtubules has clearly been demonstrated for TMV, and there is some evidence to further suggest that associations with the ER may also occur (19, 31). We do not detect co-localization of BL1 with either microtubules or actin filaments; however, we have shown an association of BL1 with ER which we also detect in transfected protoplasts (53; and B.M. Ward and S.G. Lazarowitz, unpublished). What cellular proteins or lipids may be associated with the tubules formed by other viruses in mesophyll cells and the origin of such tubules has yet to be reported. It will be of interest to see what if any associations these latter tubules make with the cytoskeletal or endomembrane systems of the cell. Do these viruses, as well as TMV and related viruses, somehow recruit the desmotubule from plasmodesmata for movement of the virus genome or virus-like particles? Are the structures formed, or lack of apparent physical alteration of the plasmodesma, a function of the form in which the viral genome is transported (virus-like particle or MP-genome complex)? Does SqLCV movement in developing phloem cells take advantage of this association to intercede during the formation of the desmotubule and plasmodesmata to create virus-specific channels through the wall? The answers to these questions will not only provide a basic understanding of the mode of action of virus-encoded MPs, but will further our understanding how interactions with the cytoskeleton, endomembrane system and wall of the plant cell are regulated such that macromolecules are directed through the cytoplasm and between cells.

4 Acknowledgments

Studies cited in this chapter were supported by grants from the National Science Foundation (no. MCB-9417664) and U.S. Department of Agriculture (no. CGRP 95-37303-1710) to S.G.L. We thank Richard Medville for superb immunogold staining and electron microscopy, and Robert Turgeon and Rich for helpful discussions.

5 References

1. **Atabekov, J. G., and Y. L. Dorokhov.** 1984. Plant virus-specific transport function and resistance of plants to viruses. Adv. Virus Res. **29**:313-363.

2. **Atabekov, J. G., and M. E. Taliansky.** 1990. Expression of a plant virus-coded transport function by different virus genomes. Adv. Virus Res. **38:**201-248.
3. **Atkins, D., R. Hull, B. Wells, K. Roberts, P. Moore, and R. N. Beachy.** 1991. The tobacco mosaic virus 30K movement protein in transgenic tobacco plants is localized to plasmodesmata. Journal of General Virology. **72:**209-211.
4. **Baulcombe, D.** 1989. Strategies for virus resistance in plants. Trends in Gen. **5:**56-60.
5. **Berna, A., R. Gafny, S. Wolf, W. J. Lucas, C. A. Holt, and R. N. Beachy.** 1991. The TMV movement protein: role of the C-terminal 73 amino acids in subcellular localization. Virology. **182:**682-689.
6. **Citovsky, V., D. Knorr, G. Schuster, and P. Zambryski.** 1990. The P30 movement protein of tobacco mosaic virus is a single-stranded nucleic acid binding protein. Cell. **60:**637-647.
7. **Citovsky, V., and P. Zambryski.** 1993. Transport of nucleic acids through membrane channels: snaking through small holes. Annu. Rev. Microbiol. **47:**167-197.
8. **Cooper, B., M. Lapidot, J. A. Heick, J. A. Dodds, and R. N. Beachy.** 1995. A defective movement protein of TMV in transgenic plants confers resistance to multiple viruses whereas the functional analog increases susceptibility. Virol. **206:**307-313.
9. **Cunningham, B. C., and J. A. Wells.** 1989. High-resolution epitope mapping of hGH-receptor interactions by alanine-scanning mutagenesis. Science. **244:**1081-1085.
10. **Dawson, W. O., and M. E. Hilf.** 1992. Host-range determinants of plant viruses. Annual Review of Plant Physiology and Plant Molecular Biology. **43:**527-555.
11. **Deom, C. M., M. J. Oliver, and R. N. Beachy.** 1987. The 30-kilodalton gene product of tobacco mosaic virus potentiates virus movement. Science. **237:**389-394.
12. **Ding, B., J. S. Haudenshield, R. J. Hull, S. Wolf, R. N. Beachy, and W. J. Lucas.** 1992. Secondary plasmodesmata are specific sites of loxcalization of the tobacco mosaic virus movement protein in transgenic tobacco plants. The Plant Cell. **4:**915-928.
13. **Ding, B., Q.-B. Li, L. Nguyen, P. Palukaitis, and W. J. Lucas.** 1995. Cucumber mosaic virus 3a protein potentiates cell-to-cell trafficking of CMV RNA in tobacco plants. Virology. **207:**345-353.
14. **Fischer, U., J. Huber, W. C. Boelens, I. W. Mattaj, and R. Luhrmann.** 1995. The HIV-1 Rev activation domain is a nuclear export signal that accesses an export pathway used by specific cellular RNAs. Cell. **82:**475-483.
15. **Fridell, R. A., U. Fischer, R. Luhrmann, B. E. Meyer, J. L. Meinkoth, M. H. Malim, and B. R. Cullen.** 1996. Amphibian transcription factor IIIA proteins contain a sequence element functionally equivalent to the nuclear export signal of human immunodeficiency virus type 1 Rev. Proc. Natl. Acad. Sci. **93:**2936-2940.
16. **Fujiwara, T., D. Giesman-Cookmeyer, B. Ding, S. A. Lommel, and W. J. Lucas.** 1993. Cell-to-cell trafficking of macromolecules through plasmodesmata

potentiated by the red clover necrotic virus movement protein. Plant Cell. 5:1783-1794.

17. **Gerace, L.** 1995. Nuclear export signals and the fast track to the cytoplasm. Cell. **82**:341-344.
18. **Gorlich, D., and I. W. Mattaj.** 1996. Nucleocytoplasmic transport. Science. **271**:1513-1518.
19. **Heinlein, M., B. L. Epel, H. S. Padgett, and R. N. Beachy.** 1996. Interaction of tobamovirus movement proteins with the plant cytoskeleton. Science. **270**:1983-1985.
20. **Hepler, P. K.** 1982. Endoplasmic reticulum in the formation of the cell plate and plasmodesmata. Protoplasma. **111**:121-133.
21. **Hicks, G. R., H. M. S. Smith, S. Lobreaux, and N. V. Raikhel.** 1996. Nuclear import in premeabilized protoplasts from higher plants has unique features. Plant Cell. **8**:1337-1352.
22. **Hicks, G. R., H. M. S. Smith, M. Shieh, and N. V. Raikhel.** 1995. Three classes of nuclear import signals bind to plant nuclei. Plant Physiol. **107**:1055-1058.
23. **Hull, R.** 1991. The movement of viruses within plants. Semin. Virol. **2**:89-95.
24. **Ingham, D. J., and S. G. Lazarowitz.** 1993. A single missense mutation in the BR1 movement protein alters the host range of the squash leaf curl geminivirus. Virology. **196**:694-772.
25. **Ingham, D. J., E. Pascal, and S. G. Lazarowitz.** 1995. Both geminivirus movement proteins define viral host range, but only BL1 determines viral pathogenicity. Virology. **207**:191-204.
26. **Kormelink, R., M. Storms, J. van Lent, D. Petters, and R. Goldbach.** 1994. Expression and subcellular localization of the NS_M protein of tomato spotted wilt virus (TSWV), a putative viral movement protein. Virology. **200**:56-65.
27. **Lazarowitz, S. G.** 1992. Geminiviruses: Genome structure and gene function. Critical Reviews in Plant Sciences. **11**:327-349.
28. **Linstead, P. J., G. J. Hills, K. A. Plaskitt, I. G. Wilson, C. L. Harker, and A. J. Maule.** 1988. The subcellular localization of the gene 1 product of Cauliflower mosaic virus is consistent with a function associated with virus spread. J. Gen. Virol. **69**:1809-1818.
29. **Lucas, W. J., B. Ding, and C. van der Schoot.** 1993. Plasmodesmata and the supracellular nature of plants. New Phytol. **125**:435-476.
30. **Lucas, W. J., and R. L. Gilbertson.** 1994. Plasmodesmata in relation to viral movement within leaf tissues. Annu. Rev. Phytopathol. **32**:387-411.
31. **McLean, B. G., J. Zupan, and P. C. Zambryski.** 1996. Tobacco mosiac virus movement protein associates with the cytoskeleton in tobacco cells. Plant Cell. **7**:2101-2114.
32. **Meshi, T., D. Hosokawa, M. Kawagishi, Y. Watanabe, and Y. Okada.** 1992. Reinvestigation of intracellular localization of the 30 kD protein in tobacco protoplasts infected with tobacco mosaic virus RNA. Virology. **187**:809-813.
33. **Meshi, T., F. Motoyoshi, T. Maeda, S. Yoshiwoka, H. Watanbe, and Y. Okada.** 1989. Mutations in the tobacco mosaic virus 30-kD protein overcome Tm-2 resistance in tomato. Plant Cell. **1**:515-522.

34. **Moore, P. J., C. A. Fenczik, C. M. Deom, and R. N. Beachy.** 1992. Developmental changes in plasmodesmata in transgenic tobacco expressing the movement protein of tobacco mosaic virus. Protoplasma. **170**:115-127.
35. **Nelson, R. S., G. Li, R. A. J. Hodgson, B. R.N., and M. H. Shintaku.** 1993. Impeded phloem-dependent accumulation of the masked strain of tobacco mosaic virus. Molecular Plant Microbe Interactions:45-54.
36. **Ohno, T., N. Takamutsu, T. Meshi, Y. Okada, M. Nishiguchi, and Y. Kiho.** 1983. Single amino acid substitution in 30 K protein of TMV defective in virus transport function. Virology. **131**:255-258.
37. **Oparka, K. J., P. Boevink, and S. Santa Cruz.** 1996. Studying the movement of plant viruses using green fluorescent protein. Trends Plant Sci. **1**:412-418.
38. **Oparka, K. J., A. G. Roberts, D. A. M. Prior, S. Chapman, D. Baulcombe, and S. Santa Cruz.** 1995. Imaging the green fluorescent protein in plants - viruses carry the torch. Protoplasma. **189**:133-141.
39. **Overall, R. L., J. Wolf, and B. E. S. Gunning.** 1982. Intercellular communication in *Azolla* roots: I. Ultrastructure of Plasmodesmata. Protoplasma. **111**:134-150.
40. **Pascal, E., P. E. Goodlove, L. C. Wu, and S. G. Lazarowitz.** 1993. Transgenic tobacco plants expressing the geminivirus BL1 protein exhibit symptoms of viral disease. Plant Cell. **5**:795-807.
41. **Pascal, E., A. A. Sanderfoot, B. M. Ward, R. Medville, R. Turgeon, and S. G. Lazarowitz.** 1994. The geminivirus BR1 movement protein binds single-stranded DNA and localizes to the cell nucleus. Plant Cell. **6**:995-1006.
42. **Perbal, M.-C., C. L. Thomas, and A. J. Maule.** 1993. Cauliflower mosaic virus gene I product (P1) forms tubular structures which extend from the surface of infected protoplasts. Virology. **195**:281-285.
43. **Sanderfoot, A. A., D. J. Ingham, and S. G. Lazarowitz.** 1996. A viral movement protein as a nuclear shuttle: The geminivirus BR1 movement protein contains domains essential for interaction with BL1 and nuclear localization. Plant Physiol. **110**:23-33.
44. **Sanderfoot, A. A., and S. G. Lazarowitz.** 1995. Cooperation in viral movement: the geminivirus BL1 movement protein interacts with BR1 and redirects it from the nucleus to the cell periphery. Plant Cell. **7**:1185-1194.
45. **Sanderfoot, A. A., and S. G. Lazarowitz.** 1996. Getting it together in plant virus movement: cooperative interactions between bipartite geminivirus movement proteins. Trends in Cell Biol. **6**:353-358.
46. **Schmidt-Zachmann, M. S., C. Dargemont, L. C. Kuhn, and E. A. Nigg.** 1993. Nuclear export of proteins: the role of nuclear retention. Cell. **74**:493-504.
47. **Stanley, J.** 1991. The molecular determinants of geminivirus pathogenesis. . **2**:139-149.
48. **Storms, M. M. H., R. Kormelink, D. Peters, J. W. M. Van Lent, and R. W. Goldbach.** 1995. The nonstructural NSm protein of tomato spotted wilt virus induces tubular structures in plant and insect cells. Virology. **214**:485-493.
49. **van Lent, J., M. Storms, F. van der Meer, J. Wellink, and R. Goldbach.** 1991. Tubular structures involved in movement of cowpea mosaic virus are also formed in infected cowpea protoplasts. J. Gen. Virol. **72**:2615-2623.

50. **van Lent, J., J. Wellink, and R. Goldbach.** 1990. Evidence for the involvement of the 58K and 48K proteins in the intercellular movement of cowpea mosaic virus. J. Gen. Virol. **71:**219-223.
51. **Waigmann, E., W. J. Lucas, V. Citovsky, and P. Zambryski.** 1994. Direct functional assay for tobacco mosaic virus cell-to-cell movement protein and identification of a domain involved in increasing plasmodesmal permeability. Proc. Natl. Acad. Sci. USA. **91:**1433-1437.
52. **Waigmann, E., and P. Zambriski.** 1995. Tobacco mosiac virus movement protein-mediated protein transport between trichome cells. Plant Cell. **7:**2069-2079.
53. **Ward, B. M., R. Medville, S. G. Lazarowitz, and R. Turgeon.** 1997. The geminivirus BL1 movement protein is associated with endoplasmic reticulum-derived tubules in developing phloem cells. J. Virology. **71:3726-3733.**
54. **Weiczorek, A., and H. Sanfacon.** 1993. Characterization and subcellular localization of tomato ringspot nepovirus putative movement protein. Virology. **194:**734-742.
55. **Wellink, J., J. W. M. van Lent, J. Verver, T. Sijen, R. W. Goldbach, and A. van Kammen.** 1993. The cowpea mosaic virus M RNA-encoded 48-kilodalton protein is responsible for induction of tubular structures in protoplasts. J. Virol. **67:**3660-3664.
56. **Wen, W., J. L. Meinkoth, R. Y. Tsien, and S. S. Taylor.** 1995. Identification of a signal for rapid export of proteins from the nucleus. Cell. **82:**463-473.
57. **Wolf, S., C. M. Deom, R. N. Beachy, and W. J. Lucas.** 1989. Movement protein of tobacco mosaic virus modifies plasmodesmatal size exclusion limit. Science. **246:**377-379.

Signaling during legume root symbiosis

Catherine Albrecht and Ton Bisseling

Dept. Molecular Biology, Agricultural University Wageningen, Dreijenlaan 3, 6703 HA Wageningen, the Netherlands.

Key words : *Rhizobium*, Mycorrhizae, Nodulins, Nod factor

1. Introduction

Legumes have the ability to establish a symbiotic interaction with different soil microorganisms. Arbuscular mycorrhiza and nodules represent the two main categories. In both systems, the infection process culminates with the colonization of host plant cells by endosymbiotic microorganisms and the creation of a highly specialized symbiotic interface. The formation of this interface leads to profound cytological modifications of the host cell, membrane biogenesis and physiological adaptation to cohabitation.

In this chapter first the signalling between rhizobia and legume roots that sets nodule formation in motion will be discussed and then a comparison with the mycorhizal symbiosis will be made

2. Legume nodulation

During root nodule formation, infection and nodule organogenesis occur simultaneously. As a start of the infection process, the rhizobia induce root hair deformation and curling. Curled root hairs form so-called Shepherd's crooks, in which rhizobia become trapped in a small confinement formed by the curls. There the rhizobia enter the root, by local hydrolysis of the root hair cell wall and invagination of the plasma membrane (van Spronsen *et al.* 1994). Around this invagination new cell wall material is deposited leading to the formation of a so-called infection thread (see Kijne *et al.* 1992). Concurrently with the infection process, root cortical cells start dividing and in this way nodule primordia are formed. The infection thread filled with dividing rhizobia grows towards the nodule primordia. Primordial cells become infected by the rhizobia by endocytosis and they become surrounded by a plant derived membrane, the peribacteroid membrane (Kijne 1992, Newcomb 1981). After infection the primordia develop into root nodules.

3. Rhizobial *nod* genes and Nod factors

The rhizobial nodulation (*nod*) genes are essential for nodule formation. The regulatory *nodD* gene is the only constitutively expressed *nod* gene and its protein

NATO ASI Series, Vol. H 104
Cellular Integration of Signalling Pathways in Plant Development
Edited by F. Lo Schiavo, R. L. Last, G. Morelli, and N. V. Raikhel

product regulates the transcription of the other *nod* genes (Mulligan and Long 1985). Signals - (iso)-flavonoids and betaines- secreted by the plant activate the NodD protein, after which it induces transcription of the other *nod* genes (Djordjevic *et al.* 1987, Peters *et al.* 1986, Phillips *et al.* 1995). These genes encode nodulation (Nod) proteins, most of which are involved in the production of signal molecules called Nod factors.

Nod factors are lipo-chitooligosaccharide with a chitin ß-1,4-linked N-acetyl-D-glucosamine backbone, varying in length between three to six sugar units and a fatty acyl chain on the C-2 position of the non-reducing sugar. Three of the *nod* genes, *nodABC*, are required for the synthesis of this basic structure. Other *nod* genes, varying in different rhizobial species, are involved in the modification of the fatty acyl chain or the addition of strain specific substitutions that are important in determining host specificity (Long 1996).

4. The role of plant compounds in nodule organogenesis

Purified Nod factors can induce early responses of nodulation, like root hair deformation, induction of expression of nodulin (nodule specific) genes and nodule primordium formation. One of the first genes that is activated by Nod factor addition is *ENOD40*. This gene is induced within a few hours in the pericycle opposite the proto-xylem pole, whereas cell divisions in the cortex first occur after 24 hours (W.C. Yang, 1993, unpublished).

Recently it was shown that the early nodulin genes *ENOD40* encode peptides of 10-13 amino acids. *ENOD40* genes have been found in legumes as well as non-legumes and appear to play a general role in regulating plant development. For example ectopic expression in alfalfa disturbed embryogenesis and regeneration (Crespi *et al.* 1994). In tobacco the ENOD40 peptide changes the response to phytohormones. Wildtype protoplasts divide with a maximum frequency of about 50 % at 0.9 μM kinetin and 5.5 μM NAA and at higher concentrations of these hormones the efficiency of division drops. The ENOD40 peptide, at picomolar levels, conferred tolerance of high auxin and cytokinin concentration (van de Sande 1996, unpublished). It is probable that the ENOD40 peptide affects the response to phytohormones in cortical cells and is involved in the mitotic reactivation of cortical cells.

5. Nod factor perception

The mechanism of Nod factor perception and transduction is still poorly understood, but studies on the relationship of Nod factor structure and its inducing activity provided some insight in the mechanisms controling Nod factor perception in the epidermis.

In the epidermis Nod factors induce within minutes depolarisation of the plasma membrane potential (Erhardt *et al.* 1992, Felle *et al.* 1995), spiking of cytoplasmic calcium levels around the nuclei in the root hairs (Erhardt *et al.* 1996) and alkalinization of the root hair cytoplasm (Felle *et al.* 1996). These responses preceed root hair deformation which is first visible one hour after addition of Nod factor

(Heidstra *et al.* 1994). However, whether these physiological responses are part of the signal tranduction cascades leading to deformation is unclear. When root hair deformation is set in motion several early nodulin genes are induced in the the epidermis of which *ENOD12* is the best studied example (Journet *et al.* 1994, Vijn *et al.* 1995a, b). For a detailed overview of the responses induced by Nod factors in the epidermis see *et al.* (1996), Long (1996) and Spaink (1996).

By using *Rhizobium* mutants and various purified Nod factors, the structure function relationship of Nod factors and epidermal responses has been studied . The most complete studies have been done with the *R. meliloti* alfalfa interaction and these will be summarised here. The major *Rhizobium meliloti* Nod factor has a tetrameric N-acetyl-glucosamine backbone, with a C16:2 acyl chain, an O-sulphate substitution at the reducing sugar moiety and an O-acetate substitution at the non-reducing sugar moiety.

The epidermal responses can be divided in three categories, based on the structural properties of *R. meliloti* Nod factors required to induce a response

1) Alkalisation of root hair cytoplasm requires a chitin backbone with an acyl chain, but the structure of the fatty acyl chain and the presence of the sulphate substitution is not important.

2) Deformation of root hairs, induction of *ENOD12* expression and membrane depolarisation require the basic Nod factor, but the sulfate substitution is essential.

3) Infection thread formation is the most stringent response since it requires both the sulfate and the acetyl substitutions and the C16:2 unsaturated acyl chain on the Nod factor.

The different Nod factor structure requirements of induced responses indicate that more than one receptor may be involved in Nod factor perception (Ardourel *et al.* 1994). This is supported by the studies of Felle and co-workers (1995, 1996) who demonstrated that sulfated as well as of non-sulfated Nod factor induce alkalinization of alfalfa root hairs. A second addition of the same Nod factor had no effect, but when first non-sulfated Nod factor was added at saturating concentration, followed by sulfated Nod factor, alkalinization took place at each addition of Nod factor. This indicates that the sulfated and non-sulfated Nod factors bind to different receptors (Felle *et al.* 1996).

6. Are elements of Nod factor activated perception and transduction recruited from the mycorrhizal symbiosis?

The ability of plant to interact with mycorrhizal fungi is wide spread in the plant kingdom and a very ancient phenomenon. 80% of the plant families are able to establish arbuscular mycorrhiza (AM) and fossil evidence suggests that this kind of symbiosis existed more than 400 millions of years ago in the tissue of the first land plant (Pirozynski and Dalpe 1989, Simon et *al.* 1993). AM is aspecific, a single species of fungus has the capacity to colonize many plant species and furthermore a given plant can interact with different fungal species (Harley and Smith 1983). In contrast, *Rhizobium* nodulation is highly specific and, with the exception of *Parasponia*, restricted to the legume family. Within this family, individual species, strains or biovar of bacteria nodulate a restricted set of host plants. The molecular mechanism of the host specificity is in part controled by the structure of Nod factors (Long 1996)

When contacting the root surface, fungal hyphae differentiate and form an appressorium. Penetration of the root occuring via hyphae that originates from the appressoria is followed by intercellular growth toward the inner cortex. On reaching the inner cortex, fungal hyphae penetrate the cells and differentiate into arbuscules. At this stage, in both endosymbiotic interactions, the endosymbiont are separated from the plant cytoplasm by a plant-derived perisymbiotic membrane (Brewin 1990).

Despite the obvious differences between the two symbioses, plant mutants have shown that the same plant genes can modulate some steps in both arbuscular mycorrhiza and nodule development. Such mutants (nod$^-$, myc$^-$) have been found in *M. truncatula*, alfalfa, faba bean, bean and pea (Sagan *et al.* 1995, Bradbury *et al.* 1991, Duc *et al.* 1989, Shirtliffe and Vessey 1996). In the latter 2 genes (*sym8* and *sym9*) have been identified that are blocked at early stages of the interaction. The *sym8* mutants show the most severe phenotype, since they are blocked upstream of root hair deformation (had$^-$, inf$^-$, noi$^-$), whereas *sym9* is blocked downstream of this response (had$^+$, inf$^-$, noi$^-$) (Markwei and LaRue 1992). In the interaction with arbuscular mycorrhiza, both mutants are blocked at the stage of appressoria formation (Gianinazzi-Pearson *et al.* 1991).

The ancestral nature of arbuscular mycorrhiza together with the involvment of common host genes suggests that part of the plant processes leading to nodulation may have evolved from those already established for the fungal endosymbiosis (Duc *et al.* 1989, LaRue and Weeden 1992). This is furthermore supported by the observation that rhizobia as well as purified Nod factors have the capacity to extend the level of mycorrhizal colonization. This effect seems to be mediated by flavonoids which are produced in increased amounts in response to specific rhizobial Nod factors. Indeed, only Nod factors that induce the production of flavonoids are active (Xie *et al.* 1995). It has been observed that the profile of flavonoids as well as the expression pattern of genes encoding enzymes specific for flavonoids biosynthesis, alters following colonization by mycorrhizal fungi (Harrisson *et al.* 1993). Some of the flavonoids that are synthetized promote spore germination, hyphal elongation and branching and probably also influence the growth of the fungus within the root (Nair *et al.* 1991; Tsai *et al.* 1991). These data suggest the existence of common steps in early signalling events between host plant and the two endosymbionts.

Recently, Van Rhijn and co-workers (1997) have reported that in addition to the genetically identified *sym* genes, the early nodulines *ENOD2* and *ENOD40* are also involved in mycorrhiza establishment. *ENOD40* is induced by Nod factors (lipochito-oligosaccharides) as well as chitin fragments (Minami *et al.* 1996). The chitin-like nature of the rhizobial Nod factor suggests that chitooligosaccharides, released from fungal cell wall by mycorrhiza-regulated chitinases, may trigger *ENOD40* induction (Albrecht *et al.* 1994, Dumas *et al.* 1992). The early nodulin gene *ENOD2* is activated at a relatively late stage of legume nodulation when the Nod factor induced primordium differentiates in a nodule and the nodule parenchyma is formed (van deWiel *et al.* 1990). Hence the activation of this gene is not the result of a simple Nod factor activated signal transduction cascade. In addition it is unclear whether Nod factor activated responses in inner layers of the root (cortex and pericycle) involve the translocation of Nod factors to these layers or whether second messengers generated in the epidermis- the tissue directly exposed to Nod factor- induce these responses. Expression of these two early nodulin genes in spontaneous nodules as well as in uninoculated roots upon application of the cytokinin 6-benzylaminopurine support the hypothesis that activation of these transcripts is related to generation of secondary signal messengers leading to altered hormone balance in the root. For these reasons the reported studies with *ENOD40* and *ENOD2* did not provide evidence that Nod factor activated signal transduction pathways are conserved (van Rhijn *et al.* 1997). To study whether such pathways are conserved we selected the early nodulin genes *ENOD12* and *ENOD5* for further studies. These genes are among the best available molecular markers to study Nod factor activated signal transduction since these genes are activated within a few hours in the epidermis, the tissue directly exposed to Nod factors. Furthermore, induction of these genes requires a Nod factor since they are not activated by chitin fragments (Horvath *et al.* 1993, Journet *et al.* 1994).

We found here that *ENOD12* as well as *ENOD5* are activated during infection with AM mycorrhizae. Analyses of the *sym*8 mutant show that in both interactions the induction of *ENOD12* and *ENOD5* is blocked. Thus *Sym*8 appears to encode a protein involved in signal perception or transduction leading to *ENOD5* and *ENOD12* induction.

References

Albrecht, C., Burgess, T., Dell, D. and Lapeyrie, F. (1994). *Eucalyptus* root chitinases and peroxidases stimulation following infection by a wide range of *Pisolithus tinctorius* ectomycorhizal strains of variable aggressivness. *The New Phytologist* **127**, 217-222

Ardourel, M., Demont, N., Debellé, F., Maillet, F., de Billy, F., Promé, J.-C., Dénarié, J. and Truchet, G. (1994) *Rhizobium meliloti* lipooligosaccharide nodulation factors: different structural requirements for bacterial entry into target root hair cells and induction of plant symbiotic developmental responses. *Plant Cell.* **6**, 1357-1374.

Bradbury , S.M., Peterson, R.L. and Bowley, S.R. (1991). Interaction between three alfalfa nodulation genotypes and two *Glomus* species. *New Phytologist* **119**, 115-120

Brewin N. J. 1990. The role of the plant plasma membrane in symbiosis. Pages 351-375 in : The Plant Plasma Membrane : Structure, Function and Molecular Biology. C. Larsson and I. M. Moller, eds. Springer-Verlag, Berlin

Crespi, M.D., Jurkevitch, E., Poiret, M., d'Aubenton-Carafa, Y., Petrovics, G., Kondorosi, E. and Kondorosi, A. (1994) *ENOD40*, a gene expressed during nodule organogenesis, codes for a non-translatable RNA involved in plant growth. *EMBO J.* **13**, 5099-5112.

Djordjevic, M.A., Redmond, J.W., Batley, M. and Rolfe, B.G. (1987) Clovers secrete specific phenolic compounds which either stimulate or repress *nod* gene expression in *Rhizobium trifolii. EMBO J.* **6**, 1173-1179.

Duc, G., Trouvelot, A., Gianinazzi-Pearson, V. and Gianinazzi, S. (1989). First report of non-mycorrhizal plant mutants (Myc-) obtained in pea (*Pisum sativum*) and Fababean (*Vicia Faba* L.). *Plant science* **60**, 215-222

Dumas-Gaudot E, Grenier J, Furlan V, Asselin A. 1992. Chitinase, chitinosane and ß-1,3-glucanase activities in *Allium* and *Pisum* roots colonized by *Glomus* species. *Plant Science* **84**, 17-24

Ehrhardt, D.W., Atkinson, E.M., and Long, S.R. (1992) Depolarization of alfalfa root hair membrane potential by *Rhizobium meliloti* Nod factors. *Science* **256**, 998-1000.

Ehrhardt, D.W., Wais, R., and Long, S.R. (1996) Calcium spiking in plant root hairs responding to *Rhizobium* nodulation signals. *Cell* **85**, 673-681.

Felle, H.H., Kondorosi, E., Kondorosi, A. and Schultze, M. (1995) Nod signal-induced plasma membrane potential changes in alfalfa root hairs are differentially sensitive to structural modifications of the lipochitooligosaccharide. *Plant J.* **7**, 939-947.

Felle, H.H., Kondorosi, E., Kondorosi, A. and Schultze, M. (1996) Rapid alkalinization in alfalfa root hairs in response to rhizobial lipochitooligosaccharide signals. *Plant J.* **10**, 295-301.

Gianinazzi-Pearson V., Gianinazzi, S, Guillemin, J.P., Trouvelot, A. and Duc, G. (1991). Genetic and cellular analysis of resistance to vesicular arbuscular (VA) mycorrhizal fungi in pea mutants. *Advances in Molecular Gentics of Plant-Microbe Interactions* (Hennecke, H. and Verma, D.P.S., eds) 336-342, Kluver

Harley, J.L. and Smith, S.E. (1983). Mycorrhizal symbiosis, Academic Press

Harrisson M.J. and Dixon R.A. (1993). Isoflavonoids accumulation and expression of defense gene transcripts during the establishment of vesicular-arbuscular mycorrhizal associations in roots of *Medicago truncatula. Mol. Plant-Microbe Interact.* **6**, 643-654

Heidstra, R., Geurts, R., Franssen, H., Spaink, H.P., van Kammen, A., and Bisseling, T. (1994) Root hair deformation activity of nodulation factors and their fate on *Vicia sativa.. Plant Physiol.* **105**, 787-797.

Horvath B., Heidstra R., Lados M., Moerman M., Spaink H. P. , Prome J. C., Van Kammen A. and Bisseling T. (1993). Lipo-oligosaccharides of *Rhizobium* induce infection-related early nodulin gene expression in pea root hairs. *The plant journal* **4**, 727-733

Journet, E.P., Pichon, M., Dedieu, A., de Billy, F., Truchet, G. and Barker, D.G. (1994) *Rhizobium meliloti* Nod factors elicit cell specific transcription of the *ENOD12* gene in transgenic alfalfa. *Plant J.* **6**, 241-249.

Kijne, J.W. (1992) The *Rhizobium* infection process, In *Biological Nitrogen Fixation* (Stacey G., Burris R.H., Evans H.J., ed), pp. 349-398, Chapman and Hall, New York.

LaRue, T. A. and Weeden, N. F. (1994) The symbiosis genes of the host in *Proceedings of the first European Nitrogen Fixation Conference*, eds. Kiss, G. B. & Endre, G. (Officina, Szeged, Hungary), 147-151

Long, S.R. (1996) *Rhizobium* symbiosis: Nod factors in perspective. *Plant Cell* **8**, 1885-1898.

Markwei C. M and LaRue T. A. (1991) Phenotypic characterization of *sym*8 and *sym*9, two genes conditioning non-nodulation in *Pisum sativum* 'Sparkle'. *Can. J. Microbiol.* **38**: 548-554.

Minami E., Kouchi H., Cohn R.J., Ogawa T. and Stacey G. (1996). Expression of the early nodulin, ENOD40, in soybean roots in response to various lipo-chitin signal molecules. *The Plant journal* **10**(1), 23-32

Mulligan, J.T. and Long, S.R. (1985) Induction of *Rhizobium meliloti nod*C expression by plant exudate requires *nod*D. *Proc. Natl. Acad. Sci. USA* **82**, 6609-6613.

Nair M. G., Safir G. R. and Siqueira J.O. (1991). Isolation and identification of vesicular-arbuscular mycorrhiza stimulatory compounds from clover (*Trifolium repens*) roots. *Appl. Environ. Microbiol.* **57**, 434-439

Newcomb, W. (1981) Nodule morphogenesis and differentiation, In *International review of cytology*, supplement 13, (Giles, K.L. and Atherly, A.G., eds.) pp. 247-297, Academic Press, New York.

Peters, N.K. and Long, S.R. (1988) Alfalfa root exudates and compounds which promote or inhibit induction of *Rhizobium meliloti* nodulation genes. *Plant Physiol.* **88**, 396-400

Phillips, D.A., Wery, J., Joseph, C.M., Jones, A.D. and Teuber, L.R. (1995) Release of flavonoids and betaines from seeds of seven medicago species. *Crop Sci.* **35**, 805-808.

Pirozynski K. A. and Y. Dalpe 1989. Geological history of the Glomaceae with particular reference to mycorrhizal symbiosis. *Symbiosis* **7**, 1-36

Sagan *et al.* (1995). Selection of nodulation and mycorrhizal mutants in the model plant *Medicago truncatula* (Gaertn.) after g-ray mutagenesis. *Plant Science* 111, 63-71

Shirtliffe, S.J. and Vessey, J.K.(1996). A nodulation (nod+/Fix-) mutant of *Phaseolus vulgaris* L. has nodule-like structures lacking peripheral vascular bundles (Pvb-) and is resistant to mycorrhizal infection (Myc-). *Plant Science* **118**, 209-220

Simon L., J. Bousquet, R. C. Levesque and M. Lalonde 1993. Origin and diversification of endomycorrhizal fungi and coincidence with vascular land plants. *Nature* **363**, 67-69

Spaink, H.P. (1996) Regulation of plant morphogenesis by lipo-chitin oligosaccharides. *Crit. Rev. Plant Sci.* **15**, 559-582.

Tsai M. and Phillips D. A. (1991). Flavonoids released naturally from alfalfa promote development of symbiotic *Glomus* spores *in vitro*. *Appl. Environ. Microbiol.* **57**, 1485-1488

Van de Sande, K., Pawlowski, K., Czaja, I., Wieneke, U., Schell, J., Schmidt, J., Walden, R., Matvienko, M., Wellink, J., van Kammen, A., Franssen, H. and Bisseling, T. (1996) Modification of phytohormone response by a peptide encoded by *ENOD40* of legumes and a non legume. *Science* **273**, 370-373.

Van de Wiel, C., Norris, J.H., Bochenek, B., Dickstein, R., Bisseling, T. and Hirsch, A.M. (1990) Nodulin gene expression and *ENOD2* localization in effective, nitrogen-fixing and ineffective, bacteria-free nodules of alfalfa. *Plant Cell* **2**, 1009-1017.

Van Rhijn, P., Fang, Y., Galili, S., Shaul, O., Atzmon, N., Wininger, S., Eshed, Y., Lum, M., Li, Y., To, V., Fujishige, N., Kapulnik, Y., and Hirsch, A. (1997). Expression of early nodulin genes in alfalfa mycorrhizae indicates that signal transduction pathways used in forming arbuscular mycorrhizae and *Rhizobium*-induced nodules may be conserved. *Proc. Natl. Sci. USA* **94**, 5467-5472.

Van Spronsen, P.C., Bakhuizen, R., van Brussel, A.A.N., and Kijne, J.W. (1994) Cell wall degradation during infection thread formation by the nodule bacterium *Rhizobium leguminosarum* is a two-step process. *Eur. J. Cell Biol.* **64**, 88-94.

Vijn, I., Martinez-Abarca, F., Yang, W.-C., das Neves, L., van Brussel, A., van Kammen, A. and Bisseling, T., (1995a). Early nodulin gene expression during Nod factor-induced processes in *Vicia sativa*.. *Plant J.* **8**, 111-119.

Vijn, I., Yang, W.-C., Pallisgård, N., Østergaard Jensen, E., Van Kammen, A. and Bisseling, T. (1995b) *VsENOD5*, *VsENOD12* and *VsENOD40* expression during *Rhizobium*-induced nodule formation on *Vicia sativa* roots. *Plant Mol. Biol.* **28**, 1111-1119.

Xie Z.P., Staehelin C., Vierheilig H., Wiemken A., Jabbouri S., Broughton W. J., Vogeli-Lange R. and Boller T. (1995) Rhizobial nodulation factors stimulate mycorrhizal colonization of nodulating and non-nodulating soybeans. *Plant Physiol.* **108**, 1519-1525

Alternate Methods of Gene Discovery -- The Candidate EST Approach and DNA Microarrays

S.C. Somerville, M. Nishimura, D. Hughes, I. Wilson, J. Vogel

Department of Plant Biology, Carnegie Institution of Washington, Stanford, CA 94305, U.S.A.

Keywords Disease resistance, DNA microarrays, expressed sequence tag (EST), gene expression, leucine-rich repeat motif

Introduction

With the advent of large scale cDNA and genome sequencing projects, gene sequence information is accumulating at an accelerating rate, especially in the model plant *Arabidopsis thaliana.* One consequence is that increasingly genes will be identified first in sequence databases rather than via traditional cloning procedures. Thus, in the future, gene discovery will be based in part on identifying genes similar in sequence to known genes or identifying clones that map to known mutations or character traits.

A second consequence is that large scale surveys of gene expression will be required to exploit this sequence information fully. No longer will it be sufficient to evaluate genes one gene at a time. From such large scale surveys, new genes that exhibit altered expression patterns in response to an environmental or developmental cue can be identified. In addition, an integrated view of how large numbers of genes are coordinately expressed in plants can be developed. DNA microarrays and related technologies permit such large scale surveys of steady-state mRNA levels.

The methods described in this report have general utility and can be applied a wide range of problems in plant biology. However, in the paragraphs below, the examples are derived from the field of plant pathology. Therefore, a brief introduction to plant disease resistance genes is provided.

Disease Resistance Genes

Flor's gene-for-gene hypothesis provides a framework for studying plant-pathogen interactions that has served the plant pathology community well (Flor 1971, Baker et al. 1997). In brief, this hypothesis states that an incompatible interaction (i.e., resistant plant + avirulent pathogen) is governed by resistance alleles at a single host disease resistance locus and avirulence allele(s) at a single pathogen virulence locus. This gene-for-gene complementarity is the basis for the gene-for-gene hypothesis. A common mechanistic interpretation of the gene-for-gene hypothesis is that host resistance loci encode receptor proteins capable of intercepting signals from pathogens (i.e., avirulence gene products) and transmitting those signals to secondary messengers leading to the activation of host defense responses. During compatible interactions, the recognition step

NATO ASI Series, Vol. H 104
Cellular Integration of Signalling Pathways in Plant Development
Edited by F. Lo Schiavo, R. L. Last, G. Morelli, and N. V. Raikhel

fails and host defenses are not activated. An underlying assumption of this model is that recognition (i.e., receptor-signal binding) is highly specific. In other words, a resistance gene product or receptor can recognize or bind only one specific avirulence gene product or signal. Although a few exceptions exist, most resistance and avirulence genes conform to this general view of the genes mediating host-pathogen interactions (Flor 1971, Staskawicz et al. 1995).

Because of their central role in host-pathogen interactions, disease resistance genes represent attractive targets for detailed studies and a number disease resistance genes have been cloned during the past five years. Among the recently cloned host resistance genes are genes encoding resistance to various viral, bacterial, fungal, and nematode pathogens (see Baker et al. 1997 for a recent listing of cloned disease resistance genes). A striking feature of most of these genes is that they contain a segment of leucine-rich repeats (LRR) (Jones and Jones 1996).

Most disease resistance genes have been cloned using either a map-based cloning (e.g., *Pto*, *RPM1*, *RPP5, RPS2)* or a transposon-tagging strategy (e.g., *Cf9*, *L6*, *N*). The major advantage of both methods is that no biochemical information about the nature of the resistance gene or its product is required. However, the sequence conservation among members of the LRR-class can be used to identify additional resistance genes. This tactic will be particularly valuable in recovering disease resistance genes for economically significant diseases of crop species such as wheat, in which it is difficult to implement map-based cloning or transposon-tagging. In the paragraphs below, two approaches are described – using degenerate oligonucleotides to amplify fragments of resistance genes and scanning the sequence databases for R-ESTs (i.e., plant ESTs similar in sequence to disease resistance genes). Both approaches are based on the underlying assumption that genetic resources exist for mapping gene fragments and EST clones relative to known disease resistance genes. Fortunately, for many crop species, plant breeders and plant pathologists have developed excellent genetic maps of disease resistance genes in conventional breeding programs.

In the discussion below, only gene fragments and ESTs that are similar in sequence to the members of the LRR-class of disease resistance genes are discussed. Although most resistance genes cloned to date fall within this class, not all resistance genes contain an LRR motif. Notably, *Hm1* encodes a reductase, which renders the pathogen-produced HC-toxin inactive (Johal and Briggs 1992). Because the HC-toxin is required for colonization by the fungal pathogen, *Cochiolobus carbonum*, maize lines of the *Hm1* genotype are resistant to *C. carbonum* infections. The tomato *Pto* gene encodes a ser/thr protein kinase (Martin et al. 1993) and a third resistance gene, the barley *ml-o* gene, codes for a novel protein of unknown biochemical activity (Büschges et al. 1997). Thus, searches for resistance gene candidates, like those described below, could be repeated using the *Hm1*, *Pto*, or *ml-o* genes.

Amplification of Resistance Gene Fragments Using Degenerate Oligonucleotides

The observation that most cloned resistance genes fall into the LRR-class suggests that additional resistance genes will also be members of this class and that sequence-based methods for discovering new disease resistance genes will be productive. Three groups have developed degenerate oligonucleotide primers based on conserved sequences of cloned disease resistance genes and have used these primers to amplify and clone candidate disease resistance genes in potato (Leister et al. 1996) and soybean (Kanzin et al. 1996, Yu et al. 1996). Nineteen soybean and twelve potato partial clones were mapped and, in several cases, the clones mapped near known disease resistance genes. In a related approach, a *Pto*-like cDNA clone was used to identify a receptor-kinase-like gene that cosegregated with the *Lr10* rust resistance gene in wheat (Feuillet et al. 1997). These successes in finding candidate clones imply this will be a promising approach for cloning disease resistance genes that should find wide applicability in crop and forestry plants. Thus, it seems safe to predict that in the near future, many disease resistance genes of the LRR-class will be cloned.

R-ESTs, ESTs with sequence similarity to disease resistance genes

Arabidopsis represents a special case because two EST sequencing projects have produced >29,000 partial sequences of >16,000 distinct genes (Cooke et al. 1996, Newman et al. 1994, Rounsley et al. 1996). More recently, full genome sequencing of *Arabidopsis* was initiated as an international collaboration among five groups. Both sources of sequence information can be scanned with BLAST (Altschul et al. 1994) or FASTA (Pearson and Lipman1988) software routines for sequences that resemble known resistance genes. It is estimated that currently ~60% of *Arabidopsis* genes are represented by at least one EST sequence (Rounsley et al. 1996) and, over time, this proportion will increase as more EST and genomic sequence is deposited in the sequence databases. Thus, there is a roughly 60% chance of identifying a specific *Arabidopsis* gene in the current EST database, if there is no bias in the representation of genes in the database.

The cDNA library used to generate the majority of *Arabidopsis* EST sequences was derived from a mixture of equal amounts of mRNA from four tissues -- roots, etiolated seedlings, flower and developing seeds, and leaves (Newman et al. 1994). Thus, genes induced under specific environmental conditions or expressed in small amounts in specialized tissues will be under-represented in the library and in the EST database. In addition, genes that are absent from Columbia but occur in other *Arabidopsis* accessions will not be represented in the EST database. However, many disease resistance genes are expressed in uninfected tissue and *Arabidopsis* ESTs corresponding to the disease resistance genes *RPS2*, *RPM1*, and *RPP5* do occur in the EST database (Botella et al. 1997).

In a collaborative project with M. Botella, M. Coleman and J. Jones (Sainsbury Laboratory, Norwich, U.K.), we screened the *Arabidopsis* EST database for ESTs with sequence similarity to resistance genes of the LRR-class (Botella et al. 1997). Ninety-one ESTs were identified. These ESTs represented 59 distinct

ESTs (TIGR tentative consensus groupings, Rounsley et al. 1996). This result suggests that the *Arabidopsis* genome contains a significant number of resistance gene-like genes (59/0.6 = ~98), which is consistent with the observation that well-characterized crop plants like wheat have >100 disease resistance loci.

Of the 59 R-ESTs, 42 were mapped on the *Arabidopsis* genetic map to 43 loci (Botella et al. 1997). A majority of these map to chromosome 5 (21/43). Notably, eight R-ESTs map to the MRC-J region of chromosome 5, defined as a ~30 cM region rich in disease resistance genes and defense response genes (Holub 1997). This clustering of disease resistance genes has been observed in other plant species and has led to speculation that intragenic and intergenic recombination contribute to the generation of new disease resistance genes and alleles (Holub 1997).

Not all LRR-containing genes are disease resistance genes. For example, *ERECTA* (Torii et al. 1996) and *CLAVATA1* (Clark et al. 1995) encode LRR-receptor kinases that are thought to play a role in development. Also, a series of LRR cell wall proteins have been described, including the polygalacturonase inhibitor proteins (PGIP) (Jones and Jones 1997). Additional characterization of the genes represented by each EST will be required to determine which ones are disease resistance genes and which ones play a role in development or some other plant process. In *Drosophila melanogaster*, five genes of the Toll pathway, which specifies dorsal/ventral polarity in the embryo, are also used in signaling defense response genes during pathogen attack in the adult fly (Lemaitre et al. 1996). Interestingly, the *Drosophila* Toll receptor shares sequence similarity to the N, L6 and RPP5 disease resistance proteins; pelle is a ser/thr kinase and similar in sequence to Pto, the tomato bacterial leaf speck resistance gene; and both cactus and NPR1 (never-induced for pathogenesis-related proteins) have an ankyrin-repeat containing domain (Wilson et al. 1997). These observations support the concept that cell-cell recognition in host-pathogen interactions and in development in multi-cellular organisms is derived from a common ancestral mechanism (Wilson et al. 1997). Although the barley *Mla* (Torp and Jørgensen 1986) and the *Arabidopsis RPM1* (Grant et al. 1995) genes appear dispensable in uninfected plants, it is possible that, like the Toll pathway genes, some disease resistance genes will serve another function in the life of the plant. To address the question of what proportion of the R-ESTs code for disease resistance genes versus development genes, we have initiated a systematic search for mutant or anti-sense plants in which the expression of R-EST-related genes is suppressed. In addition to identifying the subset of R-EST genes that play a role in development, transgenic plants can be evaluated for their disease resistance phenotype in a "reverse genetics" strategy for identifying disease resistance genes.

DNA Microarrays

The cloning and sequencing of a disease resistance gene represents but one step in developing an understanding of disease resistance and host-pathogen interactions. Detailed studies of the resistance gene product, such as determining its crystal structure, will provide insight into how the resistance gene product functions to bind the corresponding avirulence gene product and to transduce this signal to secondary messengers. However, broader questions about how groups of genes interact to effect disease resistance or susceptibility are also important.

Typically the expression patterns of 5-10 genes (e.g., pathogenesis-related proteins and genes encoding enzymes of the phenylpropanoid pathway) are surveyed to provide evidence of host responses to infection (Dixon and Harrison 1990). Thus, our understanding of how plants cope with pathogen infection is divorced from other aspects of plant biology. To understand these complex interactions fully, it will be necessary to ask broader questions such as: How are various aspects of plant biology integrated with defense responses?

With the DNA microarray technology, large scale surveys of mRNA expression patterns are feasible (Schena 1996). The DNA microarray technology is relatively new; the first paper describing this technology was published in 1995 (Schena et al. 1995). The principal steps are illustrated in Fig. 1 (see also the home page of one of our collaborators for additional photos and short videos illustrating the equipment at http://cmgm.stanford.edu/pbrown/array.html). In brief, a complex probe (e.g., poly(A)$^+$-RNA) is labeled with a fluorochrome and hybridized to an array of targets (i.e., inserts of cDNA clones immobilized on a glass slide). The hybridization signal for each target is monitored and used to assess mRNA accumulation in a given tissue sample. The expression patterns of thousands of target clones can be determined simultaneously (Schena et al. 1996, Shalon et al. 1996). In addition, quantitative output of relative mRNA levels, which is linear over a broad range, can be obtained allowing both rare and abundant mRNA species to be evaluated in the same experiment. Thus, plant responses to a variety of treatments can be evaluated more thoroughly than is possible currently.

The microarrays, which are the most notable feature of this technique, are produced by a robotic arm that spots target clones onto glass slides. By using a robotic device, the array elements can be laid out in a compact design (1 cm^2 for 10,000-element microarray; Shalon et al. 1996) and many target clones can be assayed simultaneously. An additional consequence is that hybridization volumes are very small and probe concentrations can be very high, which improves the ability to detect rare mRNA species. Another benefit of using glass slides is that the background hybridization is very low compared with the more conventional nylon or nitrocellulose membranes. Thus, weak signals can be monitored reliably. The use of fluorescence-based detection offers several benefits. Multiple probes, each labeled with a distinct fluorochrome, can be hybridized simultaneously to a given microarray slide, permitting direct comparisons between samples. Currently, two fluorochrome labels are used (e.g., Cy3 and Cy5). Emission by the fluors exhibits a broad linear range; therefore, both rare and abundant mRNA species can be evaluated in the same experiment. In addition, the microarray technique can be adapted for use in reverse Southerns in which complex DNA probes (e.g., total genomic DNA, YAC, or BAC DNA) are used in place of mRNA (Shalon et al. 1996).

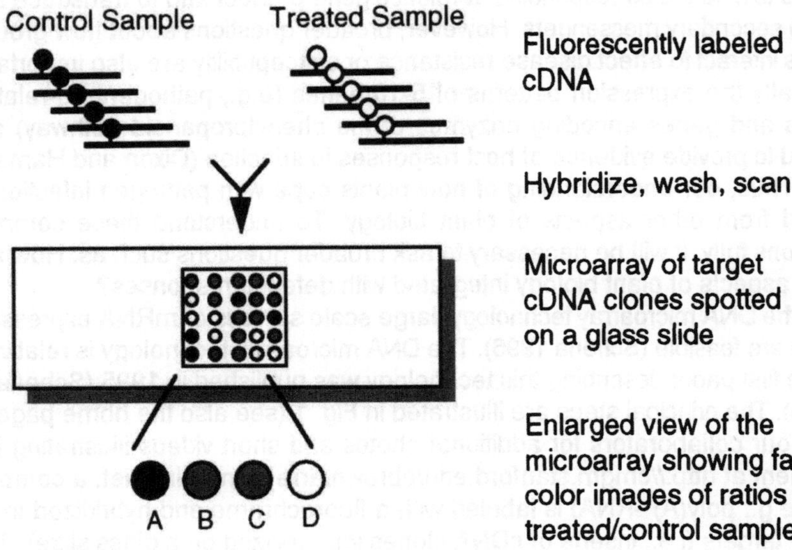

Target cDNA	Expression Levels Control	Treated	Ratio T/C
A	500	200	0.4
B	10	10	1.0
C	1,000	1,000	1.0
D	200	1,000	5.0

Fig. 1. Diagram outlining the use of DNA microarrays to assess steady-state mRNA levels in two mRNA populations. The false color image represents the ratio of the values derived from mRNA from treated versus control tissues. The color coding is as follows: *Black*, mRNA levels are higher in the control than the treated sample; *White*, mRNA levels are higher in the treated than the control sample; *Grey*, mRNA levels are similar in the control and treated samples.

Alternate methods that compete with the microarray technique for surveying thousands of discrete mRNA species simultaneously and quantitatively include high density cDNA arrays on nylon membranes (e.g., Bernard et al. 1996), the oligonucleotide-based DNA chip technology (Lipshutz et al. 1995, Southern 1996), serial analysis of gene expression (SAGE; Velculescu et al. 1995) and large scale sequencing of primary libraries (Adams et al. 1995). Probing high density arrays on nylon filters with radio-labeled complex probes is the method most directly comparable to the DNA microarray technology. The main advantage of using the microarrays over the high density arrays supported on nylon membranes is increased sensitivity and thus increased ability to monitor rare mRNAs. In a recent paper, Schena et al. (1996) estimated that an mRNA species represented at one part in 200,000 (w/w) could be detected. In contrast, Bernard et al. (1996) estimated the sensitivity of one part in 20,000 for cDNAs spotted on nylon filters. Additional advantages of DNA microarrays include the use of small hybridization volumes (i.e., 10 µL) and hence the need for smaller amounts of starting material.

The DNA chip technology is based on oligonucleotide arrays synthesized on silicon wafers using photolithography-based methods (Lipshutz et al. 1995, Southern 1996). This technology is dependent on prior knowledge of gene sequence. Therefore, the first successful applications have been in genotyping (e.g., detecting mutant alleles) and diagnostics (e.g., detecting HIV) (Lipshutz 1995). The DNA chip technology is very expensive. Currently, microarrays are less expensive to produce, more accessible, more adaptable, and thus, more suited to the academic research community.

A third method for monitoring the levels of many mRNA species is the serial analysis of gene expression (SAGE) method (Velculescu et al. 1995). This technique consists of isolating short cDNA fragments (9-20 nt) from near the 3' end of mRNAs, ligating random fragments together serially in units of 10-50 and then sequencing them. The authors claim that 9 nt of sequence is sufficient to identify the source gene (or gene family) for 95% of such fragments. At this time, the application of the SAGE method is limited in plants by a lack of gene sequence information. Although 5' sequences for about 29,000 *Arabidopsis* ESTs are available in public databases, few 3' sequences exist (Newman et al. 1994, Rounsley et al. 1996). As efforts to sequence the *Arabidopsis* genome and to sequence the 3' ends of EST clones progress, the SAGE method may represent a more viable option.

Complementary methods to those outlined above are techniques based on the differential expression of genes. Methods, such as differential screening of cDNA libraries, differential display (Liang and Pardee 1992) or subtractive hybridization (Wang and Brown 1991), identify mRNA species that are preferentially expressed in one treatment relative to a control treatment. These methods are not quantitative and by themselves give little indication of the relative abundance of the differentially expressed mRNA species. Also, these methods focus on mRNA species that exhibit dramatic differences in expression. With the microarray technology, 2-fold differences in expression can be reliably detected (Schena et al. 1996).

2A.

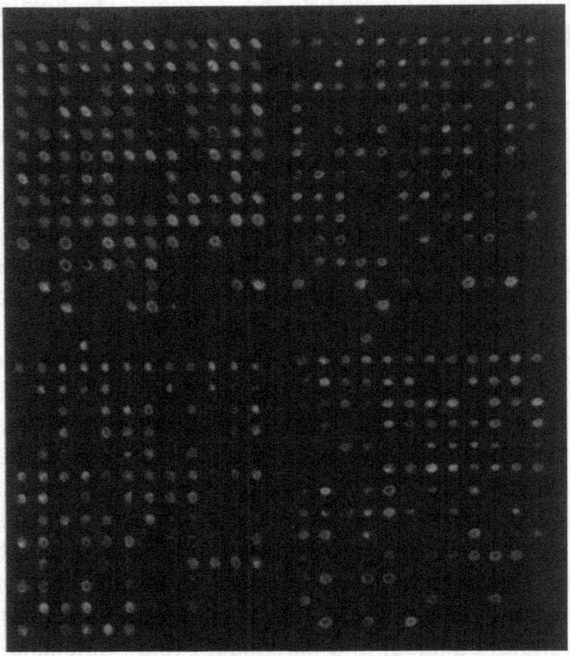

2B

Fig. 2. Greyscale images of scans of one DNA microarray, consisting of 600 cDNA inserts plus controls, that was probed with **A.** Cy-5-labeled cDNA prepared from mRNA from uninfected Columbia and **B.** Cy3-labeled cDNA prepared from mRNA from powdery mildew-infected Columbia (susceptible) plants harvested 48 hours post-inoculation. The brightness of each spot reflects the intensity of the fluorescent signal derived from a single probe and does not represent a ratioed value.

We are specifically interested in utilizing the DNA microarrays to help us evaluate host responses to powdery mildew infection in both compatible and incompatible interactions. To this end, we have generated a DNA microarray consisting of 600 elements plus controls. The greyscale images of the hybridization signal obtained by probing a DNA microarray with labeled mRNA from powdery mildew-infected susceptible *Arabidopsis* plants and with labeled mRNA from uninfected control plants are shown in Fig. 2B and 2A, respectively. From this preliminary result, it is apparent that genes previously reported to be induced in various plants infected by a variety of pathogens are also induced in powdery mildew-infected *Arabidopsis* plants (see Table 1) (Dixon and Harrison 1990, Giese et al. 1996). In addition, enhanced mRNA accumulation in infected plants is observed for several genes not previously reported to exhibit altered

Table 1: Relative steady-state mRNA levels for powdery mildew-infected versus uninfected plants. mRNA samples from Columbia (susceptible genotype) were collected 48 hours post-inoculation with the powdery mildew pathogen, *Erysiphe cichoracearum*, UCSC1. Only those genes for which the ratio was either >3.0 or <0.3 are shown.

Gene Name[a]	Ratio (I/U)[b]
Genes showing enhanced expression in infected plants	
Pathogenesis-related protein 1	37.5
Pathogenesis-related protein 2 homologue	12.0
Pathogenesis-related protein homologue	8.8
β-1,3-glucanase 2 homologue	7.5
Unknown, isolated in differential display experiment	7.2
TCH3 homologue	4.8
Glutathione-S-transferase homologue	3.9
β-fructosidase homologue	3.8
Extensin homologue	3.6
Multi-drug resistance protein homologue	3.6
Xc750 homologue, pathogen-inducible gene	3.4
Receptor-like kinase homologue	3.2
Ser/thr kinase homologue	3.0
Genes showing reduced expression in infected plants	
EST with leucine-rich repeats	0.3
Phenylalanine ammonia lyase homologue	0.2
Unknown, isolated in differential display experiment	0.2
Heat shock protein, hsp81k homologue	0.1

[a] Most of the target clones that make up the microarray are derived from cDNA clones from the EST sequencing projects and are referred to as homologues.

[b] Fluorescent signal derived from an mRNA sample from infected tissues divided by the signal derived from an mRNA sample from uninfected tissues. The average of two replicates is given.

gene expression in host-pathogen interactions (e.g., multi-drug resistance protein, β-fructosidase). Finally, several genes appear to be down-regulated in infected plants, a phenomenon that is rarely reported in the literature. Thus, based on this preliminary analysis, the DNA microarrays promise to provide detailed information about changes in host gene expression for a large number of genes in response to pathogen attack. An added benefit is that novel genes not previously reported as defense-response genes in more conventional gene expression studies were identified using this 600-element DNA microarray.

Conclusion

Although the examples cited in this paper are based on plant disease resistance genes and host-pathogen interactions, the approaches described can be utilized in any field of plant biology. Using degenerate primers to amplify gene fragments and scanning EST and genomic databases for gene sequences represent two powerful gene identification strategies that build on prior knowledge. When coupled with detailed genetic maps of mutations or natural variation affecting various developmental, physiological, biochemical or agronomic traits, these techniques provide rapid, efficient means for identifying candidate genes. Therefore, it will be important to ensure that detailed genetic maps of traits and characters are developed in parallel with large scale sequencing projects.

As we accumulate more information about the structure and function of individual genes, learning how groups of genes interact to determine a specific response to an environmental or developmental cue is one obvious next challenge in the field of genomics. DNA microarrays are one technology that permits large scale surveys of gene expression. In the future, it will be possible to evaluate changes in steady-state mRNA levels for all genes of an organism in mutants, in transgenic plants under- or over-expressing a gene of interest, and in plants treated with various growth regulators, herbicides or other regulatory compounds. Eventually comprehensive catalogues of steady-state mRNA levels for tissue types, developmental stages and plants exposed to diverse environmental conditions will accumulate. Such global knowledge will provide a basis for developing more rational strategies for modifying plant architecture and physiology to the benefit of the environment and humanity.

Acknowledgments

We are indebted to the members of the Pat Brown and Ron Davis laboratories (Stanford, U.S.A.) for their guidance and advice about DNA microarrays and for access to equipment. The project to identify and map R-ESTs was carried out in collaboration with members of the J. Jones laboratory (Norwich, U.K.). This work was supported in part by the Carnegie Institution of Washington, the U.S. Department of Agriculture, and the U.S. Department of Energy. Publication number 1345 of the Carnegie Institution of Washington.

Citations

Adams, M.E., Kerlavage, A.R., Fleischmann, R.D., Fuldner, R.A., Bult, C.J., Lee, N.H., Kirkness, E.F., Weinstock, K.G., Gocayne, J.D. et al. (1995) Initial assessment of human gene diversity and expression patterns based upon 83 million nucleotides of cDNA sequence. Nature 377:3-174.

Altschul, S.F., Boguski, M.S., Gish, W., Wootton, J.C. (1994) Issues in searching molecular sequence databases. Nature Genetics 6:119-129.

Baker, B., Zambryski, P., Staskawicz, B., Dinesh-Kumar, S.P. (1997) Signaling in plant-microbe interactions. Science 276:726-733.

Bernard, K., Auphan, N., Granjeaud, S., Victorero, G., Schmitt-Verhulst, A.-M., Jordan, B.R., Nguyen, C. (1996) Multiplex messenger assay: simultaneous, quantitative measurement of expression of many genes in the context of T cell activation. Nucl. Acids Res. 24:1435-1442.

Botella, M.A., Coleman, M.J., Hughes, D.E., Nishimura, M.T., Jones, J.D.G., Somerville, S.C. (1997) Map positions for 47 *Arabidopsis* ESTs with sequence similarity to disease resistance genes. Plant J. (accepted).

Büschges, R., Hollricher, K., Panstruga, R., Simons, G., Wolter, M., Frijters, A., van Daelen, R., van der Lee, T., Diergaarde, P., Groenendijk, J., Töpsch, S., Vos, P., Salamini, F., Schulze-Lefert, P. (1997) The barley *Mlo* gene: A novel control element of plant pathogen resistance. Cell 88:695-705.

Clark, S., Booker, M., Hao, T., Hung, C.-Y., Kayes, J., Pogany, J., Razaq, S., Schrage, K., Simon, E., Trotochaud, A., Williams, R., Yu, L. , Meyerowitz, E. (1995) A potential signaling pathway involving the CLAVATA1 receptor-kinase restricts meristem activity. Abstract S74 from the *Proceedings of the Seventh International Conference on Arabidopsis Research. June 23-27, 1996. Norwich, U.K.* (for electronic access to this abstract see http://nasc.nott.ac.uk/Norwich/sessions)

Cooke, R. et al. (33 co-authors) (1996) Further progress towards a catalogue of all *Arabidopsis* genes: analysis of a set of 5000 non-redundant ESTs. Plant J. 9:101-124.

Dixon, R.A., Harrison, M.J. (1990) Activation, structure, and organization of genes involved in microbial defense in plants. Adv. Genet. 28:165-234.

Feuillet, C., Schachermayr, G., Keller, B. (1997) Molecular cloning of a new receptor-like kinase gene encoded at the *Lr10* disease resistance locus of wheat. Plant J. 11:45-52.

Flor, H.H. (1971) Current status of the gene-for-gene concept. Annu. Rev. Phytopathol. 9:275-296.

Giese, H., Hippe-Sanwald, S., Somerville, S., Weller, J. (1996) *Erysiphe graminis.* In: Mycota, Vol. VI: Plant Relationships (Carroll, G., Tudzynski, P., eds.), Springer-Verlag, Berlin (in press).

Grant, M.R., Godiard, L., Staube, E., Ashfield, T., Lewald, J., Sattler, A., Innes, R.W., Dangl, J.L. (1995) Structure of the *Arabidopsis RPM1* gene enabling dual specificity disease resistance. Science 269:843-846.

Holub, E.B. (1997) Organization of resistance genes in *Arabidopsis*. In *The Gene-for-Gene Relationship in Host-Parasite Interactions* (Crute, I., Holub, E., Burdon, J., eds.). Wallingford, UK: CAB International (in press).

Johal, G., Briggs, S.P. (1992) Reductase activity encoded by the *Hm1* disease resistance gene in maize. Science 258:985-987.

Jones, D.A., Jones, J.D.G. (1996) The roles of leucine-rich repeat proteins in plant defenses. Adv. Plant Pathol. 24:90-167.

Kanazin, V., Marek, L.F., Shoemaker, R.C. (1996) Resistance gene analogs are conserved and clustered in soybean. Proc. Natl. Acad. Sci., USA 93:11746-11750.

Leister, D., Ballvora, A., Salamini, F., Gebhardt, C. (1996) A PCR-based approach for isolating pathogen resistance genes from potato with potential for wide application in plants. Nature Genet. 14:421-429.

Lemaitre, B., Nicolas, E., Michaut, L., Reichhart, J.-M., Hoffmann, J.A. (1996) The dorsoventral regulatory gene cassette *spätzle/Toll/cactus* controls the potent antifungal response in *Drosophila* adults. Cell 86:973-983.

Liang, P., Pardee, A.B. (1992) Differential display of eukaryotic messenger RNA by means of the polymerase chain reaction. Science 257:967-971.

Lipshutz, R.J., Morris, D., Chee, M., Hubbell, E., Kozal, M.J., Shah, N., Shen, N., Yang, R., Fodor, S.P.A. (1995) Using oligonucleotide probe arrays to access genetic diversity. BioTechniques 19:442-447.

Martin, G.B., Brommonschenkel, S.H., Chunwongse, J., Frary, A., Ganal, M.W., Spivey, R., Wu, T., Earle, E.D., Tanksley, S.D. (1993) Map-based cloning of a protein kinase gene conferring disease resistance in tomato. Science 262:1432-1436.

Newman, T., de Bruijn, F., Green, P., Keegstra, K., Kende, H., McIntosh, L., Ohlrogge, J., Raikhel, N., Somerville, S., Thomashow, M., Retzel, E., Somerville, C. (1994) Genes galore: A summary of methods for accessing results from large-scale partial sequencing of anonymous *Arabidopsis* cDNA clones. Plant Physiol. 106:1241-1255.

Pearson, W., Lipman, D. (1988) Improved tools for biological sequence comparison. Proc. Natl. Acad. Sci., USA 85:2444-2448.

Rounsley, S.D., Glodek, A., Sutton, G., Adams, M.D., Somerville, C.R., Venter, J.C., Kerlavage, A.R. (1996) The construction of *Arabidopsis* EST assemblies: A new resource to facilitate gene identification. Plant Physiol.112:1179-1183.

Schena, M. (1996) Genome analysis with gene expression microarrays. BioEssays 18:427-431.

Schena, M., Shalon, D., Davis, R.W., Brown, P.O. (1995) Quantitative monitoring of gene expression patterns with a complementary DNA microarray. Science 270:467-470.

Schena, M., Shalon, D., Heller, R., Chai, A., Brown, P.O., Davis, R.W. (1996) Parallel human genome analysis: Microarray-based expression monitoring of 1000 genes. Proc. Natl. Acad. Sci, USA 93:10614-10619.

Shalon, D., Smith, S.J., Brown, P.O. (1996) A DNA microarray system for analyzing complex DNA samples using two-color fluorescent probe hybridization. Genome Res. 6:639-645.

Southern, E.M. (1996) DNA chips: analyzing sequence by hybridization to oligonucleotides on a large scale. Trends Genet. 12:110-115.

Staskawicz, B.J., Ausubel, F.M., Baker, B.J., Ellis, J.G., Jones, J.D.G. (1995) Molecular genetics of plant disease resistance. Science 268:661-667.

Torii, K.U., Mitsukawa, N., Oosumi, T., Matsuura, Y., Yokoyama, R., Whittier, R.F., Komeda, Y. (1996) The *Arabidopsis ERECTA* gene encodes a putative receptor protein kinase with extracellular leucine-rich repeats. Plant Cell 8:735-746.

Torp, J., Jørgensen, J.H. (1986) Modification of powdery mildew resistance gene *Ml-a12* by induced mutation. Can. J. Genet. Cytol. 28:725-731.

Velculescu, V.E., Zhang, L., Vogelstein, B., Kinzler, K.W. (1995) Serial analysis of gene expression. Science 270:484-487.

Wang, Z., Brown, D.D. (1991) A gene expression screen. Proc. Natl. Acad. Sci., USA 88:11505-11509.

Wilson, I., Vogel, J., Somerville, S. (1997) Signalling pathways: a common theme in plants and animals? Curr. Biol. 7:R175-R178.

Yu, Y.G., Buss, G.R., Saghai Maroof, M.A. (1996) Isolation of a superfamily of candidate disease-resistance genes in soybean based on a conserved nucleotide-binding site. Proc. Natl. Acad. Sci., USA 93:11751-11756.

Use of transient expression in plants for the study of the "gene-for-gene" interaction

Fumiaki Katagiri and R. Todd Leister

Department of Biological Sciences, University of Maryland Baltimore County, 1000 Hilltop Circle, Baltimore, MD 21250, USA

Abstract

One of the common responses in disease resistance defined by the "gene-for-gene" interaction is the hypersensitive response (HR), which is a rapid and localized cell death of plants at the site of infection. This specific cell death was used to develop a biolistic transient expression assay for the resistance response. If the transiently expressed gene causes the HR, the expression level of a cointroduced reporter gene decreases due to the rapid death of the cell. Using this assay, it was shown that expression of the *Pseudomonas syringae* avirulence genes, *avrRpt2* or *avrB*, in *Arabidopsis thaliana* can elicit the HR when the plants carry the corresponding resistance gene, *RPS2* or *RPM1*, respectively. This indicates that the avirulence genes are the only bacterial factors that are required to elicit the specific resistance response as long as the avirulence gene products are localized properly. This observation and others strongly suggest that the molecular recognition of pathogen attack occurs inside of plant cells for these combinations of avirulence gene and resistance gene. *RPS2* and *RPM1* are members of NBS-LRR class of resistance genes. NBS-LRR proteins could be cytoplasmic receptors for specific signal molecules from avirulent pathogens. We propose a speculative model to view different classes of resistance gene products from a unified perspective.

1. Introduction

In many plant-pathogen systems, a plant that can rapidly recognize pathogen attack and effectively mount a battery of defense responses is resistant to the pathogen (3, 14). Rapid pathogen recognition is usually highly specific (9). Such a specific pathogen recognition is best described by the "gene-for-gene" relationship (10). In these relationships, a plant is resistant to a certain pathogen only when the pathogen carries an avirulence gene and the plant carries the corresponding resistance gene. Common mechanisms for pathogen recognition and the ensuing signal transduction are postulated to underlie the "gene-for-gene" relationship (11, 19, 22). One of the common responses associated with resistance is the hypersensitive response (HR). The HR is a rapid and localized plant cell death at the infection site (20), which can be used as a convenient marker for resistance. In addition to the HR, a burst of reactive oxygen species, reinforcement of cell walls, induction of defense-related genes, synthesis of

NATO ASI Series, Vol. H 104
Cellular Integration of Signalling Pathways in Plant Development
Edited by F. Lo Schiavo, R. L. Last, G. Morelli, and N. V. Raikhel

phytoalexins (3, 14), and induction of systemic acquired resistance (34) are often associated with the resistance responses. To explain the "gene-for-gene" relationship, a simple "ligand-receptor" model has been proposed (3, 11): an avirulence gene product, directly or indirectly, creates a specific molecular signal (elicitor) that is recognized by the receptor encoded by the corresponding resistance gene. The receptors then feed the signal into a common signal transduction pathway to induce a battery of defense responses. The ligand-receptor model implies the presence of a family of resistance genes that share a basic structure because the receptors are presumed to interact with components in a common signal transduction pathway.

Four classes of "gene-for-gene" resistance genes have been reported. They are: class 1, the NBS-LRR (nucleotide binding site, leucine-rich repeats) (2, 4, 13, 23, 28, 29, 35, 42); class 2, the protein serine/threonine kinase class (26); class 3, the membrane protein with extracellular LRR class (6, 8, 18); and class 4, the receptor-like serine/threonine kinase with extracellular LRR class (38). The *Arabidopsis thaliana* resistance genes, *RPS2* and *RPM1*, belong to the NBS-LRR class (4, 13, 28). *RPS2* corresponds to the avirulence gene *avrRpt2* of a bacterial pathogen *Pseudomonas syringae* (21, 44). *RPM1* corresponds to two *P. syringae* avirulence genes *avrB* and *avrRpm1* (5).

2. Development of a transient expression assay for the resistance response

When *RPS2* was isolated by a map-based approach, biolistic transient expression of *RPS2* was used to demonstrate that the *RPS2* cDNA clone can complement the mutant phenotype of rps2 plants (28). The principle of the assay is that if the HR occurs by transient expression of a certain gene, the expression level of a cointroduced reporter gene is low because the cell dies rapidly. For the actual procedure, leaves of rps2 plants were inoculated with a *P. syringae* strain carrying *avrRpt2* prior to the biolistic bombardment. A half side of each leaf was inoculated. The uninfected half side served as a reference for the transformation efficiency. The *RPS2* cDNA and the ß-glucuronidase (GUS) reporter gene (17), both of which were under the control of CaMV *35S* promoter, were cointroduced into the leaf cells by biolistic transformation. After a one-day incubation, the GUS reporter enzyme activity in the cells was visualized by histochemical staining using X-gluc (17). Decrease of GUS activity in the infected sides of the leaves, compared to the uninfected sides, was observed in a manner dependent on both *RPS2* and *avrRpt2*. Thus the gene-for-gene specificity was preserved in this assay. More recently, an assay based on the same principle - decrease of a cointroduced reporter gene expression as an indicator of cell death - has been used in the study of programmed cell death in mammals (7).

3. Expression of bacterial avirulence genes in plants

In gene-for-gene interactions between bacterial pathogens and plants, the nature of the specific elicitors generated by the bacteria has been elusive except for the case of the *P. syringae* avirulence gene *avrD* (37). One possibility is that the avirulence gene product itself serves as the specific elicitor. To investigate this possibility, we transiently expressed *avrRpt2* together with *GUS* in either RPS2 wild type or rps2 mutant plants (24). The expression of *avrRpt2* caused a lower expression level of cointroduced GUS in RPS2 plants than in rps2 plants. This observation indicates that expression of the bacterial avirulence gene *avrRpt2* in plants is sufficient to elicit the resistance response in the plants that carry the corresponding resistance gene *RPS2*. In other words, the avirulence gene product AvrRpt2 is the only bacterial factor required to elicit the specific resistance response, as long as AvrRpt2 is localized correctly. Similar results that demonstrate that expression of bacterial avirulence genes in plants leads to elicitation of the specific resistance response have been reported for *avrB* (12), for the *P. syringae* avirulence gene (36, 40), and for the *Xanthomonas campestris* avirulence gene *avrBs3* (1).

4. A quantitative transient expression assay

A problem associated with the above experiment in which the effect of *avrRpt2* expression in plants was examined is lack of a proper reference for the biolistic transformation efficiency. The original assay described in section 2 used GUS expression in uninfected sides of the leaves for this reference purpose (28). We introduced a reference in the avirulence gene assay by taking advantage of biolistic transformation, in which cells can be transformed with different sets of genes in a single transformation event (24). We mixed two different sets of biolistics and bombarded them together: one set of biolistics coated with the DNA constructs for *avrRpt2* and *GUS* and the other set of biolistics coated with a construct containing the luciferase (LUC) reporter gene (41) under the control of the *35S* promoter. Only a small number of the cells are transformed by biolistic transformation. Statistically, therefore, the cells transformed by the first set of biolistics and the ones transformed by the second set are different and apart on the leaf. Because the HR is a localized event, expression of the *LUC* reporter gene is virtually unaffected by whether or not the HR occurs in the cells transformed with the other set. We measured the GUS and LUC enzyme activities in the extracts made from the bombarded leaves and normalized the GUS activity by using the LUC activity as the reference for the transformation efficiency. Using this quantitative assay, we unequivocally demonstrated that expression of *avrRpt2* or *avrB* in plants can elicit the resistance response in the plants carrying the corresponding resistance gene, *RPS2* or *RPM1*, respectively (Fig. 1)

5. A transient expression assay for both avirulence and resistance genes

We extended our study by combining the transient expression assays for avirulence genes and resistance genes (24). In this experiment, rpm1 rps2 double mutant plants were used. An avirulence gene and a resistance gene were cointroduced with the *GUS* reporter gene. The decrease of GUS activity, indicative of the HR, was observed only when the corresponding pair of avirulence and resistance genes were expressed: either *avrRpt2* and *RPS2* or *avrB* and *RPM1*. This observation clearly demonstrates that the gene-for-gene specificity is strictly conserved in this assay system and that this assay can be used for functional analysis of both avirulence genes and resistance genes. Due to use of transient expression, this assay allows a rapid collection of data. Therefore, this assay will facilitate molecular analysis of avirulence and resistance genes.

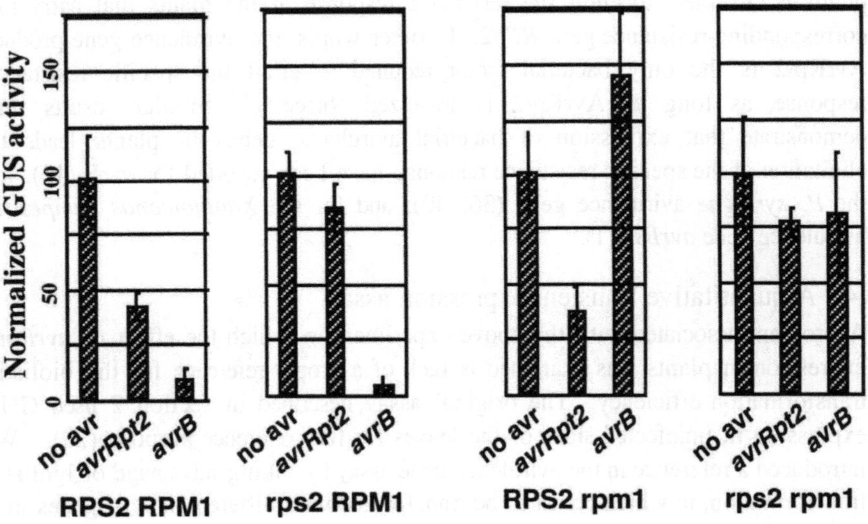

Fig. 1 *avrRpt2* and *avrB* can elicit the specific resistance response when transiently expressed in plants (24). The indicated avirulence genes were introduced together with the GUS reporter gene into plant cells with different genotypes for *RPS2* and *RPM1* resistance genes. The GUS reporter activity was normalized using the LUC reference as described in the text. The specific resistance response is viewed as decrease of the normalized GUS activity. Note that significant decrease in the normalized GUS activity was observed only when the corresponding combinations of the avirulence gene and the resistance gene were present in the assay.

Biolistic transient expression can easily be applied to many plant species. The HR is a very common response in gene-for-gene resistance (3). These two facts give this transient expression assay a large opportunity to be applied to the studies of the gene-for-gene interactions in many plant-pathogen systems. In addition, this assay will allow a quick test to see if a certain resistance gene can function in a heterologous plant species. Demonstration that a certain resistance gene can function in a heterologous plant species has so far been limited to closely related species (31, 43). To test whether a certain resistance gene can function in distantly related species is important not only to demonstrate possible applications of cloned resistance genes in agriculture but also to investigate the evolution of the resistance gene-based pathogen surveillance system of plants. This type of experiment would show how well basic components of the system are conserved through evolution and when this type of surveillance system emerged in evolution.

6. How can an avirulence gene product be transported into a plant cell?

As shown in sections 3 and 4, expression of a bacterial avirulence gene is sufficient to elicit the specific resistance response in plants (24). If the ligand-receptor model is correct, where does this molecular recognition take place? Sequence analysis of these avirulence gene products showed that when expressed in plants, they are likely to be localized in the cytoplasm (16, 39). An in vitro translation/translocation experiment suggested that RPS2 is localized in the cytoplasm (24). This notion that both avirulence and resistance gene products are localized in plant cell cytoplasm is consistent with the ligand-receptor model: both the ligand and receptor are in the cytoplasm. Then how is an avirulence gene product delivered into a plant cell in a natural situation - when the plants are infected with bacteria? The Hrp (hypersensitive response and pathogenicity) genes are in most cases required for an avirulent bacteria to cause the HR (12, 30). The Hrp genes are also required for the pathogenicity of the bacteria (25). Some of the Hrp genes encode components of the type III protein secretion system (15). In bacterial pathogens of mammals, such as *Yersinia* and *Salmonella*, the type III secretion system is apparently used by the bacteria to inject proteins important for their pathogenicity into host mammalian cells (33). By analogy, phytopathogenic bacteria may also use the type III secretion system to inject proteins important for their pathogenicity into host plant cells. Plants probably evolved a surveillance system based on resistance genes to detect some of such injected proteins as signals of pathogen attack.

7. A unified view of the gene-for-gene surveillance systems: a speculative model

There are four different classes of resistance genes. How are they organized? Do different classes of resistance gene products use different major signal transduction pathways? Here we present a speculative model that views all four classes of resistance gene products from a common perspective (Fig. 2).

The signal transduction pathway initiated by the *P. syringae* avirulence gene product AvrPto provides the basis for our model. This pathway in tomato includes two different classes of resistance gene products, Pto (class 2, protein kinase) (26) and Prf (class 1, NBS-LRR) (35). Because Pto specifically binds AvrPto, it is very likely that Pto is the direct intracellular receptor for AvrPto (36, 40). Fen is a protein kinase highly homologous to Pto, and the *Fen* gene is closely located to the *Pto* gene on the chromosome (27, 32). Although *Fen* is not a resistance gene by definition because no corresponding avirulent pathogen is known, an HR-like response caused by the insecticide fenthion is Fen-dependent. Prf is also required for this response to fenthion, whereas Pto is not. If we assume all Pto, Fen, and Prf are components of signal transduction pathways, the simplest model is to have Pto and Fen as parallel components for different stimuli and to have Prf as a common component at or after the convergence point of these two pathway. Prf seems an exception among class 1 proteins. Other class 1 NBS-LRR proteins could be direct receptors of the specific elicitors originated from pathogens. For example, RPS2 and RPM1 could be direct receptors of AvrRpt2 and AvrB, respectively. Analogously, Prf may recognize a special form of plant proteins instead of pathogen-originated proteins. Because Pto and Fen are protein kinases, the special form of the proteins may be phosphorylated proteins. Pto and Fen autophosphorylate themselves, so that phosphorylated Pto and Fen themselves may be recognized by Prf. Alternatively, protein kinase cascades initiated by Pto and Fen may lead to phosphorylation of the same (or similar enough) protein that can be recognized by Prf upon phosphorylation.

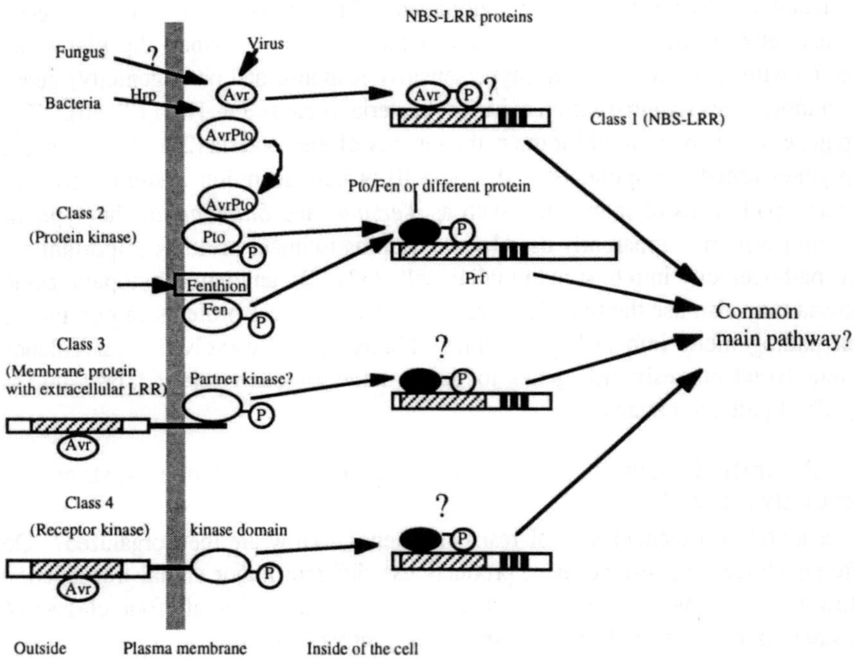

Fig. 2 A speculative model for a unified view of resistance gene functions. An assumption that NBS-LRR proteins can be receptors for phosphorylated proteins makes a unified description of all four classes of resistance gene products possible.

Several Pto-binding proteins have been identified (45). Are they consistent with our model? Pti1 is the most well characterized among them and is a protein kinase (45). Overexpression of Pti1 in plants enhanced the Pto-dependent resistance response. Pti-1 may be involved in the protein kinase cascade leading to phosphorylation of the protein that can be recognized by Prf. Because it was not shown whether the kinase activity of Pti1 is required for the enhancement of the resistance response in plants, the possibility that Pti1 enhances the resistance response simply by binding to Pto and stabilizing the phosphorylated state of Pto is not ruled out. For other Pto-binding proteins, their physiological significance has not been demonstrated. It is possible that different resistance gene products have different satellite pathways and modifying pathways in addition to the main pathway. Some of the Pto-binding proteins may be involved in such Pto-specific satellite or modifying pathways.

The notion that an NBS-LRR protein can be a receptor for a phosphorylated plant protein could be generalized more than just for the Pto- and Fen-dependent pathways. The rice *Xa21* gene (38), which belongs to class 4 resistance genes, encodes a receptor protein kinase-like protein. It is tempting to speculate, based on the protein structure, that the extracellular domain of Xa21, which contains LRR, specifically binds an extracellular elicitor and that upon binding, the kinase activity of the intracellular kinase domain is activated. The homology between the kinase domain of Xa21 and Pto (38) suggests that these protein kinases share similarity in their downstream components. According to our model, Xa21 may have an NBS-LRR protein as a downstream component, which recognizes a phosphorylated protein, either the autophosphorylated Xa21 kinase domain or a protein whose phosphorylation is initiated by Xa21 kinase. A class 3 resistance gene encodes a membrane protein that has LRR in its extracellular domain. It is postulated that this class of resistance gene product has an intracellular protein kinase partner (18) and that with the partner kinase, it functions in a similar way as Xa21 does. If this is the case, class 3 resistance gene products may also have downstream NBS-LRR proteins.

Thus all four classes of resistance gene products can be viewed from a unified viewpoint if we assume that NBS-LRR proteins can be receptors for phosphorylated plant proteins. This model predicts that NBS-LRR proteins are involved in all resistance response signal transduction pathways that are initiated by one of the four resistance gene products. There may be overlaps (e.g., Prf) and redundancy in the ligand specificity for such NBS-LRR proteins, which could make genetic identification of these factors difficult. The signal transduction pathway downstream from these NBS-LRR proteins (main pathway) would be roughly the same, so that the resistance response is similar for the resistance defined by different classes of resistance gene products. Note that as we discussed above, different resistance gene products may have their specific satellite pathways and modifying pathways. Although both RPS2 and RPM1 belong to class 1 resistance gene products, there is some difference in their downstream responses (although the possibility that quantitative and/or kinetic difference in earlier steps of the pathway is interpreted differently in downstream steps cannot be excluded).

One possibility this model suggests is that NBS-LRR proteins are in general receptors for phosphorylated proteins. This would be the case for the pathways involving classes 2, 3, and 4 resistance gene products. What about class 1 resistance gene products? Class 1 resistance gene products are postulated to be intracellular receptors for pathogen-originated proteins, such as AvrRpt2 and AvrB. These pathogen-originated proteins recognized by NBS-LRR resistance gene products may be phosphorylated in the plant cell. We have been unsuccessful in detection of direct protein-protein interactions between RPS2 and AvrRpt2 or between RPM1 and AvrB in vitro (unpublished). A simple explanation is that these protein-protein interactions may require some other plant factors. Our model suggests that such a plant factor might be protein kinases (with a relatively low substrate specificity) that phosphorylate avirulence gene products.

References

1. Ackerveken, G. v. d., E. Marois, and U. Bonas. 1996. Recognition of the bacterial avirulence protein AvrBs3 occurs inside the host plant cell. Cell 87:1307-1316.
2. Anderson, P. A., G. J. Lawrence, B. C. Morrish, M. A. Ayliffe, E. J. Finnegan, and J. G. Ellis. 1997. Inactivation of the flax rust resistance gene M associated with loss of a repeated unit within the leucine-rich repeat coding region. Plant Cell 9:641-651.
3. Bent, A. F. 1996. Plant disease resistance genes: Function meets Structure. Plant Cell 8:1757-1771.
4. Bent, A. F., B. N. Kunkel, D. Dahlbeck, K. L. Brown, R. Schmidt, J. Giraudat, J. Leung, and B. J. Staskawicz. 1994. RPS2 of Arabidopsis thaliana: A leucine-rich repeat class of plant disease resistance genes. Science 265:1856-1860.
5. Bisgrove, S. R., M. T. Simonich, N. M. Smith, A. Sattler, and R. W. Innes. 1994. A disease resistance gene in Arabidopsis with specificity for two different pathogen avirulence genes. Plant Cell 6:927-933.
6. Cai, D., M. Kleine, S. Kifle, H.-J. Harloff, N. N. Sandal, K. A. Marcker, R. M. Klein-Lankhorst, E. M. J. Salentijn, W. Lange, W. J. Stiekema, U. Wyss, F. M. W. Grundler, and C. Jung. 1997. Positional cloning of a gene for nematode resistance in sugar beet. Science 275:832-834.
7. Chittenden, T., C. Flemington, A. B. Houghton, R. G. Ebb, G. J. Gallo, B. Elangovan, G. Chinnadurai, and R. J. Lutz. 1995. A conserved domain in Bak, distinct from BH1 and BH2, mediates cell death and protein binding functions. EMBO J. 14:5589-5596.
8. Dixon, M. S., D. A. Jones, J. S. Keddie, C. M. Thomas, K. Harrison, and J. D. G. Jones. 1996. The tomato Cf-2 disease resistance locus comprises two functional genes encoding leucine-rich repeat proteins. Cell 84:451-459.
9. Ellingboe, A. H. 1981. Changing concepts in host-pathogen genetics. Ann. Rev. Phytopathol. 19:125-143.
10. Flor, H. H. 1971. Current status of gene-for-gene concept. Ann. Rev. Phytopathol. 9:275-296.
11. Gabriel, D. W., and B. G. Rolfe. 1990. Working models of specific recognition in plant-microbe interactions. Ann. Rev. Phytopathol. 28:365-391.

12. Gopalan, S., D. W. Bauer, J. R. Alfano, A. O. Loniello, S. Y. He, and A. Collmer. 1996. Expression of the Pseudomonas syringae avirulence protein AvrB in plant cells alleviates its dependence on the hypersensitive response and pathogenicity (Hrp) secretion system in eliciting genotype-specific hypersensitive cell death. Plant Cell 8:1095-1105.
13. Grant, M. R., L. Godiard, E. Straube, T. Ashfield, J. Lewald, A. Sattler, R. W. Innes, and J. L. Dangl. 1995. Structure of the Arabidopsis RPM1 gene which enables dual-specificity disease resistance. Science 269:843-846.
14. Hammond-Kosack, K. E., and J. D. G. Jones. 1996. Resistance gene-dependent plant defense responses. Plant Cell 8:1773-1791.
15. Huang, H. C., R. H. Lin, C. J. Chang, A. Collmer, and W. L. Deng. 1995. The complete hrp gene cluster of Pseudomonas syringae pv. syringae 61 includes two blocks of genes required for $harpin_{pss}$ secretion that are arranged colinearly with Yersinia ysc homologs. Mol. Plant-Microbe Interact. 8:733-746.
16. Innes, R. W., A. F. Bent, B. N. Kunkel, S. R. Bisgrove, and B. J. Staskawicz. 1993. Molecular analysis of avirulence gene avrRpt2 and identification of a putative regulatory sequence common to all known Pseudomonas syringae avirulence genes. J. Bacteriol. 175:4859-4869.
17. Jefferson, R. A., T. A. Kavanagh, and M. W. Bevan. 1987. GUS fusions: ß-glucuronidase as a sensitive and versatile gene fusion marker in higher plants. EMBO J. 13:3901-3907.
18. Jones, D. A., C. M. Thomas, K. E. Hammond-Kosack, P. J. Balint-Kurti, and J. D. G. Jones. 1994. Isolation of the Tomato Cf-9 Gene for Resistance to Cladosporium fulvum by Transposon Tagging. Science 266:789-793.
19. Keen, N. T. 1992. The molecular biology of disease resistance. Plant Molecular Biology 19:109-122.
20. Klement, Z. 1982. Hypersensitivity, p. 149-177. In M. S. Mount, and G. H. Lacy (ed.), Phytopathogenic Procaryotes, Vol. 2, vol. 2. Academic Press, New York.
21. Kunkel, B. N., A. F. Bent, D. Dahlbeck, R. W. Innes, and B. J. Staskawicz. 1993. RPS2, an Arabidopsis disease resistance locus specifying recognition of Pseudomonas syringae strains expressing the avirulence gene avrRpt2. Plant Cell 5:865-875.
22. Lamb, C. J., M. A. Lawton, M. Dron, and R. A. Dixon. 1989. Signals and transduction mechanisms for activation of plant defenses against microbial attack. Cell 56:215-224.
23. Lawrence, G. J., E. J. Finnegan, M. A. Ayliffe, and J. G. Ellis. 1995. The L6 gene for flax rust resistance is related to the Arabidopsis bacterial resistance gene RPS2 and the tobacco viral resistance gene N. The Plant Cell 7:1195-1206.
24. Leister, R. T., F. M. Ausubel, and F. Katagiri. 1996. Molecular recognition of pathogen attack occurs inside of plant cells in plant disease resistance specified by the Arabidopsis genesRPS2 and RPM1. Proc. Natl. Acad. Sci. USA 93:15497-15502.

25. Lindgren, P. B., R. C. Peet, and N. J. Panopoulos. 1986. Gene cluster of Pseudomonas syringae pv. phaseolicola controls pathogenicity on bean and hypersensitivity on non-host plants. J. Bacteriol. 168:512-522.
26. Martin, G. B., S. H. Brommonschenkel, J. Chunwongse, A. Frary, M. W. Ganal, R. Spivey, T. Wu, E. D. Earle, and S. D. Tanksley. 1993. Map-based cloning of a protein kinase gene conferring disease resistance in tomato. Science 262:1432-1436.
27. Martin, G. B., A. Frary, T. Wu, S. Brommonschenkel, J. Chunwongse, E. D. Earle, and S. D. Tanksley. 1994. A member of the tomato Pto gene family confers sensitivity to fenthion resulting in rapid cell death. Plant Cell 6:1543-1552.
28. Mindrinos, M., F. Katagiri, G.-L. Yu, and F. M. Ausubel. 1994. The A. thaliana disease resistance gene RPS2 encodes a protein containing a nucleotide-binding site and leucine-rich repeats. Cell 78:1089-1099.
29. Ori, N., Y. Eshed, I. Paran, G. Presting, D. Aviv, S. Tanksley, D. Zamir, and R. Fluhr. 1997. The I2C family from the wilt disease resistance locus I2 belongs to the nucleotide binding, leucine-rich repeat superfamily of plant resistance gene. Plant Cell 9:521-531.
30. Pirhonen, M. U., M. C. Lidell, D. L. Rowley, S. W. Lee, S. Jin, Y. Liang, S. Silverstone, N. T. Keen, and S. W. Hutcheson. 1996. Phenotypic expression of Pseudomonas syringae avr genes in E. coli linked to the activities of the hrp-encoded secretion system. Mol Plant-Microbe Interact 9:252-260.
31. Rommens, C. M., J. M. Salmeron, G. E. Oldroyd, and B. J. Staskawicz. 1995. Intergeneric transfer and functional expression of the tomato disease resistance gene Pto. Plant Cell 7:1537-1544.
32. Rommens, C. M. T., J. M. Salmeron, D. C. Baulcombe, and B. J. Staskawicz. 1995. Use of a gene expression system based on potato virus X to rapidly identify and characterize a tomato Pto homolog that controls fenthion sensitivity. Plant Cell 7:249-257.
33. Rosqvist, R., S. Hakansson, A. Forsberg, and H. Wolf-Watz. 1995. Functional conservation of the secretion and translocation machinery for virulence proteins of yersiniae, salmonellae and shigellae. EMBO J 14:4187-4195.
34. Ryals, J. A., U. H. Newenschwander, M. G. Willits, A. Molina, H.-Y. Steiner, and M. D. Hunt. 1996. Systemic acquired resistance. Plant Cell 8:1809-1819.
35. Salmeron, J. M., G. E. D. Oldroyd, C. M. T. Rommens, S. R. Scofield, H.-S. Kim, D. T. Lavelle, D. Dahlbeck, and B. J. Stackawicz. 1996. Tomato Prf is a member of the leucine-rich repeat class of plant disease resistance genes and lies embedded within the Pto kinase gene cluster. Cell 86:123-133.
36. Scofield, S. R., C. M. Tobias, J. P. Rathjen, J. H. Chang, D. T. Lavelle, R. W. Michelmore, and B. J. Staskawicz. 1996. Molecular basis of gene-for-gene specificity in bacterial speck disease of tomato. Science 274:2063-2065.

37. Smith, M., E. Mazzola, J. Sims, S. Midland, and N. Keen. 1993. The syringolides: Bacterial C-glycosyl lipids that trigger plant disease resistance. Tetrahedron Lett. 34:223-226.
38. Song, W.-Y., G.-L. Wang, L.-L. Chen, H.-S. Kim, L.-Y. Pi, T. Holsten, J. Gardner, B. Wang, W.-X. Zhai, L.-H. Zhu, C. Fauquet, and P. Ronald. 1995. A receptor kinase-like protein encoded by the rice disease resistance gene, Xa21. Science 270:1804-1806.
39. Tamaki, S., D. Dahlbeck, B. Staskawicz, and N. T. Keen. 1988. Characterization and expression of two avirulence genes cloned from Pseudomonas syringae pv. glycinea. J. Bacteriol 170:4846-4854.
40. Tang, X., R. D. Frederick, J. Zhou, D. A. Halterman, Y. Jia, and G. B. Martin. 1996. Initiation of plant disease resistance by physical interaction of AvrPto and Pto kinase. Science 274:2060-2063.
41. Wet, J. R. d., K. V. Wood, D. R. Helinski, and M. DeLuca. 1985. Cloning of firefly luciferase cDNA and the expression of active luciferase in Escherichia coli. Proc. Natl. Acad. Sci. USA 82:7870-7873.
42. Whitham, S., S. P. Dinesh-Kumar, D. Choi, R. Hehl, C. Corr, and B. Baker. 1994. The product of the tobacco mosaic resistance gene N: Similarity to toll and the interleukin-1 receptor. Cell 78:1101-1105.
43. Whitham, S., S. McCormick, and B. Baker. 1996. The N gene of tobacco confers resistance to tobacco mosaic virus in transgenic tomato. Proc. Natl. Acad. Sci. USA 93:8776-8781.
44. Yu, G.-L., F. Katagiri, and F. M. Ausubel. 1993. Arabidopsis mutations at the RPS2 locus result in loss of resistance to Pseudomonas syringae strains expressing the avirulence gene avrRpt2. Mol. Plant-Microbe Interact. 6:434-443.
45. Zhou, J., Y.-T. Loh, R. A. Bressan, and G. B. Martin. 1995. The tomato gene Pti1 encodes a serine/threonine kinase that is phosphorylated by Pto and is involved in the hypersensitive response. Cell 83:925-935.

NATO ASI Series H

Vol. 36: **Signal Molecules in Plants and Plant-Microbe Interactions.** Edited by B.J.J. Lugtenberg. 425 pages. 1989.

Vol. 37: **Tin-Based Antitumour Drugs.** Edited by M. Gielen. 226 pages. 1990.

Vol. 38: **The Molecular Biology of Autoimmune Disease.** Edited by A.G. Demaine, J-P. Banga, and A.M. McGregor. 404 pages. 1990.

Vol. 39: **Chemosensory Information Processing.** Edited by D. Schild. 403 pages. 1990.

Vol. 40: **Dynamics and Biogenesis of Membranes.** Edited by J. A. F. Op den Kamp. 367 pages. 1990.

Vol. 41: **Recognition and Response in Plant-Virus Interactions.** Edited by R. S. S. Fraser. 467 pages. 1990.

Vol. 42: **Biomechanics of Active Movement and Deformation of Cells.** Edited by N. Akkas. 524 pages. 1990.

Vol. 43: **Cellular and Molecular Biology of Myelination.** Edited by G. Jeserich, H. H. Althaus, and T. V. Waehneldt. 565 pages. 1990.

Vol. 44: **Activation and Desensitization of Transducing Pathways.** Edited by T. M. Konijn, M. D. Houslay, and P. J. M. Van Haastert. 336 pages. 1990.

Vol. 45: **Mechanism of Fertilization: Plants to Humans.** Edited by B. Dale. 710 pages. 1990.

Vol .46: **Parallels in Cell to Cell Junctions in Plants and Animals.** Edited by A. W Robards, W. J . Lucas, J . D. Pitts, H . J . Jongsma, and D. C. Spray. 296 pages. 1990.

Vol. 47: **Signal Perception and Transduction in Higher Plants.** Edited by R. Ranjeva and A. M. Boudet. 357 pages. 1990.

Vol. 48: **Calcium Transport and Intracellular Calcium Homeostasis.** Edited by D. Pansu and F. Bronner. 456 pages. 1990.

Vol. 49: **Post-Transcriptional Control of Gene Expression.** Edited by J. E. G. McCarthy and M. F. Tuite. 671 pages. 1990.

Vol. 50: **Phytochrome Properties and Biological Action.** Edited by B. Thomas and C. B. Johnson. 337 pages. 1991.

Vol. 51: **Cell to Cell Signals in Plants and Animals.** Edited by V. Neuhoff and J. Friend. 404 pages. 1991.

Vol. 52: **Biological Signal Transduction.** Edited by E. M . Ross and K . W. A. Wirtz. 560 pages. 1991.

Vol. 53: **Fungal Cell Wall and Immune Response.** Edited by J. P. Latge and D. Boucias. 472 pages. 1991.

NATO ASI Series H

Vol. 54: **The Early Effects of Radiation on DNA.**
Edited by E. M. Fielden and P. O'Neill. 448 pages. 1991.

Vol. 55: **The Translational Apparatus of Photosynthetic Organelles.**
Edited by R. Mache, E. Stutz, and A. R. Subramanian. 260 pages. 1991.

Vol. 56: **Cellular Regulation by Protein Phosphorylation.**
Edited by L. M. G. Heilmeyer, Jr. 520 pages. 1991.

Vol. 57: **Molecular Techniques in Taxonomy.**
Edited by G . M . Hewitt, A. W. B. Johnston, and J. P. W. Young .
420 pages. 1991.

Vol. 58: **Neurocytochemical Methods.**
Edited by A. Calas and D. Eugene. 352 pages. 1991.

Vol. 59: **Molecular Evolution of the Major Histocompatibility Complex.**
Edited by J. Klein and D. Klein. 522 pages. 1991.

Vol. 60: **Intracellular Regulation of Ion Channels.**
Edited by M. Morad and Z. Agus. 261 pages. 1992.

Vol. 61: **Prader-Willi Syndrome and Other Chromosome 15q Deletion Disorders.**
Edited by S. B. Cassidy. 277 pages. 1992.

Vol. 62: **Endocytosis. From Cell Biology to Health, Disease and Therapie.**
Edited by P. J. Courtoy. 547 pages. 1992.

Vol. 63: **Dynamics of Membrane Assembly.**
Edited by J. A. F. Op den Kamp. 402 pages. 1992.

Vol. 64: **Mechanics of Swelling. From Clays to Living Cells and Tissues.**
Edited by T. K. Karalis. 802 pages. 1992.

Vol. 65: **Bacteriocins, Microcins and Lantibiotics.**
Edited by R. James, C. Lazdunski, and F. Pattus. 530 pages. 1992.

Vol. 66: **Theoretical and Experimental Insights into Immunology.**
Edited by A. S. Perelson and G. Weisbuch. 497 pages. 1992.

Vol. 67: **Flow Cytometry. New Developments.**
Edited by A. Jacquemin-Sablon. 1993.

Vol. 68: **Biomarkers. Research and Application in the Assessment of Environmental Health.** Edited by D. B. Peakall and L. R. Shugart.
138 pages. 1993.

Vol. 69: **Molecular Biology and its Application to Medical Mycology.**
Edited by B. Maresca, G. S. Kobayashi, and H. Yamaguchi.
271 pages. 1993.

Vol. 70: **Phospholipids and Signal Transmission.**
Edited by R. Massarelli, L. A. Horrocks, J. N. Kanfer, and K. Löffelholz.
448 pages. 1993.

Vol. 71: **Protein Synthesis and Targeting in Yeast.**
Edited by A. J. P. Brown, M. F. Tuite, and J. E. G. McCarthy.
425 pages. 1993.

NATO ASI Series H

Vol. 72: **Chromosome Segregation and Aneuploidy.**
Edited by B. K. Vig. 425 pages. 1993.

Vol. 73: **Human Apolipoprotein Mutants III. In Diagnosis and Treatment.**
Edited by C. R. Sirtori, G. Franceschini, B. H. Brewer Jr. 302 pages. 1993.

Vol. 74: **Molecular Mechanisms of Membrane Traffic.**
Edited by D. J. Morré, K. E. Howell, and J. J. M. Bergeron.
429 pages. 1993.

Vol. 75: **Cancer Therapy. Differentiation, Immunomodulation and Angiogenesis.**
Edited by N. D'Alessandro, E. Mihich, L. Rausa, H. Tapiero, and T. R.Tritton. 299 pages. 1993.

Vol. 76: **Tyrosine Phosphorylation/Dephosphorylation and Downstream Signalling.**
Edited by L. M. G. Heilmeyer Jr. 388 pages. 1993.

Vol. 77: **Ataxia-Telangiectasia.** Edited by R. A. Gatti, R. B. Painter.
306 pages. 1993.

Vol. 78: **Toxoplasmosis.** Edited by J. E. Smith. 272 pages. 1993.

Vol. 79: **Cellular Mechanisms of Sensory Processing. The Somatosensory System.**
Edited by L. Urban. 514 pages. 1994.

Vol. 80: **Autoimmunity: Experimental Aspects.**
Edited by M. Zouali. 318 pages. 1994.

Vol. 81: **Plant Molecular Biology. Molecular Genetic Analysis of Plant Development and Metabolism.**
Edited by G. Coruzzi, P. Puigdomènech. 579 pages. 1994.

Vol. 82: **Biological Membranes: Structure, Biogenesis and Dynamics.**
Edited by Jos A. F. Op den Kamp. 367 pages. 1994.

Vol. 83: **Molecular Biology of Mitochondrial Transport Systems.**
Edited by M. Forte, M. Colombini. 420 pages. 1994.

Vol. 84: **Biomechanics of Active Movement and Division of Cells.**
Edited by N. Akkaş. 587 pages. 1994.

Vol. 85: **Cellular and Molecular Effects of Mineral and Synthetic Dusts and Fibres.**
Edited by J. M. G. Davis, M.-C. Jaurand. 448 pages. 1994.

Vol. 86: **Biochemical and Cellular Mechanisms of Stress Tolerance in Plants.**
Edited by J. H. Cherry. 616 pages. 1994.

Vol. 87: **NMR of Biological Macromolecules.**
Edited by C. I. Stassinopoulou. 616 pages. 1994.

Vol. 88: **Liver Carcinogenesis. The Molecular Pathways.**
Edited by G. G. Skouteris. 502 pages. 1994.

NATO ASI Series H

Vol. 89: **Molecular and Cellular Mechanisms of H^+ Transport.**
Edited by B. H. Hirst. 504 pages. 1994.

Vol. 90: **Molecular Aspects of Oxidative Drug Metabolizing Enzymes: Their Significance in Environmental Toxicology, Chemical Carcinogenesis and Health.**
Edited by E. Arınç, J. B. Schenkman, E. Hodgson. 623 pages. 1994.

Vol. 91: **Trafficking of Intracellular Membranes: From Molecular Sorting to Membrane Fusion.**
Edited by M. C. Pedroso de Lima, N. Düzgüneş, D. Hoekstra. 371 pages. 1995.

Vol. 92: **Signalling Mechanisms – from Transcription Factors to Oxidative Stress.**
Edited by L. Packer, K. Wirtz. 466 pages. 1995.

Vol. 93: **Modulation of Cellular Responses in Toxicity.**
Edited by C. L. Galli, A. M. Goldberg, M. Marinovich. 379 pages. 1995.

Vol. 94: **Gene Technology.**
Edited by A. R. Zander, W. Ostertag, B. V. Afanasiev, F. Grosveld. 556 pages. 1996.

Vol. 95: **Flow and Image Cytometry.**
Edited by A. Jacquemin-Sablon. 241 pages. 1996.

Vol. 96: **Molecular Dynamics of Biomembranes.**
Edited by J. A. F. Op den Kamp. 424 pages. 1996.

Vol. 97: **Post-transcriptional Control of Gene Expression.**
Edited by O. Resnekov, A. von Gabain. 283 pages. 1996.

Vol. 98: **Lactic Acid Bacteria: Current Advances in Metabolism, Genetics and Applications.**
Edited by T. F. Bozoglu, B. Ray. 412 pages. 1996.

Vol. 99: **Tumor Biology Regulation of Cell Growth, Differentiation and Genetics in Cancer.**
Edited by A. S. Tsiftsoglou, A. C. Sartorelli, D. E. Housman, T. M. Dexter. 342 pages. 1996.

Vol. 100: **Neurotransmitter Release and Uptake.**
Edited by Ş. Pöğün. 344 pages. 1997.

Vol. 101: **Molecular Mechanisms of Signalling and Membrane Transport.**
Edited by K. W. A. Wirtz. 344 pages. 1997.

Vol. 102: **Interacting Protein Domains**
Edited by L. Heilmeyer. 296 pages. 1997.

Vol. 103: **Molecular Microbiology.**
Edited by S. J. W. Busby, C. M. Thomas, N. L. Brown. 334 pages. 1998.

NATO ASI Series H

Vol. 104: **Cellular Integration of Signalling Pathways in Plant Development.** Edited by F. Lo Schiavo, R. L. Last, G. Morelli, N. V. Raikhel. 330 pages. 1998.

Vol. 105: **Gene Therapy.** Edited by Kleanthis Xanthopoulos. 238 pages. 1998.

Vol. 106: **Lipid and Protein Traffic. Pathways and Molecular Mechanisms.** Edited by J. A. F. Op den Kamp. 374 pages. 1998.